Xiao Bang Xing

W Green's thm $\oint_C pdx + qdy = \iint_A (\frac{\partial q}{\partial x} - \frac{\partial P}{\partial y}) \, dy \, dx$ $\quad R^2 \quad\quad R^2$

Divergence. $\iint^{R^2} F \cdot n d\sigma \longrightarrow \iiint^{R^3}$

VV Stokes $\nearrow$ orientable. $\quad$ (extension of Green thm)

$\searrow$ positive

Stoke's Thm shows a relationship between a line intergral in R^3 around a closed curve and a surface integral over a surface that has the closed curve as a boundary.

$\oint F \cdot dR$ is equal to the surface integral

$\iint curl \, F \cdot n d\sigma$

CALCULUS
OF
SEVERAL
VARIABLES

CALCULUS OF SEVERAL VARIABLES

SERGE LANG
Yale University

SECOND EDITION

ADDISON-WESLEY PUBLISHING COMPANY

Reading, Massachusetts · Menlo Park, California
London · Amsterdam · Don Mills, Ontario · Sydney

This book is in the
ADDISON-WESLEY SERIES IN MATHEMATICS

LYNN H. LOOMIS
Consulting Editor

Library of Congress Cataloging in Publication Data

Lang, Serge, 1927–
Calculus of several variables.

Includes index.
1. Calculus. 2. Functions of several real
variables. I. Title.
QA303.L256 1979   515'.84   78-55823
ISBN 0-201-04299-1

Reprinted with correcions, January 1983

ISBN 0-201-04299-1
GHIJK-MA-8987654

FOREWORD

The present course on calculus of several variables is meant as a text, either for one semester following *A First Course in Calculus*, or for a year if the calculus sequence is so structured.

For a one-semester course, the first eight chapters provide an appropriate amount of material. If some time is left over, one can cover some topics in maxima-minima in Chapter XI, or the beginnings of higher derivatives and Taylor's formula in Chapter XII, depending on the taste of the instructor.

This first part has considerable unity of style. Many of the results are immediate corollaries of the chain rule. The main idea is that given a function of several variables, if we want to look at its values at two points P and Q, we join these points by a curve (often a straight line segment), and then look at the values of the function on that curve. By this device, we are able to reduce a large number of problems in several variables to problems and techniques in one variable. For instance, the tangent plane, the directional derivative, the law of conservation of energy, Taylor's formula, are all handled in this manner.

Green's theorem is more important to include in a one-semester course than other topics, because it provides a very elegant mixing of integration and differentiation techniques in one and two variables. This mixing is used frequently in applications, and also serves to fix these techniques in the mind because of the way they are used.

For a year's course, the rest of the book provides an adequate amount of material to be covered during the second semester. It consists of three topics, which are logically independent of each other and could be covered in any order. Some order must be chosen because it is necessary to project the course in a totally ordered way on the page axis (and the time axis), but logically, the choice is arbitrary. Pedagogically, the order chosen here seemed the one best suited for most people. These three topics are:

a) Triple integration and surface integrals, which continue ideas of Chapters VII and VIII.

b) Maxima-minima and the Taylor formula, which continue the ideas of differentiating curves, perpendicularity, and analyzing a function of two or more variables by looking at its values on curves or line segments, thereby reducing the study of some properties to functions of one variable.

c) Matrices and determinants, which constitute the linear part of a function, and affect some properties like those of the inverse mapping theorem and the change of variables formula.

Different instructors will cover these three topics in whatever order they prefer. For applications to economics, it would make sense to cover the chapters on maxima-minima and the quadratic form in Taylor's formula before doing triple integration and surface integrals. The methods used depend only on the techniques developed as corollaries of the chain rule.

I think it is important that even at this early stage, students acquire the idea that one can operate with differentiation just as with polynomials. Thus §4 of Chapter XII could be covered early.

I have included only that part of linear algebra which is immediately useful for the applications to calculus. My *Introduction to Linear Algebra* provides an appropriate text when a whole semester is devoted to the subject. Many courses are still structured to give primary emphasis to the analytic aspects, and only a few notions involving matrices and linear maps are needed to cover, say, the chain rule for mappings of one space into another, and to emphasize the importance of linear approximations. These, it seems to me, are the essential ingredients of a second semester of calculus for students who want to become acquainted rapidly with the most important basic notions and how they are used in practice. Many years ago, there was no linear algebra introduced in calculus courses. Intermediate years have probably seen an excessive amount—more than was needed. I try to strike a proper balance here.

Some proofs have been included. On the whole, our policy has been to include those proofs which illustrate fundamental principles and are free of technicalities. Such proofs, which are also short, should be learned by students without difficulty. Examples are the uniqueness of the potential function, the law of conservation of energy, the independence of an integral on the path if a potential function exists, Green's theorem in the simplest cases, etc.

Other proofs, like those of the chain rule, or the local existence of a potential function, can be given in class or omitted, depending on the level of interest of a class and the taste of the instructor. For convenience, such proofs have usually been placed at the end of each section.

Many worked-out examples have been added since previous editions, and answers to some exercises have been expanded to include more comprehensive solutions. I have done this to lighten the text on occasion. Such expanded solutions can also be viewed as worked-out examples simply placed differently, allowing students to think before they look up the answer if they have troubles with the problem.

I include two appendices on series and Fourier series, for the convenience of courses structured so that it is desirable to give an inkling of these topics some time during the second-year calculus, without waiting for a course in advanced calculus.

I would like to express my appreciation for the helpful guidance provided by the reviewers: M. B. Abrahamse (University of Virginia), Sherwood F. Ebey (University of the South), and William F. Keigher (The University of Tennessee).

I thank Anthony Petrello for working out the answers and helping with the proofreading. I also thank Mr. Gimli Khazad for communicating to me a number of misprints.

New Haven, Connecticut   S. L.
January 1979

CONTENTS

CHAPTER I

Vectors

The concept of a vector is basic for the study of functions of several variables. It provides geometric motivation for everything that follows. Hence the properties of vectors, both algebraic and geometric, will be discussed in full.

One significant feature of all the statements and proofs of this part is that they are neither easier nor harder to prove in 3-space than they are in 2-space. Since we have to deal with $n = 2$ and $n = 3$, it is just as easy to state some things just with a neutral n. Also for physics and economics, it is useful to get used to n rather than 2 or 3. However, for purposes of pedagogy, throughout the book we always give first the definitions and formulas for the special cases of $n = 2$ and $n = 3$ so that the reader can omit any reference to higher n if he wishes.

§1. DEFINITION OF POINTS IN SPACE

We know that a number can be used to represent a point on a line, once a unit length is selected.

A pair of numbers (i.e. a couple of numbers) (x, y) can be used to represent a point in the plane.

These can be pictured as follows:

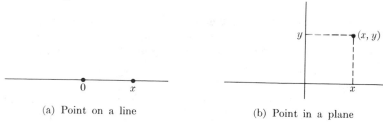

(a) Point on a line   (b) Point in a plane

Figure 1

1

We now observe that a triple of numbers (x, y, z) can be used to represent a point in space, that is 3-dimensional space, or 3-space. We simply introduce one more axis.

Figure 2 illustrates this.

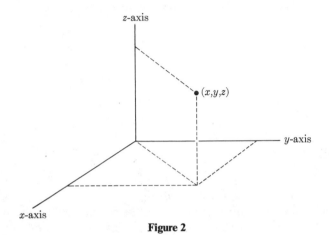

Figure 2

Instead of using x, y, z we could also use (x_1, x_2, x_3). The line could be called 1-space, and the plane could be called 2-space.

Thus we can say that a single number represents a point in 1-space. A couple represents a point in 2-space. A triple represents a point in 3-space.

Although we cannot draw a picture to go further, there is nothing to prevent us from considering a quadruple of numbers

$$(x_1, x_2, x_3, x_4)$$

and decreeing that this is a point in 4-space. A quintuple would be a point in 5-space, then would come a sextuple, septuple, octuple, ...

We let ourselves be carried away and **define a point in n-space** to be an n-tuple of numbers

$$(x_1, x_2, \ldots, x_n),$$

if n is a positive integer. We shall denote such an n-tuple by a capital letter X, and try to keep small letters for numbers and capital letters for points. We call the numbers $x_1, \ldots, x_n$ the **coordinates** of the point X. For example, in 3-space, 2 is the first coordinate of the point $(2, 3, -4)$, and -4 is its third coordinate. We denote n-space by $\mathbf{R}^n$.

Most of our examples will take place when $n = 2$ or $n = 3$. Thus the reader may visualize either of these two cases throughout the book. However, three comments must be made.

First, we have to handle $n = 2$ and $n = 3$, so that in order to avoid a lot of repetitions, it is useful to have a notation which covers both these cases simultaneously, even if we often repeat the formulation of certain results separately for both cases.

Second, no theorem or formula is simpler by making the assumption that $n = 2$ or 3.

Third, the case $n = 4$ does occur in physics.

Example 1. One classical example of 3-space is of course the space we live in. After we have selected an origin and a coordinate system, we can describe the position of a point (body, particle, etc.) by 3 coordinates. Furthermore, as was known long ago, it is convenient to extend this space to a 4-dimensional space, with the fourth coordinate as time, the time origin being selected, say, as the birth of Christ—although this is purely arbitrary (it might be more convenient to select the birth of the solar system, or the birth of the earth as the origin, if we could determine these accurately). Then a point with negative time coordinate is a BC point, and a point with positive time coordinate is an AD point.

Don't get the idea that "time is *the* fourth dimension," however. The above 4-dimensional space is only one possible example. In economics, for instance, one uses a very different space, taking for coordinates, say, the number of dollars expended in an industry. For instance, we could deal with a 7-dimensional space with coordinates corresponding to the following industries:

1. Steel   2. Auto   3. Farm products   4. Fish
5. Chemicals   6. Clothing   7. Transportation

We agree that a megabuck per year is the unit of measurement. Then a point

$$(1,000, 800, 550, 300, 700, 200, 900)$$

in this 7-space would mean that the steel industry spent one billion dollars in the given year, and that the chemical industry spent 700 million dollars in that year.

We shall now define how to add points. If A, B are two points, say in 3-space,

$$A = (a_1, a_2, a_3) \quad \text{and} \quad B = (b_1, b_2, b_3)$$

then we define $A + B$ to be the point whose coordinates are

$$A + B = (a_1 + b_1, a_2 + b_2, a_3 + b_3).$$

Example 2. In the plane, if $A = (1, 2)$ and $B = (-3, 5)$, then

$$A + B = (-2, 7).$$

In 3-space, if $A = (-1, \pi, 3)$ and $B = (\sqrt{2}, 7, -2)$, then

$$A + B = (\sqrt{2} - 1, \pi + 7, 1).$$

Using a neutral n to cover both the cases of 2-space and 3-space, the points would be written

$$A = (a_1, \ldots, a_n), \qquad B = (b_1, \ldots, b_n),$$

and we define $A + B$ to be the point whose coordinates are

$$(a_1 + b_1, \ldots, a_n + b_n).$$

We observe that the following rules are satisfied:

1. $(A + B) + C = A + (B + C)$.
2. $A + B = B + A$.
3. If we let

$$O = (0, 0, \ldots, 0)$$

be the point all of whose coordinates are 0, then

$$O + A = A + O = A$$

for all A.

4. Let $A = (a_1, \ldots, a_n)$ and let $-A = (-a_1, \ldots, -a_n)$. Then

$$A + (-A) = O.$$

All these properties are very simple, and are true because they are true for numbers, and addition of n-tuples is defined in terms of addition of their components, which are numbers.

Note. Do not confuse the number 0 and the n-tuple $(0, \ldots, 0)$. We usually denote this n-tuple by O, and also call it zero, because no difficulty can occur in practice.

We shall now interpret addition and multiplication by numbers geometrically in the plane (you can visualize simultaneously what happens in 3-space).

Example 3. Let $A = (2, 3)$ and $B = (-1, 1)$. Then

$$A + B = (1, 4).$$

The figure looks like a **parallelogram** (Fig. 3).

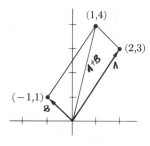

Figure 3

Example 4. Let $A = (3, 1)$ and $B = (1, 2)$. Then

$$A + B = (4, 3).$$

We see again that the geometric representation of our addition looks like a **parallelogram** (Fig. 4).

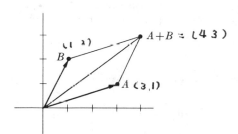

Figure 4

The reason why the figure looks like a **parallelogram** can be given in terms of plane geometry as follows. We obtain $B = (1, 2)$ by starting from the origin $O = (0, 0)$, and moving 1 unit to the right and 2 up. To get $A + B$, we start from A, and again move 1 unit to the right and 2 up. Thus the line segments between O and B, and between A and $A + B$ are the hypotenuses of right triangles whose corresponding legs are of the same length, and parallel. The above segments are therefore parallel and of the same length, as illustrated in Fig. 5.

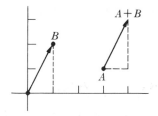

Figure 5

Example 5. If $A = (3, 1)$ again, then $-A = (-3, -1)$. If we plot this point, we see that $-A$ has opposite direction to A. We may view $-A$ as the reflection of A through the origin.

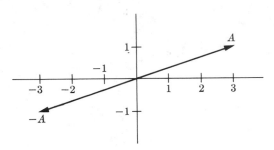

Figure 6

We shall now consider multiplication of A by a number. If c is any number, we **define** cA to be the point whose coordinates are

$$(ca_1, \ldots, ca_n).$$

Example 6. If $A = (2, -1, 5)$ and $c = 7$, then $cA = (14, -7, 35)$.

It is easy to verify the rules:

5. $c(A + B) = cA + cB$.
6. If c_1, c_2 are numbers, then

$$(c_1 + c_2)A = c_1A + c_2A \qquad \text{and} \qquad (c_1 c_2)A = c_1(c_2 A).$$

Also note that

$$(-1)A = -A.$$

What is the geometric representation of multiplication by a number?

Example 7. Let $A = (1, 2)$ and $c = 3$. Then

$$cA = (3, 6)$$

as in Fig. 7 (a).

Multiplication by 3 amounts to stretching A by 3. Similarly, $\frac{1}{2}A$ amounts to stretching A by $\frac{1}{2}$, i.e. shrinking A to half its size. In general, if t is a number,

$t > 0$, we interpret tA as a point in the same direction as A from the origin, but t times the distance. In fact, we define A and B to have the **same direction** if there exists a number $c > 0$ such that $A = cB$. We emphasize that this means A and B have the same direction **with respect to the origin**. For simplicity of language, we omit the words "with respect to the origin."

Multiplication by a negative number reverses the direction. Thus $-3A$ would be represented as in Fig. 7 (b).

We define two vectors A, B (neither of which is zero) to have **opposite directions** if there is a number $c < 0$ such that $cA = B$. Thus when $B = -A$, then A, B have opposite direction.

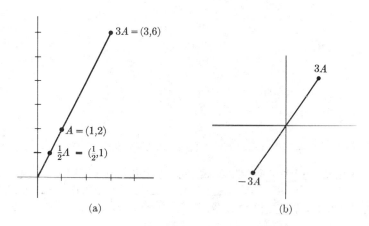

(a)   (b)

Figure 7

EXERCISES

Find $A + B$, $A - B$, $3A$, $-2B$ in each of the following cases. Draw the points of Exercises 1 and 2 on a sheet of graph paper.

1. $A = (2, -1)$, $B = (-1, 1)$ 2. $A = (-1, 3)$, $B = (0, 4)$

3. $A = (2, -1, 5)$, $B = (-1, 1, 1)$ 4. $A = (-1, -2, 3)$, $B = (-1, 3, -4)$

5. $A = (\pi, 3, -1)$, $B = (2\pi, -3, 7)$ 6. $A = (15, -2, 4)$, $B = (\pi, 3, -1)$

7. Let $A = (1, 2)$ and $B = (3, 1)$. Draw $A + B$, $A + 2B$, $A + 3B$, $A - B$, $A - 2B$, $A - 3B$ on a sheet of graph paper.

8. Let A, B be as in Exercise 1. Draw the points $A + 2B$, $A + 3B$, $A - 2B$, $A - 3B$, $A + \frac{1}{2}B$ on a sheet of graph paper.

9. Let A and B be as drawn in Fig. 8. Draw the point $A - B$.

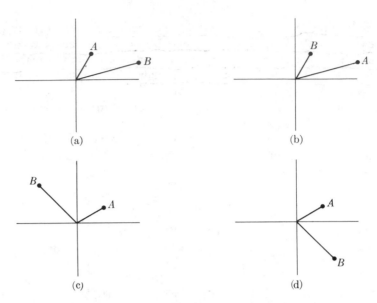

Figure 8

§2. LOCATED VECTORS

We define a **located vector** to be an ordered pair of points which we write $\overrightarrow{AB}$. (This is *not* a product.) We visualize this as an arrow between A and B. We call A the **beginning point** and B the **end point** of the located vector (Fig. 9).

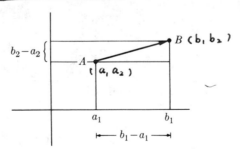

Figure 9

We observe that in the plane,

$$b_1 = a_1 + (b_1 - a_1).$$

Similarly,

$$b_2 = a_2 + (b_2 - a_2).$$

This means that

$$B = A + (B - A).$$

Let $\overrightarrow{AB}$ and $\overrightarrow{CD}$ be two located vectors. We shall say that they are **equivalent** if $B - A = D - C$. Every located vector $\overrightarrow{AB}$ is equivalent to one whose beginning point is the origin, because $\overrightarrow{AB}$ is equivalent to $\overrightarrow{O(B - A)}$. Clearly this is the only located vector whose beginning point is the origin and which is equivalent to $\overrightarrow{AB}$. If you visualize the parallelogram law in the plane, then it is clear that equivalence of two located vectors can be interpreted geometrically by saying that the lengths of the line segments determined by the pair of points are equal, and that the "directions" in which they point are the same.

In the next figures, we have drawn the located vectors $\overrightarrow{O(B - A)}$, $\overrightarrow{AB}$, and $\overrightarrow{O(A - B)}$, $\overrightarrow{BA}$.

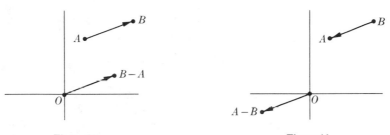

Figure 10   **Figure 11**

Example 1. Let $P = (1, -1, 3)$ and $Q = (2, 4, 1)$. Then $\overrightarrow{PQ}$ is equivalent to $\overrightarrow{OC}$, where $C = Q - P = (1, 5, -2)$. If $A = (4, -2, 5)$ and $B = (5, 3, 3)$, then $\overrightarrow{PQ}$ is equivalent to $\overrightarrow{AB}$ because

$$Q - P = B - A = (1, 5, -2).$$

Given a located vector $\overrightarrow{OC}$ whose beginning point is the origin, we shall say that it is **located at the origin**. Given any located vector $\overrightarrow{AB}$, we shall say that it is **located at** A.

A located vector at the origin is entirely determined by its end point. In view of this, we shall call an n-tuple either a **point** or a **vector**, depending on the interpretation which we have in mind.

Two located vectors $\overrightarrow{AB}$ and $\overrightarrow{PQ}$ are said to be **parallel** if there is a number $c \neq 0$ such that $B - A = c(Q - P)$. They are said to have the **same direction** if there is a number $c > 0$ such that $B - A = c(Q - P)$, and to have **opposite direction** if there is a number $c < 0$ such that $B - A = c(Q - P)$. In the next pictures, we illustrate parallel located vectors.

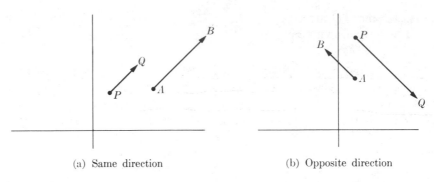

(a) Same direction   (b) Opposite direction

Figure 12

Example 2. Let

$$P = (3, 7) \quad \text{and} \quad Q = (-4, 2).$$

Let

$$A = (5, 1) \quad \text{and} \quad B = (-16, -14).$$

Then

$$Q - P = (-7, -5) \quad \text{and} \quad B - A = (-21, -15).$$

Hence $\overrightarrow{PQ}$ is parallel to $\overrightarrow{AB}$, because $B - A = 3(Q - P)$. Since $3 > 0$, we even see that $\overrightarrow{PQ}$ and $\overrightarrow{AB}$ have the same direction.

In a similar manner, any definition made concerning n-tuples can be carried over to located vectors. For instance, in the next section, we shall define what it means for n-tuples to be perpendicular. Then we can say that two located

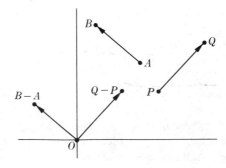

Figure 13

vectors $\overrightarrow{AB}$ and $\overrightarrow{PQ}$ are **perpendicular** if $B - A$ is perpendicular to $Q - P$. In Fig. 13, we have drawn a picture of such vectors in the plane.

EXERCISES

In each case, determine which located vectors $\overrightarrow{PQ}$ and $\overrightarrow{AB}$ are equivalent.

1. $P = (1, -1)$, $Q = (4, 3)$, $A = (-1, 5)$, $B = (5, 2)$.

2. $P = (1, 4)$, $Q = (-3, 5)$, $A = (5, 7)$, $B = (1, 8)$.

3. $P = (1, -1, 5)$, $Q = (-2, 3, -4)$, $A = (3, 1, 1)$, $B = (0, 5, 10)$.

4. $P = (2, 3, -4)$, $Q = (-1, 3, 5)$, $A = (-2, 3, -1)$, $B = (-5, 3, 8)$.

In each case, determine which located vectors $\overrightarrow{PQ}$ and $\overrightarrow{AB}$ are parallel.

5. $P = (1, -1)$, $Q = (4, 3)$, $A = (-1, 5)$, $B = (7, 1)$.

6. $P = (1, 4)$, $Q = (-3, 5)$, $A = (5, 7)$, $B = (9, 6)$.

7. $P = (1, -1, 5)$, $Q = (-2, 3, -4)$, $A = (3, 1, 1)$, $B = (-3, 9, -17)$.

8. $P = (2, 3, -4)$, $Q = (-1, 3, 5)$, $A = (-2, 3, -1)$, $B = (-11, 3, -28)$.

9. Draw the located vectors of Exercises 1, 2, 5, and 6 on a sheet of paper to illustrate these exercises. Also draw the located vectors $\overrightarrow{OP}$ and $\overrightarrow{BA}$. Draw the points $Q - P$, $B - A$, $P - Q$, and $A - B$.

§3. SCALAR PRODUCT

It is understood that throughout a discussion we select vectors always in the same n-dimensional space. You may think of the cases $n = 2$ and $n = 3$ only.

In 2-space, let $A = (a_1, a_2)$ and $B = (b_1, b_2)$. We define their **scalar product** to be

$$A \cdot B = a_1 b_1 + a_2 b_2.$$

In 3-space, let $A = (a_1, a_2, a_3)$ and $B = (b_1, b_2, b_3)$. We define their **scalar product** to be

$$A \cdot B = a_1 b_1 + a_2 b_2 + a_3 b_3.$$

In n-space, covering both cases with one notation, let $A = (a_1, \ldots, a_n)$ and $B = (b_1, \ldots, b_n)$ be two vectors. We define their **scalar** or **dot product** $A \cdot B$ to be

$$a_1 b_1 + \cdots + a_n b_n.$$

This product is a **number**. For instance, if

$$A = (1, 3, -2) \quad \text{and} \quad B = (-1, 4, -3),$$

then

$$A \cdot B = -1 + 12 + 6 = 17.$$

For the moment, we do not give a geometric interpretation to this scalar product. We shall do this later. We derive first some important properties. The basic ones are:

SP 1. *We have $A \cdot B = B \cdot A$.*

SP 2. *If A, B, C are three vectors, then*

$$A \cdot (B + C) = A \cdot B + A \cdot C = (B + C) \cdot A.$$

SP 3. *If x is a number, then*

$$(xA) \cdot B = x(A \cdot B) \quad \text{and} \quad A \cdot (xB) = x(A \cdot B).$$

SP 4. *If $A = O$ is the zero vector, then $A \cdot A = 0$, and otherwise $A \cdot A > 0$.*

We shall now prove these properties.
Concerning the first, we have

$$a_1 b_1 + \cdots + a_n b_n = b_1 a_1 + \cdots + b_n a_n,$$

because for any two numbers a, b, we have $ab = ba$. This proves the first property.

For **SP 2**, let $C = (c_1, \ldots, c_n)$. Then

$$B + C = (b_1 + c_1, \ldots, b_n + c_n)$$

and

$$A \cdot (B + C) = a_1(b_1 + c_1) + \cdots + a_n(b_n + c_n)$$
$$= a_1 b_1 + a_1 c_1 + \cdots + a_n b_n + a_n c_n.$$

Reordering the terms yields

$$a_1 b_1 + \cdots + a_n b_n + a_1 c_1 + \cdots + a_n c_n,$$

which is none other than $A \cdot B + A \cdot C$. This proves what we wanted.

We leave property **SP 3** as an exercise.

Finally, for **SP 4**, we observe that if one coordinate a_i of A is not equal to 0, then there is a term $a_i^2 \neq 0$ and $a_i^2 > 0$ in the scalar product

$$A \cdot A = a_1^2 + \cdots + a_n^2.$$

Since every term is ≥ 0, it follows that the sum is > 0, as was to be shown.

In much of the work which we shall do concerning vectors, we shall use only the ordinary properties of addition, multiplication by numbers, and the four properties of the scalar product. We shall give a formal discussion of these later. For the moment, observe that there are other objects with which you are familiar and which can be added, subtracted, and multiplied by numbers, for instance the continuous functions on an interval $[a, b]$ (cf. Exercise 6).

Instead of writing $A \cdot A$ for the scalar product of a vector with itself, it will be convenient to write also A^2. (This is the only instance when we allow ourselves such a notation. Thus A^3 has no meaning.) As an exercise, verify the following identities:

$$(A + B)^2 = A^2 + 2A \cdot B + B^2,$$
$$(A - B)^2 = A^2 - 2A \cdot B + B^2.$$

A dot product $A \cdot B$ may very well be equal to 0 without either A or B being the zero vector. For instance, let $A = (1, 2, 3)$ and $B = (2, 1, -\frac{4}{3})$. Then

$$A \cdot B = 0.$$

We define two vectors A, B to be **perpendicular** (or as we shall also say, orthogonal) if $A \cdot B = 0$. For the moment, it is not clear that in the plane, this definition coincides with our intuitive geometric notion of perpendicularity. We shall convince you that it does in the next section. Here we merely note an example. Say in $\mathbf{R}^3$, let

$$E_1 = (1, 0, 0), \qquad E_2 = (0, 1, 0), \qquad E_3 = (0, 0, 1)$$

be the three unit vectors, as shown on the diagram (Fig. 14).

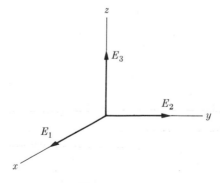

Figure 14

Then we see that $E_1 \cdot E_2 = 0$, and similarly $E_i \cdot E_j = 0$ if $i \neq j$. And these vectors look perpendicular. If $A = (a_1, a_2, a_3)$, then we observe that the i-th component of A, namely

$$a_i = A \cdot E_i$$

is the dot product of A with the i-th unit vector. We see that A is perpendicular to E_i (according to our definition of perpendicularity with the dot product) if and only if its i-th component is equal to 0.

EXERCISES

1. Find $A \cdot A$ for each of the following n-tuples.

    (a) $A = (2, -1), B = (-1, 1)$ (b) $A = (-1, 3), B = (0, 4)$
    (c) $A = (2, -1, 5), B = (-1, 1, 1)$ (d) $A = (-1, -2, 3), B = (-1, 3, -4)$
    (e) $A = (\pi, 3, -1), B = (2\pi, -3, 7)$ (f) $A = (15, -2, 4), B = (\pi, 3, -1)$

2. Find $A \cdot B$ for each of the above n-tuples.

3. Using only the four properties of the scalar product, verify in detail the identities given in the text for $(A + B)^2$ and $(A - B)^2$.

4. Which of the following pairs of vectors are perpendicular?
    (a) $(1, -1, 1)$ and $(2, 1, 5)$ (b) $(1, -1, 1)$ and $(2, 3, 1)$
    (c) $(-5, 2, 7)$ and $(3, -1, 2)$ (d) $(\pi, 2, 1)$ and $(2, -\pi, 0)$

5. Let A be a vector perpendicular to every vector X. Show that $A = O$.

Scalar product for functions.

If you want to see the scaler product in a more abstract context, applicable to other situations, e.g. to functions, see Appendix 1.

§4. THE NORM OF A VECTOR

We define the **norm** of a vector A, and denote by $\|A\|$, the number

$$\|A\| = \sqrt{A \cdot A} .$$

Since $A \cdot A \geq 0$, we can take the square root. The norm is also sometimes called the **magnitude** of A.

When $n = 2$ and $A = (a, b)$, then

$$\|A\| = \sqrt{a^2 + b^2} ,$$

as in the following picture (Fig. 15).

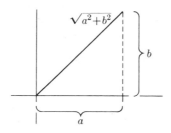

Figure 15

Example 1. If $A = (1, 2)$, then

$$\|A\| = \sqrt{1 + 4} = \sqrt{5} .$$

When $n = 3$ and $A = (a_1, a_2, a_3)$, then

$$\|A\| = \sqrt{a_1^2 + a_2^2 + a_3^2} .$$

Example 2. If $A = (-1, 2, 3)$, then

$$\|A\| = \sqrt{1 + 4 + 9} = \sqrt{14} .$$

If $n = 3$, then the picture looks like Fig. 16, with $A = (x, y, z)$.

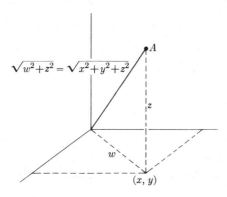

Figure 16

If we first look at the two components (x, y), then the length of the segment
between $(0, 0)$ and (x, y) is equal to $w = \sqrt{x^2 + y^2}$, as indicated.

宇さ于拉斯

Then again the norm of A by the Pythagoras theorem would be

$$\sqrt{w^2 + z^2} = \sqrt{x^2 + y^2 + z^2}.$$

Thus when $n = 3$, our definition of norm is compatible with the geometry of the Pythagoras theorem.

In terms of coordinates, $A = (a_1, \ldots, a_n)$ we see that

$$\|A\| = \sqrt{a_1^2 + \cdots + a_n^2}.$$

If $A \neq O$, then $\|A\| \neq 0$ because some coordinate $a_i \neq 0$, so that $a_i^2 > 0$, and hence $a_1^2 + \cdots + a_n^2 > 0$, so $\|A\| \neq 0$.

Observe that for any vector A we have

$$\boxed{\|A\| = \| - A\|.}$$

This is due to the fact that

$$(-a_1)^2 + \cdots + (-a_n)^2 = a_1^2 + \cdots + a_n^2,$$

because $(-1)^2 = 1$. Of course, this is as it should be from the picture:

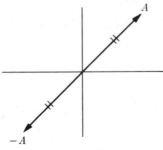

Figure 17

Recall that A and $-A$ are said to have **opposite direction**. However, they have the same norm (magnitude, as is sometimes said when speaking of vectors).

Let A, B be two points. We define the **distance** between A and B to be

$$\|A - B\| = \sqrt{(A - B) \cdot (A - B)}.$$

相距

This definition coincides with our geometric intuition when A, B are points in the plane (Fig. 18). It is the same thing as the length of the located vector $\overrightarrow{AB}$ or the located vector $\overrightarrow{BA}$.

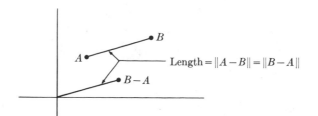

Figure 18

Example 3. Let $A = (-1, 2)$ and $B = (3, 4)$. Then the length of the located vector $\overrightarrow{AB}$ is $\|B - A\|$. But $B - A = (4, 2)$. Thus

$$\|B - A\| = \sqrt{16 + 4} = \sqrt{20} .$$

In the picture, we see that the horizontal side has length 4 and the vertical side has length 2. Thus our definitions reflect our geometric intuition derived from Pythagoras.

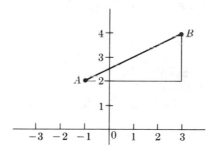

Figure 19

Let P be a point in the plane, and let a be a number > 0. The set of points X such that

$$\|X - P\| < a$$

will be called the **open disc** of radius a centered at P. The set of points X such that

$$\|X - P\| \leqq a$$

will be called the **closed disc** of radius a and center P. The set of points X such that

$$\|X - P\| = a$$

is called the circle of radius a and center P. These are illustrated in Fig. 20.

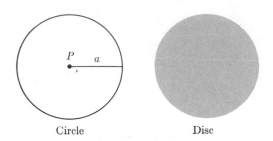

Circle   Disc

Figure 20

In 3-dimensional space, the set of points X such that

$$\|X - P\| < a$$

will be called the **open ball** of radius a and center P. The set of points X such that

$$\|X - P\| \leqq a$$

will be called the **closed ball** of radius a and center P. The set of points X such that

$$\|X - P\| = a$$

will be called the **sphere** of radius a and center P. In higher dimensional space, one uses this same terminology of ball and sphere.

Figure 21 illustrates a sphere and a ball in 3-space.

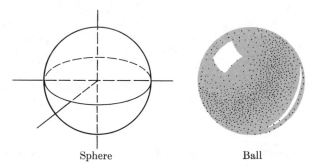

Sphere   Ball

Figure 21

The sphere is the outer shell, and the ball consists of the region inside the shell. The open ball consists of the region inside the shell excluding the shell itself. The closed ball consists of the region inside the shell *and* the shell itself.

From the geometry of the situation, it is also reasonable to expect that if $c > 0$, then $\|cA\| = c\|A\|$, i.e. if we stretch a vector A by multiplying by a positive number c, then the length stretches also by that amount. We verify this formally using our definition of the length.

Theorem 1. *Let x be a number. Then*

$$\|xA\| = |x|\,\|A\|$$

(*absolute value of x times the norm of A*).

Proof. By definition, we have

$$\|xA\|^2 = (xA) \cdot (xA),$$

which is equal to

$$x^2(A \cdot A)$$

by the properties of the scalar product. Taking the square root now yields what we want.

Let S_1 be the sphere of radius 1, centered at the origin. Let a be a number > 0. If X is a point of the sphere S_1, then aX is a point of the sphere of radius a, because

$$\|aX\| = a\|X\| = a.$$

In this manner, we get all points of the sphere of radius a. (Proof?) Thus the sphere of radius a is obtained by stretching the sphere of radius 1, through multiplication by a.

A similar remark applies to the open and closed balls of radius a, they being obtained from the open and closed balls of radius 1 through multiplication by a.

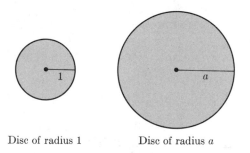

Disc of radius 1   Disc of radius a

Figure 22

We shall say that a vector E is a **unit** vector if $\|E\| = 1$. Given any vector A, let $a = \|A\|$. If $a \neq 0$, then

$$\frac{1}{a}A$$

is a unit vector, because

$$\left\| \frac{1}{a} A \right\| = \frac{1}{a} a = 1.$$

We say that two vectors A, B (neither of which is O) have the **same direction** if there is a number $c > 0$ such that $cA = B$. In view of this definition, we see that the vector

$$\frac{1}{\|A\|} A$$

is a unit vector in the direction of A (provided $A \neq O$).

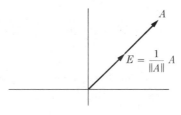

Figure 23

If E is the unit vector in the direction of A, and $\|A\| = a$, then

$$A = aE.$$

Example 4. Let $A = (1, 2, -3)$. Then $\|A\| = \sqrt{14}$. Hence the unit vector in the direction of A is the vector

$$E = \left(\frac{1}{\sqrt{14}}, \frac{2}{\sqrt{14}}, \frac{-3}{\sqrt{14}} \right).$$

Warning. There are as many unit vectors as there are directions. The three **standard unit vectors** in 3-space, namely

$$E_1 = (1, 0, 0), \qquad E_2 = (0, 1, 0), \qquad E_3 = (0, 0, 1)$$

are merely the three unit vectors in the directions of the coordinate axes.

We are also in the position to justify our definition of perpendicularity. Given A, B in the plane, the condition that

$$\|A + B\| = \|A - B\|$$

(illustrated in Fig. 24(b)) coincides with the geometric property that A should be perpendicular to B.

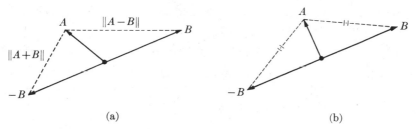

(a)   (b)

Figure 24

We shall prove:

$$\|A + B\| = \|A - B\| \text{ if and only if } A \cdot B = 0.$$

Let $\Leftrightarrow$ denote "if and only if". Then:

$$\|A + B\| = \|A - B\| \Leftrightarrow \|A + B\|^2 = \|A - B\|^2$$

$$\Leftrightarrow A^2 + 2A \cdot B + B^2 = A^2 - 2A \cdot B + B^2$$

$$\Leftrightarrow 4A \cdot B = 0$$

$$\Leftrightarrow A \cdot B = 0.$$

This proves what we wanted.

General Pythagoras Theorem. *If A and B are perpendicular, then*

$$\|A + B\|^2 = \|A\|^2 + \|B\|^2.$$

The theorem is illustrated on Fig. 25.

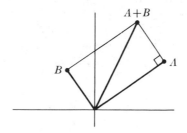

Figure 25

To prove this, we use the definitions, namely

$$\|A + B\|^2 = (A + B) \cdot (A + B) = A^2 + 2A \cdot B + B^2$$

$$= \|A\|^2 + \|B\|^2,$$

because $A \cdot B = 0$, and $A \cdot A = \|A\|^2$, $B \cdot B = \|B\|^2$ by definition.

Remark. If A is perpendicular to B, and x is any number, then A is also perpendicular to xB because

$$A \cdot xB = xA \cdot B = 0.$$

We shall now use the notion of perpendicularity to derive the notion of **projection.** Let A, B be two vectors and $B \neq O$. Let P be the point on the line through $\overrightarrow{OB}$ such that $\overrightarrow{PA}$ is perpendicular to $\overrightarrow{OB}$, as shown on Fig. 26(a).

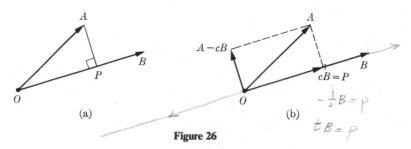

(a)   (b)

Figure 26

We can write

$$P = cB$$

for some number c. We want to find this number c explicitly in terms of A and B. The condition $\overrightarrow{PA} \perp \overrightarrow{OB}$ means that

$$A - P \text{ is perpendicular to } B,$$

and since $P = cB$ this means that

$$(A - cB) \cdot B = 0,$$

in other words,

$$A \cdot B - cB \cdot B = 0.$$

We can solve for c, and we find $A \cdot B = cB \cdot B$, so that

$$\boxed{c = \frac{A \cdot B}{B \cdot B}.}$$

Conversely, if we take this value for c, and then use distributivity, dotting $A - cB$ with B yields 0, so that $A - cB$ is perpendicular to B. Hence we have seen that there is a unique number c such that $A - cB$ is perpendicular to B, and c is given by the above formula.

We call this number c the **component** of A along B, and we call cB the **projection of A along B.**

Example 5. Suppose

$$B = E_i = (0, \ldots, 0, 1, 0, \ldots, 0)$$

is the i-th unit vector, with 1 in the i-th component and 0 in all other components.

$$\text{If } A = (a_1, \ldots, a_n), \text{ then } A \cdot E_i = a_i.$$

Thus $A \cdot E_i$ is the ordinary i-th component of A.

More generally, if B is a unit vector, not necessarily one of the E_i, then we have simply

$$c = A \cdot B,$$

because $B \cdot B = 1$ by definition of a unit vector.

Example 6. Let $A = (1, 2, -3)$ and $B = (1, 1, 2)$. Then the component of A along B is the **number**

$$c = \frac{A \cdot B}{B \cdot B} = \frac{-3}{6} = -\frac{1}{2}.$$

Hence the projection of A along B is the **vector**

$$cB = \left(-\tfrac{1}{2}, -\tfrac{1}{2}, -1\right).$$

Our construction gives an immediate geometric interpretation for the scalar product. Namely, assume $A \neq O$ and look at the angle θ between A and B (Fig. 27). Then from plane geometry we see that

$$\cos \theta = \frac{c\|B\|}{\|A\|},$$

or substituting the value for c obtained above,

$$\boxed{A \cdot B = \|A\|\,\|B\|\cos \theta} \qquad \text{and} \qquad \boxed{\cos \theta = \frac{A \cdot B}{|A|\,|B|}.}$$

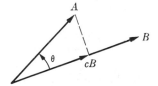

Figure 27

In some treatments of vectors, one takes the relation

$$A \cdot B = \|A\|\,\|B\|\cos \theta$$

as definition of the scalar product. This is subject to the following disadvantages, not to say objections:

(a) The four properties of the scalar product **SP 1** through **SP 4** are then by no means obvious.

(b) Even in 3-space, one has to rely on geometric intuition to obtain the cosine of the angle between A and B, and this intuition is less clear than in the plane. In higher dimensional space, it fails even more.

(c) It is extremely hard to work with such a definition to obtain further properties of the scalar product.

Thus we prefer to lay obvious algebraic foundations, and then recover very simply all the properties. We used plane geometry to see the expression $A \cdot B = |A| |B| \cos \theta$. If you want to see a proof that

$$\frac{|A \cdot B|}{|A| |B|} \leqq 1$$

based only on the algebraic definitions, and applicable to wider contexts, see Theorem 1 in Appendix 1.

Example 7. Let $A = (1, 2, -3)$ and $B = (2, 1, 5)$. Find the cosine of the angle θ between A and B.

By definition,

$$\cos \theta = \frac{A \cdot B}{\|A\| \|B\|} = \frac{2 + 2 - 15}{\sqrt{14} \sqrt{30}} = \frac{-11}{\sqrt{420}}.$$

Example 8. Find the cosine of the angle between the two located vectors $\overrightarrow{PQ}$ and $\overrightarrow{PR}$ where

$$P = (1, 2, -3), \qquad Q = (-2, 1, 5), \qquad R = (1, 1, -4).$$

The picture looks like this:

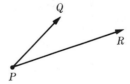

Figure 28

We let

$$A = Q - P = (-3, -1, 8) \qquad \text{and} \qquad B = R - P = (0, -1, -1).$$

Then the angle between $\overrightarrow{PQ}$ and $\overrightarrow{PR}$ is the same as that between A and B. Hence its cosine is equal to

$$\cos \theta = \frac{A \cdot B}{\|A\| \|B\|} = \frac{0 + 1 - 8}{\sqrt{74} \sqrt{2}} = \frac{-7}{\sqrt{74} \sqrt{2}}.$$

EXERCISES

1. Find the norm of the vector A in the following cases.
   (a) $A = (2, -1), B = (-1, 1)$ (b) $A = (-1, 3), B = (0, 4)$
   (c) $A = (2, -1, 5), B = (-1, 1, 1)$ (d) $A = (-1, -2, 3), B = (-1, 3, -4)$
   (e) $A = (\pi, 3, -1), B = (2\pi, -3, 7)$ (f) $A = (15, -2, 4), B = (\pi, 3, -1)$

2. Find the norm of vector B in the above cases.

3. Find the projection of A along B in the above cases.

4. Find the projection of B along A in the above cases.

5. Find the cosine between the following vectors A and B.
   (a) $A = (1, -2)$ and $B = (5, 3)$
   (b) $A = (-3, 4)$ and $B = (2, -1)$
   (c) $A = (1, -2, 3)$ and $B = (-3, 1, 5)$
   (d) $A = (-2, 1, 4)$ and $B = (-1, -1, 3)$
   (e) $A = (-1, 1, 0)$ and $B = (2, 1, -1)$

6. Determine the cosine of the angles of the triangle whose vertices are
   (a) $(2, -1, 1), (1, -3, -5), (3, -4, -4)$.
   (b) $(3, 1, 1), (-1, 2, 1), (2, -2, 5)$.

7. Let $A_1, \ldots, A_r$ be non-zero vectors which are mutually perpendicular, in other words $A_i \cdot A_j = 0$ if $i \neq j$. Let $c_1, \ldots, c_r$ be numbers such that

$$c_1 A_1 + \cdots + c_r A_r = 0.$$

   Show that all $c_i = 0$.

8. For any vectors A, B, prove the following relations:
   (a) $\|A + B\|^2 + \|A - B\|^2 = 2\|A\|^2 + 2\|B\|^2$.
   (b) $\|A + B\|^2 = \|A\|^2 + \|B\|^2 + 2A \cdot B$.
   (c) $\|A + B\|^2 - \|A - B\|^2 = 4A \cdot B$.
   Interpret (a) as a relation between lengths of the sides of a parallelogram and the length of the diagonals.

9. Show that if θ is the angle between A and B, then

$$\|A - B\|^2 = \|A\|^2 + \|B\|^2 - 2\|A\| \, \|B\| \cos \theta.$$

10. Let A, B, C be three non-zero vectors. If $A \cdot B = A \cdot C$, show by an example that we do not necessarily have $B = C$.

§5. PARAMETRIC LINES

We define the **parametric equation** or **parametric representation** of a straight line passing through a point P in the direction of a vector $A \neq O$ to be

$$X = P + tA,$$

where t runs through all numbers (Fig. 29).

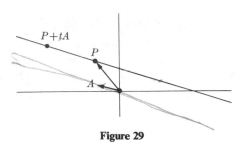

Figure 29

When we give such a parametric representation, we may think of a bug 业,是出 starting from a point P at time $t = 0$, and moving in the direction of A. At time t, the bug is at the position $P + tA$. Thus we may interpret physically the parametric representation as a description of motion, in which A is interpreted as the velocity of the bug. At a given time t, the bug is at the point

$$X(t) = P + tA,$$

which is called the **position** of the bug at time t.

This parametric representation is also useful to describe the set of points lying on the line segment between two given points. Let P, Q be two points. Then the **segment** between P and Q consists of all the points

$$S(t) = P + t(Q - P) \qquad \text{with} \qquad 0 \leq t \leq 1.$$

Indeed, $\overrightarrow{O(Q - P)}$ is a vector having the same direction as $\overrightarrow{PQ}$, as shown on Fig. 30.

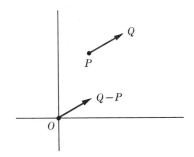

Figure 30

When $t = 0$, we have $S(0) = P$, so at time $t = 0$ the bug is at P. When $t = 1$, we have

$$S(1) = P + (Q - P) = Q,$$

so when $t = 1$ the bug is at Q. As t goes from 0 to 1, the bug goes from P to Q.

Example 1. Let $P = (1, -3, 4)$ and $Q = (5, 1, -2)$. Find the coordinates of the point which lies one third of the distance from P to Q.

Let $S(t)$ as above be the parametric representation of the segment from P to Q. The desired point is $S(1/3)$, that is:

$$S\left(\frac{1}{3}\right) = P + \frac{1}{3}(Q - P) = (1, -3, 4) + \frac{1}{3}(4, 4, -6)$$

$$= \left(\frac{7}{3}, \frac{-5}{3}, 2\right).$$

Warning. The desired point in the above example is *not* given by

$$\frac{P + Q}{3}.$$

Example 2. Find a parametric representation for the line passing through the two points $P = (1, -3, 1)$ and $Q = (-2, 4, 5)$.
We first have to find a vector in the direction of the line. We let

$$A = P - Q,$$

so

$$A = (3, -7, -4).$$

The parametric equation of the line is therefore

$$X(t) - P + tA - (1, -3, 1) + t(3, -7, -4).$$

Remark. It would be equally correct to give a parametric representation of the line as

$$Y(t) = P + tB \qquad \text{where} \qquad B = Q - P.$$

Interpreted in terms of the moving bug, however, one parametrization gives the position of a bug moving in one direction along the line, starting from P at time $t = 0$, while the other parametrization gives the position of another bug moving in the **opposite** direction along the line, also starting from P at time $t = 0$.

We shall now discuss the relation between a parametric representation and the ordinary equation of a line in the plane.
Suppose that we work in the plane, and write the coordinates of a point X as (x, y). Let $P = (p, q)$ and $A = (a, b)$. Then in terms of the coordinates, we can write

$$x = p + ta, \qquad y = q + tb.$$

We can then eliminate t and obtain the usual equation relating x and y.

Example 3. Let $P = (2, 1)$ and $A = (-1, 5)$. Then the parametric equation of the line through P in the direction of A gives us

(*) $x = 2 - t, \qquad y = 1 + 5t.$

Multiplying the first equation by 5 and adding yields

(**) $5x + y = 11,$

which is the familiar equation of a line.

This elimination of t shows that every pair (x, y) which satisfies the parametric equation (*) for some value of t also satisfies equation (**). Conversely, suppose we have a pair of numbers (x, y) satisfying (**). Let $t = 2 - x$. Then

$$y = 11 - 5x = 11 - 5(2 - t) = 1 + 5t.$$

Hence there exists some value of t which satisfies equation (*). Thus we have proved that the pairs (x, y) which are solutions of (**) are exactly the same pairs of numbers as those obtained by giving arbitrary values for t in (*). Thus the straight line can be described parametrically as in (*) or in terms of its usual equation (**). Starting with the ordinary equation

$$5x + y = 11,$$

we let $t = 2 - x$ in order to recover the specific parametrization of (*).

When we parametrize a straight line in the form

$$X = P + tA,$$

we have of course infinitely many choices for P on the line, and also infinitely many choices for A, differing by a scalar multiple. We can always select at least one. Namely, given an equation

$$ax + by = c$$

with numbers a, b, c, suppose that $a \neq 0$. We use y as parameter, and let

$$y = t.$$

Then we can solve for x, namely

$$x = \frac{c}{a} - \frac{b}{a} t.$$

Let $P = (c/a, 0)$ and $A = (-b/a, 1)$. We see that an arbitrary point (x, y) satisfying the equation

$$ax + by = c$$

can be expressed parametrically, namely

$$(x, y) = P + tA.$$

In higher dimension, starting with a parametric equation

$$X = P + tA,$$

we cannot eliminate t, and thus the parametric equation is the only one available to describe a straight line.

EXERCISES

1. Find a parametric representation for the line passing through the following pairs of points.
   (a) $P_1 = (1, 3, -1)$ and $P_2 = (-4, 1, 2)$
   (b) $P_1 = (-1, 5, 3)$ and $P_2 = (-2, 4, 7)$

Find a parametric equation for the line passing through the following points.

2. $(1, 1, -1)$ and $(-2, 1, 3)$ 3. $(-1, 5, 2)$ and $(3, -4, 1)$

4. Let $P = (1, 3, -1)$ and $Q = (-4, 5, 2)$. Determine the coordinates of the following points:
   (a) The midpoint of the line segment between P and Q.
   (b) The two points on this line segment lying one-third and two-thirds of the way from P to Q.
   (c) The point lying one-fifth of the way from P to Q.
   (d) The point lying two-fifths of the way from P to Q.

5. If P, Q are two arbitrary points in n-space, give the general formula for the midpoint of the line segment between P and Q.

§6. PLANES

We can describe planes in 3-space by an equation analogous to the single equation of the line. We proceed as follows.

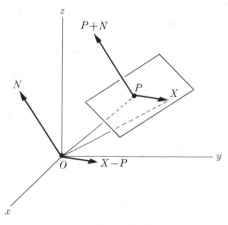

Figure 31

Let P be a point in 3-space and consider a located vector $\overrightarrow{ON}$. We define the **plane** passing through P perpendicular to $\overrightarrow{ON}$ to be the collection of all points X such that the located vector $\overrightarrow{PX}$ is perpendicular to $\overrightarrow{ON}$. According to our definitions, this amounts to the condition

$$(X - P) \cdot N = 0,$$

which can also be written as

$$\boxed{X \cdot N = P \cdot N.}$$

We shall also say that this plane is the one perpendicular to N, and consists of all vectors X such that $X - P$ is perpendicular to N. We have drawn a typical situation in 3-space in Fig. 31.

Instead of saying that N is **perpendicular** to the plane, one also says that N is **normal** to the plane.

Let t be a number $\neq 0$. Then the set of points X such that

$$(X - P) \cdot N = 0$$

coincides with the set of points X such that

$$(X - P) \cdot tN = 0.$$

Thus we may say that our plane is the plane passing through P and perpendicular to the **line** in the direction of N. To find the equation of the plane, we could use any vector tN (with $t \neq 0$) instead of N.

Example 1. Let

$$P = (2, 1, -1) \qquad \text{and} \qquad N = (-1, 1, 3).$$

Let $X = (x, y, z)$. Then

$$X \cdot N = (-1)x + y + 3z.$$

Therefore the equation of the plane passing through P and perpendicular to N is

$$-x + y + 3z = -2 + 1 - 3$$

or

$$-x + y + 3z = -4.$$

Observe that in 2-space, with $X = (x, y)$, the formulas lead to the equation of the line in the ordinary sense.

Example 2. The equation of the line in the (x, y)-plane, passing through $(4, -3)$ and perpendicular to $(-5, 2)$ is

$$-5x + 2y = -20 - 6 = -26.$$

We are now in position to interpret the coefficients $(-5, 2)$ of x and y in this equation. They give rise to a vector perpendicular to the line. **In any** equation

$$ax + by = c$$

the vector (a, b) **is perpendicular to the line determined by the equation**. Similarly, in 3-space, the vector (a, b, c) is perpendicular to the plane determined by the equation

$$ax + by + cz = d.$$

Example 3. The plane determined by the equation

$$2x - y + 3z = 5$$

is perpendicular to the vector $(2, -1, 3)$. If we want to find a point in that plane, we of course have many choices. We can give arbitrary values to x and y, and then solve for z. To get a concrete point, let $x = 1, y = 1$. Then we solve for z, namely

$$3z = 5 - 2 + 1 = 4,$$

so that $z = \frac{4}{3}$. Thus

$$\left(1, 1, \tfrac{4}{3}\right)$$

is a point in the plane.

In n-space, the equation $X \cdot N = P \cdot N$ is said to be the equation of a **hyperplane**. For example,

$$3x - y + z + 2w = 5$$

is the equation of a hyperplane in 4-space, perpendicular to $(3, -1, 1, 2)$.

Two vectors A, B are said to be parallel if there exists a number $c \neq 0$ such that $cA = B$. Two lines are said to be **parallel** if, given two distinct points P_1, Q_1 on the first line and P_2, Q_2 on the second, the vectors

$$P_1 - Q_1$$

and

$$P_2 - Q_2$$

are parallel.

Two planes are said to be **parallel** (in 3-space) if their normal vectors are parallel. They are said to be **perpendicular** if their normal vectors are perpendicular. The **angle** between two planes is defined to be the angle between their normal vectors.

Example 4. Find the cosine of the angle between the planes.

$$2x - y + z = 0,$$
$$x + 2y - z = 1.$$

This cosine is the cosine of the angle between the vectors.

$$A = (2, -1, 1) \quad \text{and} \quad B = (1, 2, -1).$$

It is therefore equal to

$$\frac{A \cdot B}{\|A\| \, \|B\|} = -\frac{1}{6}.$$

Example 5. Let

$$Q = (1, 1, 1) \quad \text{and} \quad P = (1, -1, 2).$$

Let

$$N = (1, 2, 3).$$

Find the point of intersection of the line through P in the direction of N, and the plane through Q perpendicular to N.

The parametric equation of the line through P in the direction of N is

(1) $$X = P + tN.$$

The equation of the plane through Q perpendicular to N is

(2) $$(X - Q) \cdot N = 0.$$

We visualize the line and plane as follows:

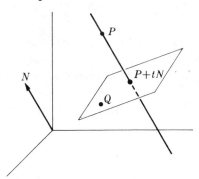

Figure 32

We must find the value of t such that the vector X in (1) also satisfies (2), that is

$$(P + tN - Q) \cdot N = 0,$$

or after using the rules of the dot product,

$$(P - Q) \cdot N + tN \cdot N = 0.$$

Solving for t yields

$$t = \frac{(Q - P) \cdot N}{N \cdot N} = \frac{1}{14}.$$

Thus the desired point of intersection is

$$P + tN = (1, -1, 2) + \tfrac{1}{14}(1, 2, 3) = \left(\tfrac{15}{14}, -\tfrac{12}{14}, \tfrac{31}{14}\right).$$

Example 6. Find the equation of the plane passing through the three points

$$P_1 = (1, 2, -1), \qquad P_2 = (-1, 1, 4), \qquad P_3 = (1, 3, -2).$$

We visualize schematically the three points as follows:

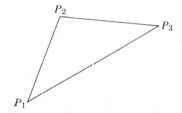

Figure 33

Then we find a vector N perpendicular to $\overrightarrow{P_1P_2}$ and $\overrightarrow{P_1P_3}$, or in other words, perpendicular to $P_2 - P_1$ and $P_3 - P_1$. We have

$$P_2 - P_1 = (-2, -1, +5),$$
$$P_3 - P_1 = (0, 1, -1).$$

Let $N = (a, b, c)$. We must solve

$$N \cdot (P_2 - P_1) = 0 \qquad \text{and} \qquad N \cdot (P_3 - P_1) = 0,$$

in other words,

$$-2a - b + 5c = 0,$$
$$b - c = 0.$$

We take $b = c = 1$ and solve for a, getting $a = 2$. Then

$$N = (2, 1, 1)$$

satisfies our requirements. The plane perpendicular to N, passing through P_1 is the desired plane. Its equation is therefore $X \cdot N = P_1 \cdot N$, that is

$$2x + y + z = 2 + 2 - 1 = 3.$$

EXERCISES

1. Show that the lines $2x + 3y = 1$ and $5x - 5y = 7$ are not perpendicular.

2. Let $y = mx + b$ and $y = m'x + c$ be the equations of two lines in the plane. Write down vectors perpendicular to these lines. Show that these vectors are perpendicular to each other if and only if $mm' = -1$.

Find the equation of the line in 2-space, perpendicular to N and passing through P, for the following values of N and P.

3. $N = (1, -1), P = (-5, 3)$ 4. $N = (-5, 4), P = (3, 2)$

5. Show that the lines

$$3x - 5y = 1, \qquad 2x + 3y = 5 \qquad \bullet$$

are not perpendicular.

6. Which of the following pairs of lines are perpendicular?
   (a) $3x - 5y = 1$ and $2x + y = 2$
   (b) $2x + 7y = 1$ and $x - y = 5$
   (c) $3x - 5y = 1$ and $5x + 3y = 7$
   (d) $-x + y = 2$ and $x + y = 9$

7. Find the equation of the plane perpendicular to the given vector N and passing through the given point P.
   (a) $N = (1, -1, 3), P = (4, 2, -1)$
   (b) $N = (-3, -2, 4), P = (2, \pi, -5)$
   (c) $N = (-1, 0, 5), P = (2, 3, 7)$

8. Find the equation of the plane passing through the following three points.
   (a) $(2, 1, 1), (3, -1, 1), (4, 1, -1)$
   (b) $(-2, 3, -1), (2, 2, 3), (-4, -1, 1)$
   (c) $(-5, -1, 2), (1, 2, -1), (3, -1, 2)$

9. Find a vector perpendicular to $(1, 2, -3)$ and $(2, -1, 3)$, and another vector perpendicular to $(-1, 3, 2)$ and $(2, 1, 1)$.

10. Let P be the point $(1, 2, 3, 4)$ and Q the point $(4, 3, 2, 1)$. Let A be the vector $(1, 1, 1, 1)$. Let L be the line passing through P and parallel to A.
    (a) Given a point X on the line L, compute the distance between Q and X (as a function of the parameter t).
    (b) Show that there is precisely one point X_0 on the line such that this distance achieves a minimum, and that this minimum is $2\sqrt{5}$.
    (c) Show that $X_0 - Q$ is perpendicular to the line.

11. Let P be the point $(1, -1, 3, 1)$ and Q the point $(1, 1, -1, 2)$. Let A be the vector $(1, -3, 2, 1)$. Solve the same questions as in the preceding problem, except that in this case the minimum distance is $\sqrt{146/15}$.

12. Find a vector parallel to the line of intersection of the two planes

$$2x - y + z = 1, \qquad 3x + y + z = 2.$$

13. Same question for the planes,

$$2x + y + 5z = 2, \qquad 3x - 2y + z = 3.$$

14. Find a parametric equation for the line of intersection of the planes of Exercises 12 and 13.

15. Find the cosine of the angle between the following planes:

(a) $x + y + z = 1$
$x - y - z = 5$

(b) $2x + 3y - z = 2$
$x - y + z = 1$

(c) $x + 2y - z = 1$
$-x + 3y + z = 2$

(d) $2x + y + z = 3$
$-x - y + z = \pi$

16. (a) Let $P_{\cdot} = (1, 3, 5)$ and $A = (-2, 1, 1)$. Find the intersection of the line through P
be careful. in the direction of A, and the plane $2x + 3y - z = 1$.

(b) Let $P = (1, 2, -1)$. Find the point of intersection of the plane

$$3x - 4y + z = 2,$$

with the line through P, perpendicular to that plane.

17. Let $Q = (1, -1, 2)$, $P = (1, 3, -2)$, and $N = (1, 2, 2)$. Find the point of the intersection of the line through P in the direction of N, and the plane through Q perpendicular to N.

18. Let P, Q be two points and N a vector in 3-space. Let P' be the point of intersection of the line through P, in the direction of N, and the plane through Q, perpendicular to N. We define the **distance** from P to that plane to be the distance between P and P'. Find the distance when

$$P = (1, 3, 5), \qquad Q = (-1, 1, 7), \qquad N = (-1, 1, -1).$$

19. In the notation of Exercise 18, show that the general formula for the distance is given by

$$\frac{|(Q - P) \cdot N|}{\|N\|}.$$

in the plane

$P' = P + tN$

$(P' - Q) \cdot N = 0$

$\|P' - P\| = t \|N\|$

20. Find the distance between the indicated point and plane.

(a) $(1, 1, 2)$ and $3x + y - 5z = 2$

(b) $(-1, 3, 2)$ and $2x - 4y + z = 1$

$(P + tN - Q) \cdot N = 0$

$(P - Q) \cdot N + tN \cdot N = 0$

§7. THE CROSS PRODUCT

$tN \cdot N = (P - Q) \cdot N$

$tN \cdot N = (Q - P) \cdot N$

$t = \dfrac{(Q - P)}{N}$

This section will not be used until either Chapter X, on surface integrals, or Chapter XVII, on the change of variables formula. Consequently, it can be omitted until then. We include it here because as a matter of taste, some people like to see immediately how to construct a perpendicular vector to a plane by means of the cross product. Also this section is completely elementary, not depending on anything much, and a reader might want to use it independently.

$\|P' - P\| = t\|N\| = \dfrac{(Q - P) \cdot N}{N \cdot N}$

$\|P' - P\| = \dfrac{(Q - P) \cdot N}{\|N\|}$

Hence we do not want to make it appear as if it is tied up with the more elaborate material of the later chapters.

Let $A = (a_1, a_2, a_3)$ and $B = (b_1, b_2, b_3)$ be two vectors in 3-space. We define their **cross product**

$$A \times B = (a_2b_3 - a_3b_2, a_3b_1 - a_1b_3, a_1b_2 - a_2b_1).$$

For instance, if $A = (2, 3, -1)$ and $B = (-1, 1, 5)$, then

$$A \times B = (16, -9, 5).$$

Remark. At first sight, the pattern of indices for the components of $A \times B$ seems rather random and hard to remember. It is possible to give a more easily remembered form to this cross product by using the expansion rule for a determinant according to the pattern of Chapter XV, §2. Indeed, let

$$E_1 = (1, 0, 0), \qquad E_2 = (0, 1, 0), \qquad E_3 = (0, 0, 1).$$

If we follow the above-mentioned pattern, we may write symbolically the cross product in the form of a determinant

$$A \times B = \begin{vmatrix} E_1 & E_2 & E_3 \\ a_1 & a_2 & a_3 \\ b_1 & b_2 & b_3 \end{vmatrix}.$$

The right-hand side, by definition, is supposed to be:

$$E_1(a_2b_3 - a_3b_2) - E_2(a_1b_3 - a_3b_1) + E_3(a_1b_2 - a_2b_2),$$

which gives precisely the expression for the cross product $A \times B$.

We leave the following assertions as exercises:

CP 1. $A \times B = -(B \times A)$.

CP 2. $A \times (B + C) = (A \times B) + (A \times C)$, *and*

$$(B + C) \times A = B \times A + C \times A.$$

CP 3. *For any number a, we have*

$$(aA) \times B = a(A \times B) = A \times (aB).$$

CP 4. $(A \times B) \times C = (A \cdot C)B - (B \cdot C)A$.

CP 5. $A \times B$ *is perpendicular to both A and B.*

As an example, we carry out this computation. We have

$$A \cdot (A \times B) = a_1(a_2b_3 - a_3b_2) + a_2(a_3b_1 - a_1b_3) + a_3(a_1b_2 - a_2b_1)$$
$$= 0$$

because all terms cancel. Similarly for $B \cdot (A \times B)$. This perpendicularity may be drawn as follows.

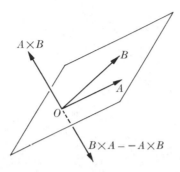

Figure 34

The vector $A \times B$ is perpendicular to the plane spanned by A and B. So is $B \times A$, but $B \times A$ points in the opposite direction.

Finally, as a last property, we have

CP 6. $(A \times B)^2 = (A \cdot A)(B \cdot B) - (A \cdot B)^2.$

Again, this can be verified by a computation on the coordinates. Namely, we have

$$(A \times B) \cdot (A \times B) = (a_2 b_3 - a_3 b_2)^2 + (a_3 b_1 - a_1 b_3)^2 + (a_1 b_2 - a_2 b_1)^2,$$

$$(A \cdot A)(B \cdot B) - (A \cdot B)^2$$
$$= (a_1^2 + a_2^2 + a_3^2)(b_1^2 + b_2^2 + b_3^2) - (a_1 b_1 + a_2 b_2 + a_3 b_3)^2.$$

Expanding everything out, we find that **CP 6** drops out.

From our interpretation of the dot product, and the definition of the norm, we can rewrite **CP 6** in the form

$$\|A \times B\|^2 = \|A\|^2 \|B\|^2 - \|A\|^2 \|B\|^2 \cos^2 \theta,$$

where θ is the angle between A and B. Hence we obtain

$$\|A \times B\|^2 = \|A\|^2 \|B\|^2 \sin^2 \theta$$

or

$$\boxed{\|A \times B\| = \|A\| \, \|B\| \, |\sin \theta|.}$$

This is analogous to the formula which gave us the absolute value of $A \cdot B$.

This formula can be used to make another interpretation of the cross product. Indeed, we see that $\|A \times B\|$ is the area of the parallelogram spanned by A and B, as shown on Fig. 35.

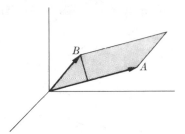

Figure 35

If we consider the plane containing the located vectors $\overrightarrow{OA}$ and $\overrightarrow{OB}$, then the picture looks like that in Fig. 36, and our assertion amounts simply to the statement that the area of a parallelogram is equal to the base times the altitude.

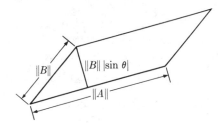

Figure 36

Example. Let $A = (3, 1, 4)$ and $B = (-2, 5, 3)$. Then the area of the parallelogram spanned by A and B is easily computed. First we get the cross product,

$$A \times B = (3 - 20, -8 - 9, 15 + 2) = (-17, -17, 17).$$

The area of the parallelogram spanned by A and B is therefore equal to the norm of this vector, and that is

$$\|A \times B\| = \sqrt{3 \cdot 17^2} = 17\sqrt{3}\,.$$

These considerations will be used especially in Chapter X, when we discuss surface area, and in Chapter XVII, when we deal with the change of variables formula.

EXERCISES

Find $A \times B$ for the following vectors.

1. $A = (1, -1, 1)$ and $B = (-2, 3, 1)$

2. $A = (-1, 1, 2)$ and $B = (1, 0, -1)$

3. $A = (1, 1, -3)$ and $B = (-1, -2, -3)$

4. Find $A \times A$ and $B \times B$, in Exercises 1 through 3.

5. Let $E_1 = (1, 0, 0)$, $E_2 = (0, 1, 0)$, and $E_3 = (0, 0, 1)$. Find $E_1 \times E_2$, $E_2 \times E_3$, $E_3 \times E_1$.

6. Show that for any vector A in 3-space we have $A \times A = O$.

7. Compute $E_1 \times (E_1 \times E_2)$ and $(E_1 \times E_1) \times E_2$. Are these vectors equal to each other?

8. Carry out the proofs of **CP 1** through **CP 4**.

9. Compute the area of the parallelogram spanned by the following vectors.
   (a) $A = (3, -2, 4)$ and $B = (5, 1, 1)$
   (b) $A = (3, 1, 2)$ and $B = (-1, 2, 4)$
   (c) $A = (4, -2, 5)$ and $B = (3, 1, -1)$
   (d) $A = (-2, 1, 3)$ and $B = (2, -3, 4)$

Do the next exercises after you have read Chapter II, §1.

10. Using coordinates, prove that if $X(t)$ and $Y(t)$ are two differentiable curves (defined for the same values of t), then

$$\frac{d[X(t) \times Y(t)]}{dt} = X(t) \times \frac{dY(t)}{dt} + \frac{dX(t)}{dt} \times Y(t).$$

11. Show (using only Exercise 10) that

$$\frac{d}{dt}[X(t) \times X'(t)] = X(t) \times X''(t).$$

12. Let $Y(t) = X(t) \cdot (X'(t) \times X''(t))$. Show that

$$Y'(t) = X(t) \cdot (X'(t) \times X'''(t)).$$

CHAPTER II

Differentiation of Vectors

§1. DERIVATIVE

Consider a bug moving along some curve in 3-dimensional space. The position of the bug at time t is given by the three coordinates

$$(x(t), y(t), z(t)),$$

which depend on t. We abbreviate these by $X(t)$. For instance, the position of a bug moving along a straight line was seen in the preceding chapter to be given by

$$X(t) = P + tA,$$

where P is the starting point, and A gives the direction of the bug. However, we can give examples when the bug does not move on a straight line. First we look at an example in the plane.

Example 1. Let $X(\theta) = (\cos \theta, \sin \theta)$. Then the bug moves around a circle of radius 1 in counterclockwise direction.

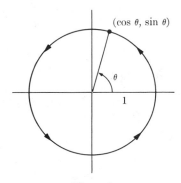

Figure 1

41

Here we used θ as the variable, corresponding to the angle as shown on the figure. Let ω be the angular speed of the bug, and assume ω constant. Thus $d\theta/dt = \omega$ and

$$\theta = \omega t + \text{a constant}.$$

For simplicity, assume that the constant is 0. Then we can write the position of the bug as

$$X(\theta) = X(\omega t) = (\cos \omega t, \sin \omega t).$$

If the angular speed is 1, then we have simply the representation

$$X(t) = (\cos t, \sin t).$$

Example 2. If the bug moves around a circle of radius 2 with angular speed equal to 1, then its position at time t is given by

$$X(t) = (2 \cos t, 2 \sin t).$$

More generally, if the bug moves around a circle of radius r, then the position is given by

$$X(t) = (r \cos t, r \sin t).$$

In these examples, we assume of course that at time $t = 0$ the bug starts at the point $(r, 0)$, that is

$$X(0) = (r, 0),$$

where r is the radius of the circle.

Example 3. Suppose the position of the bug is given in 3-space by

$$X(t) = (\cos t, \sin t, t).$$

Then the bug moves along a spiral. Its coordinates are given as functions of t by

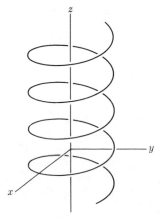

$$x(t) = \cos t,$$
$$y(t) = \sin t,$$
$$z(t) = t.$$

The position at time t is obtained by plugging in the special value of t. Thus:

$$X(\pi) = (\cos \pi, \sin \pi, \pi) = (-1, 0, \pi)$$
$$X(1) = (\cos 1, \sin 1, 1).$$

We may now give the definition of a curve in general.

Figure 2

Let I be an interval. A parametrized **curve** (defined on this interval) is an association which to each point of I associates a vector. If X denotes a curve defined on I, and t is a point of I, then $X(t)$ denotes the vector associated to t by X. We often write the association $t \mapsto X(t)$ as an arrow

$$X : I \to \mathbf{R}^n.$$

We call $X(t)$ the **position vector** at time t. It can be written in terms of coordinates,

$$X(t) = \left(x_1(t), \ldots, x_n(t) \right),$$

each $x_i(t)$ being a function of t. We say that this curve is **differentiable** if each function $x_i(t)$ is a differentiable function of t.

Remark. We take the intervals of definition for our curves to be open, closed, or also half-open or half-closed. When we define the derivative of a curve, it is understood that the interval of definition contains more than one point. In that case, at an end point the usual limit of

$$\frac{f(a + h) \quad f(a)}{h}$$

is taken for those h such that the quotient makes sense, i.e. $a + h$ lies in the interval. If a is a left end point, the quotient is considered only for $h > 0$. If a is a right end point, the quotient is considered only for $h < 0$. Then the usual rules for differentiation of functions are true in this greater generality, and thus Rules 1 through 4 below, and the chain rule of §2 remain true also. [An example of a statement which is not always true for curves defined over closed intervals is given in Exercise 11(b).]

Let us try to differentiate vectors. We consider the Newton quotient

$$\frac{X(t + h) - X(t)}{h}.$$

Its numerator looks like Fig. 3.

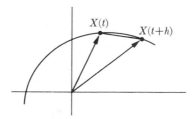

Figure 3

As h approaches 0, we see geometrically that

$$\frac{X(t + h) - X(t)}{h}$$

should approach a vector pointing in the direction of the curve. We can write
the Newton quotient in terms of coordinates,

$$\frac{X(t + h) - X(t)}{h} = \left(\frac{x_1(t + h) - x_1(t)}{h}, \ldots, \frac{x_n(t + h) - x_n(t)}{h} \right)$$

and see that each component is a Newton quotient for the corresponding
coordinate. We assume that each $x_i(t)$ is differentiable. Then each quotient

$$\frac{x_i(t + h) - x_i(t)}{h}$$

approaches the derivative dx_i/dt. For this reason, we define the derivative
dX/dt to be

$$\frac{dX}{dt} = \left(\frac{dx_1}{dt}, \ldots, \frac{dx_n}{dt} \right).$$

In fact, we could also say that the vector

$$\left(\frac{dx_1}{dt}, \ldots, \frac{dx_n}{dt} \right)$$

is the limit of the Newton quotient

$$\frac{X(t + h) - X(t)}{h}$$

as h approaches 0. Indeed, as h approaches 0, each component

$$\frac{x_i(t + h) - x_i(t)}{h}$$

approaches dx_i/dt. Hence the Newton quotient approaches the vector

$$\left(\frac{dx_1}{dt}, \ldots, \frac{dx_n}{dt} \right).$$

Example 4. If $X(t) = (\cos t, \sin t, t)$ then

$$\frac{dX}{dt} = (-\sin t, \cos t, 1).$$

Physicists often denote dX/dt by $\dot{X}$; thus in the previous example, we could also
write

$$\dot{X}(t) = (-\sin t, \cos t, 1) = X'(t).$$

We define the **velocity vector** of the curve at time t to be the vector $X'(t)$.

Example 5. When $X(t) = (\cos t, \sin t, t)$, then

$$X'(t) = (-\sin t, \cos t, 1);$$

the velocity vector at $t = \pi$ is

$$X'(\pi) = (0, -1, 1),$$

and for $t = \pi/4$ we get

$$X'(\pi/4) = (-1/\sqrt{2}, 1/\sqrt{2}, 1).$$

The velocity vector is located at the origin, but when we translate it to the point $X(t)$, then we visualize it as tangent to the curve, as in the next figure.

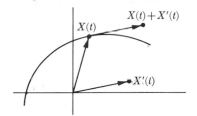

Figure 4

We define the **tangent line** to a curve X at time t to be the line passing through $X(t)$ in the direction of $X'(t)$, provided that $X'(t) \neq O$. Otherwise, we don't define a tangent line. We have therefore given two interpretations for $X'(t)$:

> $X'(t)$ is the velocity at time t;
>
> $X'(t)$ is parallel to a tangent vector at time t.

By abuse of language, we sometimes call $X'(t)$ a tangent vector, although strictly speaking, we should refer to the located vector $X(t)(X(t) + X'(t))$ as the tangent vector. However, to write down this located vector each time is cumbersome.

Example 6. Find a parametric equation of the tangent line to the curve $X(t) = (\sin t, \cos t)$ at $t = \pi/3$.

We have $X'(t) = (\cos t, -\sin t)$, so that at $t = \sqrt{3}$ we get

$$X'\left(\frac{\pi}{3}\right) = \left(\frac{1}{2}, -\frac{\sqrt{3}}{2}\right) \quad \text{and} \quad X\left(\frac{\pi}{3}\right) = \left(\frac{\sqrt{3}}{2}, \frac{1}{2}\right).$$

Let $P = X(\pi/3)$ and $A = X'(\pi/3)$. Then a parametric equation of the tangent line at the required point is

$$L(t) = P + tA = \left(\frac{\sqrt{3}}{2}, \frac{1}{2} \right) + \left(\frac{1}{2}, -\frac{\sqrt{3}}{2} \right) t.$$

(We use another letter L because X is already occupied.) In terms of the coordinates $L(t) = (x(t), y(t))$, we can write the tangent line as

$$x(t) = \frac{\sqrt{3}}{2} + \frac{1}{2} t,$$

$$y(t) = \frac{1}{2} - \frac{\sqrt{3}}{2} t.$$

Example 7. Find the equation of the plane perpendicular to the spiral

$$X(t) = (\cos t, \sin t, t)$$

when $t = \pi/3$.

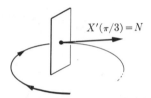

$$X'(\pi/3) = N$$

Figure 5

Let the given point be

$$P = X\left(\frac{\pi}{3} \right) = \left(\cos \frac{\pi}{3}, \sin \frac{\pi}{3}, \frac{\pi}{3} \right),$$

so that more simply,

$$P = \left(\frac{1}{2}, \frac{\sqrt{3}}{2}, \frac{\pi}{3} \right).$$

We must then find a vector N perpendicular to the plane at the given point P. We have $X'(t) = (-\sin t, \cos t, 1)$, so

$$X'\left(\frac{\pi}{3} \right) = \left(-\frac{\sqrt{3}}{2}, \frac{1}{2}, 1 \right) = N.$$

The equation of the plane through P perpendicular to N is $X \cdot N = P \cdot N$, so the equation of the desired plane is

$$-\frac{\sqrt{3}}{2} x + \frac{1}{2} y + z = -\frac{\sqrt{3}}{4} + \frac{\sqrt{3}}{4} + \frac{\pi}{3}$$

$$= \frac{\pi}{3}.$$

We define the **speed** of the curve $X(t)$ to be the norm of the velocity vector. If we denote the speed by $v(t)$, then by definition we have

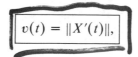

$$v(t) = \|X'(t)\|,$$

and thus

$$v(t)^2 = X'(t)^2 = X'(t) \cdot X'(t).$$

We can also omit the t from the notation, and write

$$v^2 = X' \cdot X' = X'^2.$$

Example 8. The speed of the bug moving on the circle

$$X(t) = (\cos t, \sin t)$$

is the norm of the velocity $X'(t) = (-\sin t, \cos t)$, and so is

$$v(t) = \sqrt{(-\sin t)^2 + (\cos^2 t)} = 1.$$

Example 9. The speed of the bug moving on the spiral

$$X(t) = (\cos t, \sin t, t)$$

is the norm of the velocity $X'(t) = (-\sin t, \cos t, 1)$, and so is

$$v(t) = \sqrt{(-\sin t)^2 + (\cos^2 t) + 1}$$
$$= \sqrt{2}.$$

We define the **acceleration vector** to be the derivative

$$\frac{dX'(t)}{dt} = X''(t),$$

provided of course that X' is differentiable. We shall also denote the acceleration vector by $X''(t)$ as above.

We define the **acceleration scalar** to be the norm of the acceleration vector, and denote it by $a(t)$.

Example 10. Let $X(t) = (\cos t, \sin t, t)$. We find that

$$X''(t) = (-\cos t, -\sin t, 0).$$

Therefore $\|X''(t)\| = 1$ and we see that the spiral has a constant acceleration scalar, but not a constant acceleration vector.

Warning. $a(t)$ is not necessarily the derivative of $v(t)$. Almost any example shows this. For instance, let

$$X(t) = (\sin t, \cos t).$$

Then $v(t) = \|X'(t)\| = 1$ so that $dv/dt = 0$. However, a simple computation shows that $X''(t) = (-\sin t, -\cos t)$ and hence $a(t) = 1$.

We shall list the rules for differentiation. These will concern sums, products, and the chain rule which is postponed to the next section.

The derivative of a curve is defined componentwise. Thus the rules for the derivative will be very similar to the rules for differentiating functions.

Rule 1. *Let $X(t)$ and $Y(t)$ be two differentiable curves (defined for the same values of t). Then the sum $X(t) + Y(t)$ is differentiable, and*

$$\frac{d(X(t) + Y(t))}{dt} = \frac{dX}{dt} + \frac{dY}{dt}.$$

Rule 2. *Let c be a number, and let $X(t)$ be differentiable. Then $cX(t)$ is differentiable, and*

$$\frac{d(cX(t))}{dt} = c\frac{dX}{dt}.$$

Rule 3. *Let $X(t)$ and $Y(t)$ be two differentiable curves (defined for the same values of t). Then $X(t) \cdot Y(t)$ is a differentiable function whose derivative is*

$$\frac{d}{dt}[X(t) \cdot Y(t)] = X'(t) \cdot Y(t) + X(t) \cdot Y'(t).$$

(This is formally analogous to the derivative of a product of functions, namely the first times the derivative of the second plus the second times the derivative of the first, except that the product is now a scalar product.)

As an example of the proofs we shall give the third one in detail, and leave the others to you as exercises.

Let for simplicity

$$X(t) = (x_1(t), x_2(t)) \qquad \text{and} \qquad Y(t) = (y_1(t), y_2(t)).$$

Then

$$\frac{d}{dt}X(t) \cdot Y(t) = \frac{d}{dt}[x_1(t)y_1(t) + x_2(t)y_2(t)]$$

$$= x_1(t)\frac{dy_1(t)}{dt} + \frac{dx_1}{dt}y_1(t) + x_2(t)\frac{dy_2}{dt} + \frac{dx_2}{dt}y_2(t)$$

$$= X(t) \cdot Y'(t) + X'(t) \cdot Y(t),$$

by combining the appropriate terms.

The proof for 3-space or n-space is obtained by replacing 2 by 3 or n, and inserting ... in the middle to take into account the other coordinates.

Example 11. The square $X(t)^2 = X(t) \cdot X(t)$ comes up frequently in applications, for instance because it can be interpreted as the square of the distance of $X(t)$ from the origin. Using the rule for the derivative of a product, we find the formula

$$\frac{d}{dt} X(t)^2 = 2X(t) \cdot X'(t).$$

You should memorize this formula by repeating it out loud.

Suppose that $\|X(t)\|$ is constant. This means that $X(t)$ lies on a sphere of constant radius k. Taking the square yields

$$X(t)^2 = k^2$$

that is, $X(t)^2$ is also constant. Differentiate both sides with respect to t. Then we obtain

$$2X(t) \cdot X'(t) = 0$$

and therefore

$$X(t) \cdot X'(t) = 0.$$

Interpretation. *Suppose a bug moves along a curve $X(t)$ which remains at constant distance from the origin, i.e. $\|X(t)\| = k$ is constant. Then the position vector $X(t)$ is perpendicular to the velocity $X'(t)$.*

If $X(t)$ is a curve and $f(t)$ is a function, defined for the same values of t, then we may also form the product

$$f(t)X(t)$$

of the number $f(t)$ by the vector $X(t)$.

Example 12. Let $X(t) = (\cos t, \sin t, t)$ and $f(t) = e^t$, then

$$f(t)X(t) = (e^t \cos t, e^t \sin t, e^t t),$$

and

$$f(\pi)X(\pi) = (e^\pi(-1), e^\pi(0), e^\pi \pi) = (-e^\pi, 0, e^\pi \pi).$$

If $X(t) = (x(t), y(t), z(t))$, then

$$f(t)X(t) = (f(t)x(t), f(t)y(t), f(t)z(t)).$$

We have a rule for such differentiation analogous to Rule 3.

Rule 4. *If both $f(t)$ and $X(t)$ are defined over the same interval, and are differentiable, then so is $f(t)X(t)$, and*

$$\frac{d}{dt}f(t)X(t) = f(t)X'(t) + f'(t)X(t).$$

The proof is just the same as for Rule 3.

Example 13. Let A be a fixed vector, and let f be an ordinary differentiable function of one variable. Let $F(t) = f(t)A$. Then $F'(t) = f'(t)A$. For instance, if $F(t) = (\cos t)A$ and $A = (a, b)$ where a, b are fixed numbers, then

$$F(t) = (a \cos t, b \cos t)$$

and thus

$$F'(t) = (-a \sin t, -b \sin t) = (-\sin t)A.$$

Similarly, if A, B are fixed vectors, and

$$G(t) = (\cos t)A + (\sin t)B,$$

then

$$G'(t) = (-\sin t)A + (\cos t)B.$$

EXERCISES

Find the velocity vector of the following curves.

1. $(e^t, \cos t, \sin t)$

2. $(\sin 2t, \log(1 + t), t)$

3. $(\cos t, \sin t)$

4. $(\cos 3t, \sin 3t)$

5. In Exercises 3 and 4, show that the velocity vector is perpendicular to the position vector. Is this also the case in Exercises 1 and 2?

6. In Exercises 3 and 4, show that the acceleration vector is in the opposite direction from the position vector.

7. Let A, B be two constant vectors. What is the velocity vector of the curve

$$X = A + tB? \qquad \llcorner B \,)$$

8. Let $X(t)$ be a differentiable curve. A plane or line which is perpendicular to the velocity vector $X'(t)$ at the point $X(t)$ is said to be **normal** to the curve at the point t or also at the point $X(t)$. Find the equation of a line normal to the curves of Exercises 3 and 4 at the point $\pi/3$.

9. Find the equation of a plane normal to the curve

$$(e^t, t, t^2)$$

at the point $t = 1$.

10. Same question at the point $t = 0$.

11. Let $X(t)$ be a differentiable curve defined on an open interval. Let Q be a point which is not on the curve.
    (a) Write down the formula for the distance between Q and an arbitrary point on the curve.
    (b) If t_0 is a value of t such that the distance between Q and $X(t_0)$ is at a minimum, show that the vector $Q - X(t_0)$ is normal to the curve, at the point $X(t_0)$. [*Hint:* Investigate the minimum of the square of the distance.]
    (c) If $X(t)$ is the parametric equation of a straight line, show that there exists a unique value t_0 such that the distance between Q and $X(t_0)$ is a minimum.

12. Let A be a non-zero vector, c a number, and Q a point. Let P_0 be the point of intersection of the line passing through Q, in the direction of N, and the plane $X \cdot N = c$. Show that for all points P of the plane, we have

$$\|Q - P_0\| \leqq \|Q - P\|.$$

13. Prove that if the speed is constant, then the acceleration is perpendicular to the velocity.

14. Prove that if the acceleration of a curve is always perpendicular to its velocity, then its speed is constant.

★ ●. Let B be a non-zero vector, and let $X(t)$ be such that $X(t) \cdot B = t$ for all t. Assume also that the angle between $X'(t)$ and B is constant. Show that $X''(t)$ is perpendicular to $X'(t)$.

16. Write a parametric equation for the tangent line to the given curve at the given point in each of the following cases.
    (a) $(\cos 4t, \sin 4t, t)$ at the point $t = \pi/8$
    (b) $(t, 2t, t^2)$ at the point $(1, 2, 1)$
    (c) $(e^{3t}, e^{-3t}, 3\sqrt{2}\,t)$ at $t = 1$
    (d) (t, t^3, t^4) at the point $(1, 1, 1)$

17. Let A, B be fixed non-zero vectors. Let

$$X(t) = e^{2t}A + e^{-2t}B.$$

Show that $X''(t)$ has the same direction as $X(t)$.

18. Show that the two curves $(e^t, e^{2t}, 1 - e^{-t})$ and $(1 - \theta, \cos\theta, \sin\theta)$ intersect at the point $(1, 1, 0)$. What is the angle between their tangents at that point?

19. At what points does the curve $(2t^2, 1 - t, 3 + t^2)$ intersect the plane

$$3x - 14y + z - 10 = 0?$$

●. Let $X(t)$ be a differentiable curve.
    ★ (a) Suppose that $X'(t) = O$ for all t throughout its interval of definition I. What can you say about the curve?

$$X(t) = (X_1(t) \overset{0}{\cdot} \quad X_i(t) \overset{0}{\cdots} \quad X_n(t))\overset{0}{}$$

$$X'(t) = (\frac{dX_1}{dt} \overset{0}{\cdot} \quad \frac{dX_i}{dt} \overset{0}{\cdots} \quad \frac{dX_n}{dt})\overset{0}{}$$

(b) Suppose $X'(t) \neq O$ but $X''(t) = O$ for all t in the interval. What can you say about the curve?

21. Let $X(t) = (a \cos t, a \sin t, bt)$, where a, b are constant. Let $\theta(t)$ be the angle which the tangent line at a given point of the curve makes with the z-axis. Show that $\cos \theta(t)$ is the constant $b/\sqrt{a^2 + b^2}$.

22. Show that the velocity and acceleration vectors of the curve in Exercise 21 have constant norms (magnitudes).

23. Let B be a fixed unit vector, and let $X(t)$ be a curve such that $X(t) \cdot B = e^{2t}$ for all t. Assume also that the velocity vector of the curve has a constant angle θ with the vector B, with $0 < \theta < \pi/2$.
(a) Show that the speed is $2e^{2t}/\cos \theta$.
(b) Determine the dot product $X'(t) \cdot X''(t)$ in terms of t and θ.

24. Let

$$X(t) = \left(\frac{2t}{1 + t^2}, \frac{1 - t^2}{1 + t^2}, 1 \right).$$

Show that the cosine of the angle between $X(t)$ and $X'(t)$ is constant.

25. Suppose that a bug moves along a differentiable curve $B(t) = (x(t), y(t), z(t))$, lying in the surface $z^2 = 1 + x^2 - y^2$. (This means that the coordinates (x, y, z) of the curve satisfy this equation.) Show that

$$2x(t)x'(t) = B(t) \cdot B'(t).$$

Assume that the cosine of the angle between the position vector $B(t)$ and the velocity vector $B'(t)$ is always positive. Show that the distance of the bug to the yz-plane increases whenever its x-coordinate is positive.

§2. LENGTH OF CURVES

Suppose a bug travels along a curve $X(t)$. The rate of change of the distance traveled is equal to the speed, so we may write the equation

$$\frac{ds(t)}{dt} = v(t).$$

Consequently it is reasonable to make the following definition.

We define the **length** of a curve X between two values a, b of t ($a \leq b$) in the interval of definition of the curve to be the integral of the speed:

$$\int_a^b v(t)\, dt = \int_a^b \|X'(t)\|\, dt.$$

By definition, we can rewrite this integral in the form

$$\int_a^b \sqrt{\left(\frac{dx}{dt}\right)^2 + \left(\frac{dy}{dt}\right)^2}\; dt \qquad\qquad \text{when}\quad X(t) = (x(t), y(t))$$

$$\int_a^b \sqrt{\left(\frac{dx}{dt}\right)^2 + \left(\frac{dy}{dt}\right)^2 + \left(\frac{dz}{dt}\right)^2}\; dt \quad \text{when}\quad X(t) = (x(t), y(t), z(t))$$

$$\int_a^b \sqrt{\left(\frac{dx_1}{dt}\right)^2 + \cdots + \left(\frac{dx_n}{dt}\right)^2}\; dt \quad \text{when}\quad X(t) = (x_1(t), \ldots, x_n(t)).$$

Example 1. Let the curve be defined by

$$X(t) = (\sin t, \cos t).$$

Then $X'(t) = (\cos t, -\sin t)$ and $v(t) = \sqrt{\cos^2 t + \sin^2 t} = 1$. Hence the length of the curve between $t = 0$ and $t = 1$ is

$$\int_0^1 v(t)\, dt = t\Big|_0^1 = 1.$$

In this case, of course, the integral is easy to evaluate. There is no reason why this should always be the case.

Example 2. Set up the integral for the length of the curve

$$X(t) = (e^t, \sin t, t)$$

between $t = 1$ and $t = \pi$.

We have $X'(t) = (e^t, \cos t, 1)$. Hence the desired integral is

$$\int_1^\pi \sqrt{e^{2t} + \cos^2 t + 1}\; dt.$$

In this case, there is no easy formula for the integral. In the exercises, however, the functions are adjusted in such a way that the integral can be evaluated by elementary techniques of integration. Don't expect this to be the case in real life, though.

EXERCISES

1. Find the length of the spiral $(\cos t, \sin t, t)$ between $t = 0$ and $t = 1$.
2. Find the length of the spirals.
   (a) $(\cos 2t, \sin 2t, 3t)$ between $t = 1$ and $t = 3$.
   (b) $(\cos 4t, \sin 4t, t)$ between $t = 0$ and $t = \pi/8$.

3. Find the length of the indicated curve for the given interval:

    (a) $(t, 2t, t^2)$ between $t = 1$ and $t = 3$. [*Hint:* You will get at some point the integral $\int \sqrt{1 + u^2}\ du$. The easiest way of handling that is to let

$$u = \frac{e^t - e^{-t}}{2} = \sinh t, \text{ so } 1 + \sinh^2 t = \cosh^2 t$$

where

$$\cosh t = \frac{e^t + e^{-t}}{2}.$$

This makes the expression under the square root sign into a perfect square. This method will in fact prove the general formula

$$\boxed{\int \sqrt{a^2 + x^2}\ dx = \frac{1}{2}\left[x\sqrt{a^2 + x^2} + a^2 \log(x + \sqrt{a^2 + x^2}\,) \right].}$$

Of course, you can check the formula by differentiating the right-hand side, and just use it for the exercise.

    (b) $(e^{3t}, e^{-3t}, 3\sqrt{2}\ t)$ between $t = 0$ and $t = \frac{1}{3}$.

4. Find the length of the curve defined by

$$X(t) = (t - \sin t, 1 - \cos t)$$

between (a) $t = 0$ and $t = 2\pi$, (b) $t = 0$ and $t = \pi/2$.

5. Find the length of the curve $X(t) = (t, \log t)$ between (a) $t = 1$ and $t = 2$, (b) $t = 3$ and $t = 5$. [*Hint:* Substitute $u^2 = 1 + t^2$ to evaluate the integral. Use partial fractions.]

6. Find the length of the curve defined by $X(t) = (t, \log \cos t)$ between $t = 0$ and $t = \pi/4$.

CHAPTER III

Functions of Several Variables

We view functions of several variables as functions of points in space. This appeals to our geometric intuition, and also relates such functions more easily with the theory of vectors. The gradient will appear as a natural generalization of the derivative. In this chapter we are mainly concerned with basic definitions and notions. We postpone the important theorems to the next chapter.

§1. GRAPHS AND LEVEL CURVES

In order to conform with usual terminology, and for the sake of brevity, a collection of objects will simply be called a **set**. In this chapter, we are mostly concerned with sets of points in space.

Let S be a set of points in n-space. A **function** (defined on S) is an association which to each element of S associates a **number**. For instance, if to each point we associate the numerical value of the temperature at that point, we have the temperature function.

Remark. In the previous chapter, we considered parametrized curves, associating a vector to a point. We do **not** call these functions. Only when the values of the association are **numbers** do we use the word **function**. We find this to be the most useful convention for this course.

In practice, we sometimes omit mentioning explicitly the set S, since the context usually makes it clear for which points the function is defined.

Example 1. In 2-space (the plane) we can define a function f by the rule

$$f(x, y) = x^2 + y^2.$$

It is defined for all points (x, y) and can be interpreted geometrically as the square of the distance between the origin and the point.

Example 2. Again in 2-space, we can define a function f by the formula

$$f(x, y) = \frac{x^2 - y^2}{x^2 + y^2} \qquad \text{for all} \qquad (x, y) \neq (0, 0).$$

We do not define f at $(0, 0)$ (also written O).

Example 3. In 3-space, we can define a function f by the rule

$$f(x, y, z) = x^2 - \sin(xyz) + yz^3.$$

Since a point and a vector are represented by the same thing (namely an n-tuple), we can think of a function such as the above also as a function of vectors. When we do not want to write the coordinates, we write $f(X)$ instead of $f(x_1, \ldots, x_n)$. As with numbers, we call $f(X)$ the **value** of f at the point (or vector) X.

Just as with functions of one variable, one can define the **graph** of a function f of n variables $x_1, \ldots, x_n$ to be the set of points in $(n + 1)$-space of the form

$$(x_1, \ldots, x_n, f(x_1, \ldots, x_n)),$$

the $(x_1, \ldots, x_n)$ being in the domain of definition of f.

When $n = 1$, the graph of a function f is a set of points $(x, f(x))$. Thus the graph itself is in 2-space.

When $n = 2$, the graph of a function f is the set of points $(x, y, f(x, y))$. When $n = 2$, it is already difficult to draw the graph since it involves a figure in 3-space. The graph of a function of two variables may look like this:

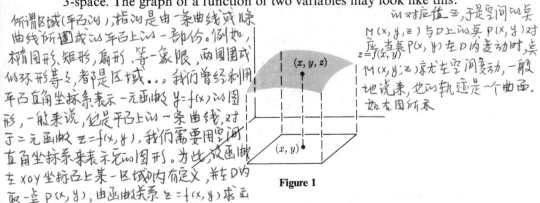

Figure 1

For each number c, the equation $f(x, y) = c$ is the equation of a curve in the plane. We have considerable experience in drawing the graphs of such curves, and we may therefore assume that we know how to draw this graph in principle. This curve is called the **level curve** of f at c. It gives us the set of points (x, y) where f takes on the value c. By drawing a number of such level curves, we can get a good description of the function.

Example 4. Let $f(x, y) = x^2 + y^2$. The level curves are described by equations

$$x^2 + y^2 = c.$$

These have a solution only when $c \geq 0$. In that case, they are circles (unless $c = 0$ in which case the circle of radius 0 is simply the origin). In Fig. 2, we have drawn the level curves for $c = 1$ and 4.

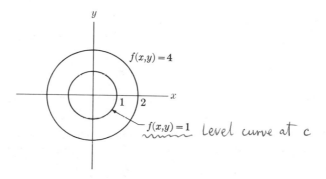

$f(x,y) = 4$

1 2

$f(x,y) = 1$ Level curve at c

Figure 2

The graph of the function $z = f(x, y) = x^2 + y^2$ is then a figure in 3-space, which we may represent as follows.

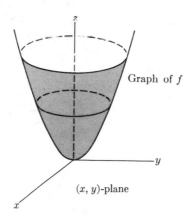

Graph of f

(x, y)-plane

Figure 3

Example 5. Let the elevation of a mountain in meters be given by the formula

$$f(x, y) = 4{,}000 - 2x^2 - 3y^4.$$

We see that $f(0, 0) = 4{,}000$ is the highest point of the mountain, As x, y increase, the altitude decreases. The mountain and its level curves might look like this.

Figure 4

In this case, the highest point is at the origin, and the level curves indicate decreasing altitude as they move away from the origin.

Example 6. We wish to find the level curves of the function

$$f(x, y) = \frac{x^2 - y^2}{x^2 + y^2}, \qquad \text{defined for} \qquad (x, y) \neq (0, 0).$$

Instead of using the previous techniques, we use another idea. We observe that the numerator and denominator in the expression for $f(x, y)$ have exact degree 2. We then let a be some fixed number, and consider the line passing through the origin

$$y = ax.$$

For $x \neq 0$ we find:

$$f(x, ax) = \frac{x^2 - a^2 x^2}{x^2 + a^2 x^2} = \frac{x^2(1 - a^2)}{x^2(1 + a^2)}.$$

We can cancel x^2 to obtain

$$f(x, ax) = \frac{1 - a^2}{1 + a^2}.$$

Thus we see that on this straight line, the function has the constant value

$$\frac{1 - a^2}{1 + a^2}.$$

Therefore this line is a level curve for the function. Since every point of the plane (other than the origin $(0, 0)$) lies on a unique line passing through the origin, we have found all level curves by this method.

We have drawn some of them in Fig. 5. (The numbers indicate the value of the function on the corresponding line.)

It would of course be technically much more disagreeable to draw the level curves in Example 3, and we shall not do so.

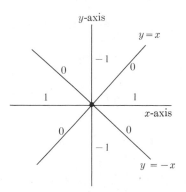

Figure 5

We see that the level lines are based on the same principle as the contour lines of a map. Each line describes, so to speak, the altitude of the function. If the graph is interpreted as a mountainous region, then each level curve gives the set of points of constant altitude. In Example 1, a person wanting to stay at a given altitude need but walk around in circles. In Example 6 such a person should walk on a straight line toward or away from the origin.

If we deal with a function of three variables, say $f(x, y, z)$, then $(x, y, z) = X$ is a point in 3-space. In that case, the set of points satisfying the equation

$$f(x, y, z) = c$$

for some constant c is a surface. The notion analogous to that of level curve is that of level surface.

The graph of a function of three variables is the set of points

$$(x, y, z, f(x, y, z))$$

in 4-dimensional space. Not only is this graph hard to draw, it is impossible to draw. It is, however, possible to define it as we have done by writing down coordinates of points.

In physics, a function f might be a potential function, giving the value of the potential energy at each point of space. The level surfaces are then sometimes called surfaces of **equipotential**. The function f might also give a temperature distribution (i.e. its value at a point X is the temperature at X). In that case, the level surfaces are called **isothermal** surfaces.

EXERCISES

Sketch the level curves for the functions $z = f(x, y)$, where $f(x, y)$ is given by the following expressions.

1. $x^2 + 2y^2$ 2. $y - x^2$ 3. $y - 3x^2$

4. $x - y^2$　　　　　　5. $3x^2 + 3y^2$　　　　　6. xy

7. $(x - 1)(y - 2)$　　　8. $(x + 1)(y + 3)$　　9. $\dfrac{x^2}{4} + \dfrac{y^2}{16}$

10. $2x - 3y$　　　　　　11. $\dfrac{xy}{x^2 + y^2}$　　　　12. $\dfrac{xy^2}{x^2 + y^4}$

13. $\dfrac{4xy(x^2 - y^2)}{x^2 + y^2}$ (try polar coordinates)　　　14. $\dfrac{x}{\sqrt{x^2 + y^2}}$

15. $\dfrac{x + y}{x - y}$　　　　　16. $\dfrac{x^2 + y^2}{x^2 - y^2}$　　　　17. $\sqrt{x^2 + y^2}$

(In Exercises 11, 12, 13, and 14 the function is not defined at $(0, 0)$. In Exercise 15, it is not defined for $y = x$, and in Exercise 16 it is not defined for $y = x$ or $y = -x$.)

18. $(x - 1)^2 + (y + 3)^2$　　19. $x^2 - y^2$　　　　20. $(x + 1)^2 + y^2$

§2. PARTIAL DERIVATIVES

In this section and the next, we discuss the notion of differentiability for functions of several variables. When we discussed the derivative of functions of one variable, we assumed that such a function was defined on an interval. We shall have to make a similar assumption in the case of several variables, and for this we need to introduce a new notion.

Let U be a set in the plane. We shall say that U is an **open set** if the following condition is satisfied. Given a point P in U, there exists an open disc D of radius $a > 0$ which is centered at P and such that D is contained in U.

Let U be a set in space. We shall say that U is an **open set** in space if given a point P in U, there exists an open ball B of radius $a > 0$ which is centered at P and such that B is contained in U.

A similar definition is given of an open set in n-space.

Given a point P in an open set, we can go in all directions from P by a small distance and still stay within the open set.

Example 1. In the plane, the set consisting of the first quadrant, excluding the x- and y-axes, is an open set.

The x-axis is not open in the plane (i.e. in 2-space). Given a point on the x-axis, we cannot find an open disc centered at the point and contained in the x-axis.

Example 2. Let U be the open ball of radius $a > 0$ centered at the origin. Then U is an open set. This is illustrated on Fig. 6.

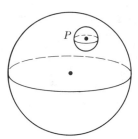

Figure 6

In the next picture we have drawn an open set in the plane, consisting of the region inside the curve, but not containing any point of the boundary. We have also drawn a point P in U, and a sphere (disc) around P contained in U.

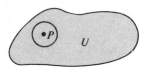

Figure 7

When we defined the derivative as a limit of

$$\frac{f(x + h) - f(x)}{h},$$

we needed the function f to be defined in some open interval around the point x.

Now let f be a function of n variables, defined on an open set U. Then for any point X in U, the function f is also defined at all points which are close to X, namely all points which are contained in an open ball centered at X and contained in U. We shall obtain the partial derivative of f by keeping all but one variable fixed, and taking the ordinary derivative with respect to the one variable.

Let us start with two variables. Given a function $f(x, y)$ of two variables x, y, let us keep y constant and differentiate with respect to x. We are then led to consider the limit of

$$\frac{f(x + h, y) - f(x, y)}{h}$$

as h approaches 0. If this limit exists, we call it the **derivative** of f **with respect to the first variable**, or also the **first partial derivative** of f, and denote it by

$$(D_1 f)(x, y).$$

This notation allows us to use any letters to denote the variables. For instance,

$$\lim_{h \to 0} \frac{f(u + h, v) - f(u, v)}{h} = D_1 f(u, v).$$

Note that $D_1 f$ is a single function. We often omit the parentheses, writing

$$D_1 f(u, v) = (D_1 f)(u, v)$$

for simplicity.

Also, if the variables x, y are agreed upon, then we write

$$D_1 f(x, y) = \frac{\partial f}{\partial x}.$$

Similarly, we define

$$D_2 f(x, y) = \lim_{h \to 0} \frac{f(x, y + k) - f(x, y)}{h}$$

and also write

$$D_2 f(x, y) = \frac{\partial f}{\partial y}.$$

Example 3. Let $f(x, y) = x^2 y^3$. Then

$$\frac{\partial f}{\partial x} = 2xy^3 \qquad \text{and} \qquad \frac{\partial f}{\partial y} = 3x^2 y^2.$$

We observe that the partial derivatives are themselves functions. This is the reason why the notation $D_i f$ is sometimes more useful than the notation $\partial f / \partial x_i$. It allows us to write $D_i f(P)$ for any point P in the set where the partial is defined. There cannot be any ambiguity or confusion with a (meaningless) symbol $D_i(f(P))$, since $f(P)$ is a number. Thus $D_i f(P)$ means $(D_i f)(P)$. It is the value of the function $D_i f$ at P.

Example 4. Let $f(x, y) = \sin xy$. To find $D_2 f(1, \pi)$, we first find $\partial f / \partial y$, or $D_2 f(x, y)$, which is simply

$$D_2 f(x, y) = (\cos xy)x.$$

Hence

$$D_2 f(1, \pi) = (\cos \pi) \cdot 1 = -1.$$

Also,

$$D_2 f\left(3, \frac{\pi}{4}\right) = \left(\cos \frac{3\pi}{4}\right) \cdot 3 = -\frac{1}{\sqrt{2}} \cdot 3 = -\frac{3}{\sqrt{2}}.$$

A similar definition of the partial derivatives is given in 3-space. Let f be a function of three variables (x, y, z), defined on an open set U in 3-space. We define, for instance,

$$(D_3 f)(x, y, z) = \frac{\partial f}{\partial z} = \lim_{h \to 0} \frac{f(x, y, z + h) - f(x, y, z)}{h},$$

and similarly for the other variables.

Example 5. Let $f(x, y, z) = x^2 y \sin(yz)$. Then

$$D_3 f(x, y, z) = \frac{\partial f}{\partial z} = x^2 y \cos(yz) y = x^2 y^2 \cos(yz).$$

Let $X = (x, y, z)$ for abbreviation. Let

$$E_1 = (1, 0, 0), \qquad E_2 = (0, 1, 0), \qquad E_3 = (0, 0, 1)$$

be the three standard unit vectors in the directions of the coordinate axes. Then we can abbreviate the Newton quotient for the partial derivatives by writing

$$D_i f(X) = \frac{\partial f}{\partial x_i} = \lim_{h \to 0} \frac{f(X + hE_i) - f(X)}{h}.$$

Indeed, observe that

$$hE_1 = (h, 0, 0) \qquad \text{so} \qquad f(X + hE_1) = f(x + h, y, z),$$

and similarly for the other two variables.

In a similar fashion we can define the partial derivatives in n-space, by a definition which applies simultaneously to 2-space and 3-space. Let f be a function defined on an open set U in n-space. Let the variables be $(x_1, \ldots, x_n)$. For small values of h, the point

$$(x_1 + h, x_2, \ldots, x_n)$$

is contained in U. Hence the function is defined at that point, and we may form the quotient

$$\frac{f(x_1 + h, x_2, \ldots, x_n) - f(x_1, \ldots, x_n)}{h}.$$

If the limit exists as h tends to 0, then we call it the **first partial derivative** of f and denote it by

$$D_1 f(x_1, \ldots, x_n), \text{ or } D_1 f(X), \text{ or also by } \frac{\partial f}{\partial x_1}.$$

Similarly, we let

$$D_i f(X) = \frac{\partial f}{\partial x_i}$$

$$= \lim_{h \to 0} \frac{f(x_1, \ldots, x_i + h, \ldots, x_n) - f(x_1, \ldots, x_n)}{h}$$

if it exists, and call it the i-th **partial derivative**.
Let

$$E_i = (0, \ldots, 0, 1, 0, \ldots, 0)$$

be the i-th unit vector in the direction of the i-th coordinate axis, having components equal to 0 except for the i-th component which is 1. Then we have

$$(D_i f)(X) = \lim_{h \to 0} \frac{f(X + hE_i) - f(X)}{h}.$$

This is a very useful brief notation which applies simultaneously to 2-space, 3-space, or n-space.

If f is a function of two variables (x, y), then we define the **gradient** of f, written **grad** f, to be the vector

$$\boxed{\operatorname{grad} f(x, y) = \left(\frac{\partial f}{\partial x}, \frac{\partial f}{\partial y} \right).}$$

Example 6. Let $f(x, y) = x^2 y^3$. Then

$$\operatorname{grad} f(x, y) = (2xy^3, 3x^2 y^2),$$

so that in this case,

$$\operatorname{grad} f(1, 2) = (16, 12).$$

Thus the gradient of a function f associates a **vector** to a point X.

If f is a function of three variables (x, y, z), then we define the gradient to be

$$\boxed{\operatorname{grad} f(x, y, z) = \left(\frac{\partial f}{\partial x}, \frac{\partial f}{\partial y}, \frac{\partial f}{\partial z} \right).}$$

Example 7. Let $f(x, y, z) = x^2 y \sin(yz)$. Find grad $f(1, 1, \pi)$.
First we find the three partial derivatives, which are:

$$\frac{\partial f}{\partial x} = 2xy \sin(yz)$$

$$\frac{\partial f}{\partial y} = x^2 [y \cos(yz)z + \sin(yz)]$$

$$\frac{\partial f}{\partial z} = x^2 y \cos(yz)y = x^2 y^2 \cos(yz).$$

We then substitute $(1, 1, \pi)$ for (x, y, z) in these partials, and get

$$\operatorname{grad} f(1, 1, \pi) = (0, \pi, -1).$$

Let f be defined in an open set U and assume that the partial derivatives of f exist at each point X of U. The **vector**

$$\left(\frac{\partial f}{\partial x_1}, \ldots, \frac{\partial f}{\partial x_n}\right) = (D_1 f(X), \ldots, D_n f(X)),$$

whose components are the partial derivatives, will be called the **gradient** of f at X and will be denoted by $\operatorname{grad} f(X)$. One must read this

$$(\operatorname{grad} f)(X),$$

but we shall usually omit the parentheses around $\operatorname{grad} f$. Sometimes one also writes ∇f instead of $\operatorname{grad} f$. Thus in 2-space we also write

$$\nabla f(x, y) = (\nabla f)(x, y) = (D_1 f(x, y), D_2 f(x, y)),$$

and similarly in 3-space,

$$\nabla f(x, y, z) = (\nabla f)(x, y, z) = (D_1 f(x, y, z), D_2 f(x, y, z), D_3 f(x, y, z)).$$

We shall give a geometric interpretation for the gradient in Chapter IV, §3. There we shall see that it gives the direction of maximal increase of the function, and that its magnitude is the rate of increase in that direction.

Using the formula for the derivative of a sum of two functions, and the derivative of a constant times a function, we conclude at once that the gradient satisfies the following properties:

Theorem 1. *Let f, g be two functions defined on an open set U, and assume that their partial derivatives exist at every point of U. Let c be a number. Then*

$$\operatorname{grad}(f + g) = \operatorname{grad} f + \operatorname{grad} g$$
$$\operatorname{grad}(cf) = c \operatorname{grad} f.$$

We shall give later several geometric and physical interpretations for the gradient.

EXERCISES

Find the partial derivatives

$$\frac{\partial f}{\partial x}, \quad \frac{\partial f}{\partial y}, \quad \text{and} \quad \frac{\partial f}{\partial z},$$

for the following functions $f(x, y)$ or $f(x, y, z)$.

1. $xy + z$ 2. $x^3 y^5 + 1$ 3. $\sin(xy) + \cos z$

4. $\cos(xy)$ 5. $\sin(xyz)$ 6. e^{xyz}

7. $x^2 \sin(yz)$ 8. xyz 9. $xz + yz + xy$

10. $x \cos(y - 3z) + \arcsin(xy)$

11. Find grad $f(P)$ if P is the point $(1, 2, 3)$ in Exercises 1, 2, 6, 8, and 9.

12. Find grad $f(P)$ if P is the point $(1, \pi, \pi)$ in Exercises 4, 5, 7.

13. Find grad $f(P)$ if

$$f(x, y, z) = \log(z + \sin(y^2 - x))$$

and

$$P = (1, -1, 1).$$

14. Find the partial derivatives of x^y. [*Hint*: $x^y = e^{y \log x}$.]

Find the gradient of the following functions at the given point.

15. $f(x, y, z) = e^{-2x} \cos(yz)$ at $(1, \pi, \pi)$

16. $f(x, y, z) = e^{3x+y} \sin(5z)$ at $(0, 0, \pi/6)$

§3. DIFFERENTIABILITY AND GRADIENT

Let f be a function defined on an open set U. Let X be a point of U. For all vectors H such that $\|H\|$ is small (and $H \neq O$), the point $X + H$ also lies in the open set. However we **cannot** form a quotient

$$\frac{f(X + H) - f(X)}{H}$$

because it is **meaningless to divide by a vector**. In order to define what we mean for a function f to be differentiable, we must therefore find a way which does not involve dividing by H.

We reconsider the case of functions of one variable. Let us fix a number x. We had defined the derivative to be

$$f'(x) = \lim_{h \to 0} \frac{f(x + h) - f(x)}{h}.$$

Let

$$\varphi(h) = \frac{f(x + h) - f(x)}{h} - f'(x).$$

Then $\varphi(h)$ is not defined when $h = 0$, but

$$\lim_{h \to 0} \varphi(h) = 0.$$

We can write

$$f(x + h) - f(x) = f'(x)h + h\varphi(h).$$

This relation has meaning so far only when $h \neq 0$. However, we observe that if we define $\varphi(0)$ to be 0, then the preceding relation is obviously true when $h = 0$ (because we just get $0 = 0$).

Let

$$g(h) = \varphi(h) \quad \text{if} \quad h > 0$$
$$g(h) = -\varphi(h) \quad \text{if} \quad h < 0.$$

Then we have shown that if f is differentiable, there exists a function g such that

(1) $$f(x + h) - f(x) = f'(x)h + |h| g(h),$$

and $$\lim_{h \to 0} g(h) = 0.$$

Conversely, suppose that there exists a number a and a function $g(h)$ such that

(1a) $$f(x + h) - f(x) = ah + |h| g(h).$$

and $$\lim_{h \to 0} g(h) = 0.$$

We find for $h \neq 0$,

$$\frac{f(x + h) - f(x)}{h} = a + \frac{|h|}{h} g(h).$$

Taking the limit as h approaches 0, we observe that

$$\lim_{h \to 0} \frac{|h|}{h} g(h) = 0.$$

Hence the limit of the Newton quotient exists and is equal to a. Hence f is differentiable, and its derivative $f'(x)$ is equal to a.

Therefore, the existence of a number a and a function g satisfying (1a) above could have been used as the definition of differentiability in the case of functions of one variable. The great advantage of (1) is that no h appears in the denominator. It is this relation which will suggest to us how to define differentiability for functions of several variables, and how to prove the chain rule for them.

Let us begin with two variables. We let

$$X = (x, y) \quad \text{and} \quad H = (h, k).$$

Then the notion corresponding to $x + h$ in one variable is here

$$X + H = (x + h, y + k).$$

We wish to compare the values of a function f at X and $X + H$, i.e. we wish to investigate the difference

$$f(X + H) - f(X) = f(x + h, y + k) - f(x, y).$$

We say that f is **differentiable** at X if the partial derivatives

$$\frac{\partial f}{\partial x} \quad \text{and} \quad \frac{\partial f}{\partial y}$$

exist, and if there exists a function g (defined for small H) such that

$$\lim_{H \to o} g(H) = 0$$

and

(2) $$\boxed{f(x + h, y + k) - f(x, y) = \frac{\partial f}{\partial x} h + \frac{\partial f}{\partial y} k + \|H\| g(H).}$$

We view the term

$$\frac{\partial f}{\partial x} h + \frac{\partial f}{\partial y} k$$

as an approximation to $f(X + H) - f(X)$, depending in a particularly simple way on h and k.

If we use the abbreviation

$$\text{grad } f = \nabla f,$$

then formula (2) can be written

$$\boxed{f(X + H) - f(X) = \nabla f(X) \cdot H + \|H\| g(H).}$$

As with grad f, one must read $(\nabla f)(X)$ and not the meaningless $\nabla(f(X))$ since $f(X)$ is a number for each value of X, and thus it makes no sense to apply ∇ to a number. The symbol ∇ is applied to the function f, and $(\nabla f)(X)$ is the value of ∇f at X.

We now consider a function of n variables.

Let f be a function defined on an open set U. Let X be a point of U. If $H = (h_1, \ldots, h_n)$ is a vector such that $\|H\|$ is small enough, then $X + H$ will also be a point of U and so $f(X + H)$ is defined. Note that

$$X + H = (x_1 + h_1, \ldots, x_n + h_n).$$

This is the generalization of the $x + h$ with which we dealt previously in one variable, or the $(x + h, y + k)$ in two variables. For three variables, we already run out of convenient letters, so we may as well write n instead of 3.

We say that f is **differentiable** at X if the partial derivatives $D_1 f(X), \ldots, D_n f(X)$ exist, and if there exists a function g (defined for small H) such that

$$\lim_{H \to O} g(H) = 0 \qquad \left(\text{also written} \quad \lim_{\|H\| \to 0} g(H) = 0 \right)$$

and

$$f(X + H) - f(X) = D_1 f(X) h_1 + \cdots + D_n f(X) h_n + \|H\| g(H).$$

With the other notation for partial derivatives, this last relation reads:

$$f(X + H) - f(X) = \frac{\partial f}{\partial x_1} h_1 + \cdots + \frac{\partial f}{\partial x_n} h_n + \|H\| g(H).$$

We say that f is **differentiable** in the open set U if it is differentiable at every point of U, so that the above relation holds for every point X in U.

In view of the definition of the gradient in §2, we can rewrite our fundamental relation in the form

(3)
$$\boxed{f(X + H) - f(X) = (\text{grad } f(X)) \cdot H + \|H\| g(H).}$$

The term $\|H\| g(H)$ has an order of magnitude smaller than the previous term involving the dot product. This is one advantage of the present notation. We know how to handle the formalism of dot products and are accustomed to it, and its geometric interpretation. This will help us later in interpreting the gradient geometrically.

Example 1. Suppose that we consider values for H pointing only in the direction of the standard unit vectors. In the case of two variables, consider for instance $H = (h, 0)$. Then for such H, the condition for differentiability reads:

$$f(X + H) = f(x + h, y) = f(x, y) + \frac{\partial f}{\partial x} h + |h| g(H).$$

In higher dimensional space, let $E_i = (0, \ldots, 0, 1, 0, \ldots, 0)$ be the i-th unit vector. Let $H = hE_i$ for some number h, so that

$$H = (0, \ldots, 0, h, 0, \ldots, 0).$$

Then for such H,

$$f(X + H) = f(X + hE_i) = f(X) + \frac{\partial f}{\partial x_i} h + |h| g(H),$$

and therefore if $h \neq 0$, we obtain

$$\frac{f(X + H) - f(X)}{h} = D_i f(X) + \frac{|h|}{h} g(H).$$

Because of the special choice of H, we can divide by the *number h*, but we are *not* dividing by the vector H.

The functions which we meet in practice are differentiable. The next theorem gives a criterion which shows that this is true. A function $\varphi(X)$ is said to be **continuous** if

$$\lim_{H \to O} \varphi(X + H) = \varphi(X),$$

for all X in the domain of definition of the function.

Theorem 2. *Let f be a function defined on some open set U. Assume that its partial derivatives exist for every point in this open set, and that they are continuous. Then f is differentiable.*

We shall omit the proof. Observe that in practice, the partial derivatives of a function are given by formulas from which it is clear that they are continuous.

EXERCISES

1. Let $f(x, y) = 2x - 3y$. What is $\partial f / \partial x$ and $\partial f / \partial y$?

2. Let $A = (a, b)$ and let f be the function on $\mathbf{R}^2$ such that $f(X) = A \cdot X$.
   Let $X = (x, y)$. In terms of the coordinates of A, determine $\partial f / \partial x$ and $\partial f / \partial y$.

3. Let $A = (a, b, c)$ and let f be the function on $\mathbf{R}^3$ such that $f(X) = A \cdot X$.
   Let $X = (x, y, z)$. In terms of the coordinates of A, determine $\partial f / \partial x$, $\partial f / \partial y$, and $\partial f / \partial z$.

4. Generalize the above two exercises to n-space.

5. Let f be defined on an open set U. Let X be a point of U. Let A be a vector, and let g be a function defined for small H, such that

$$\lim_{H \to O} g(H) = 0.$$

   Assume that

$$f(X + H) - f(X) = A \cdot H + \|H\| g(H).$$

   Prove that $A = \text{grad } f(X)$. You may do this exercise in 2 variables first and then in 3 variables, and let it go at that. Use coordinates, e.g. let $A = (a,b)$ and $X = (x, y)$. Use special values of H, as in Example 1.

§4. REPEATED PARTIAL DERIVATIVES

Let f be a function of two variables, defined on an open set U in 2-space. Assume that its first partial derivative exists. Then $D_1 f$ (which we also write $\partial f / \partial x$ if x is the first variable) is a function defined on U. We may then ask for its first or second partial derivatives, i.e. we may form $D_2 D_1 f$ or $D_1 D_1 f$ if these exist. Similarly, if $D_2 f$ exists, and if the first partial derivative of $D_2 f$ exists, we may form $D_1 D_2 f$.

Suppose that we write f in terms of the two variables (x, y). Then we can write

$$D_1 D_2 f(x, y) = \frac{\partial}{\partial x} \left(\frac{\partial f}{\partial y} \right) = (D_1(D_2 f))(x, y),$$

and

$$D_2 D_1 f(x, y) = \frac{\partial}{\partial y} \left(\frac{\partial f}{\partial x} \right) = (D_2(D_1 f))(x, y).$$

Example 1. Let $f(x, y) = \cos(xy)$. Then

$$\frac{\partial f}{\partial x} = -y \sin(xy) \qquad \text{and} \qquad \frac{\partial f}{\partial y} = -x \sin(xy).$$

Using the rule for the derivative of a product, we can then obtain the second order (or iterated) partial derivatives as follows:

$$D_2 D_1 f(x, y) = -xy \cos(xy) - \sin(xy).$$

But differentiating $\partial f / \partial y$ with respect to x, we see that

$$D_1 D_2 f(x, y) = -xy \cos(xy) - \sin(xy).$$

These two repeated partial derivatives are equal!

The next theorem tells us that in practice, this will always happen.

Theorem 3. *Let f be a function of two variables, defined on an open set U of 2-space. Assume that the partial derivatives $D_1 f$, $D_2 f$, $D_1 D_2 f$, and $D_2 D_1 f$ exist and are continuous. Then*

$$D_1 D_2 f = D_2 D_1 f.$$

The proof will be omitted.

Consider a function of three variables $f(x, y, z)$. We can then take three kinds of partial derivatives: D_1, D_2, or D_3 (in other notation, $\partial / \partial x$, $\partial / \partial y$, and $\partial / \partial z$). Let us assume throughout that all the partial derivatives which we shall consider exist and are continuous, so that we may form as many repeated partial derivatives as we please. Then using Theorem 3, we can show that it does not matter in which order we take these partials.

For instance, we see that

$$D_3 D_1 f = D_1 D_3 f.$$

This is simply an application of Theorem 3, keeping the second variable fixed. We may take a further partial derivative, for instance

$$D_1 D_3 D_1 f.$$

Here D_1 occurs twice and D_3 once. Then this expression will be equal to any other repeated partial derivative of f in which D_1 occurs twice and D_3 once. For example, we apply the theorem to the function $(D_1 f)$. Then the theorem allows us to interchange D_1 and D_3 in front of $(D_1 f)$ (always assuming that all partials we want to take exist and are continuous). We obtain

$$D_1 D_3 (D_1 f) = D_3 D_1 (D_1 f).$$

As another example, consider

(4) $$D_2 D_1 D_3 D_2 f.$$

We wish to show that it is equal to $D_1 D_2 D_2 D_3 f$. By Theorem 3, we have $D_3 D_2 f = D_2 D_3 f$. Hence:

(5) $$D_2 D_1 (D_3 D_2 f) = D_2 D_1 (D_2 D_3 f).$$

We then apply Theorem 3 again, and interchange D_2 and D_1 to obtain the desired expression.

Instead of writing $D_1 D_1 f$, we shall write more briefly

$$D_1^2 f,$$

and similarly $D_2^2 f$ instead of $D_2 D_2 f$.

Example 2. Let $f(x, y, z) = x^2 y z^3$. Then

$$D_1 f(x, y, z) = 2xyz^3$$

$$D_2 D_1 f(x, y, z) = 2xz^3$$

$$D_3 D_2 D_1 f(x, y, z) = 6xz^2.$$

On the other hand,

$$D_3 f(x, y, z) = 3x^2 y z^2$$

$$D_2 D_3 f(x, y, z) = 3x^2 z^2$$

$$D_1 D_2 D_3 f(x, y, z) = 6xz^2.$$

Thus we see experimentally that $D_3 D_2 D_1 f = D_1 D_2 D_3 f$.

Let $f(x, y)$ be a function of two variables x, y. We shall use the notation

$$\boxed{D_1 D_2 f(x, y) = \frac{\partial^2 f}{\partial x\, \partial y}.}$$

We could also write

$$D_1 D_2 f(x, y) = \frac{\partial^2 f}{\partial y\, \partial x}.$$

In this notation, one would thus have

$$\left(\frac{\partial}{\partial x}\right)^2 f = \frac{\partial^2 f}{\partial x^2} = D_1^2 f(x, y)$$

and

$$\frac{\partial}{\partial x}\left(\frac{\partial f}{\partial y}\right) = \frac{\partial^2 f}{\partial x\, \partial y} = D_1 D_2 f(x, y).$$

All the above notations are used in the scientific literature, and this is the reason for including them here.

Warning. **Do not confuse the two expressions**

$$\left(\frac{\partial}{\partial x}\right)^2 f = \frac{\partial^2 f}{\partial x^2} \qquad \text{and} \qquad \left(\frac{\partial f}{\partial x}\right)^2,$$

which are usually *not* equal. For instance, if $f(x, y) = x^2 y$, then

$$\frac{\partial^2 f}{\partial x^2} = 2y \qquad \text{and} \qquad \left(\frac{\partial f}{\partial x}\right)^2 = 4x^2 y^2.$$

Observe that

$$(D_1 f)^2 = \left(\frac{\partial f}{\partial x}\right)^2$$

is the *square* of the function $D_1 f$, whereas

$$D_1^2 f = \left(\frac{\partial}{\partial x}\right)^2 f$$

is obtained from f by *differentiating twice* with respect to x. Similarly,

$$D_1 D_2 f \neq (D_1 f)(D_2 f).$$

Example 3. Let $f(x, y) = \cos(xy)$. Then we already computed $\partial f/\partial x$ and $\partial f/\partial y$ in Example 1. Taking one more partial derivative, we find:

$$\frac{\partial^2 f}{\partial x^2} = \frac{\partial}{\partial x}(-y \sin xy) = -y^2 \cos xy$$

$$\frac{\partial^2 f}{\partial y^2} = \frac{\partial}{\partial y}(-x \sin xy) = -x^2 \cos xy.$$

EXERCISES

Find the partial derivatives of order 2 for the following functions and verify explicitly in each case that $D_1D_2f = D_2D_1f$.

1. e^{xy} 2. $\sin(xy)$

3. $x^2y^3 + 3xy$ 4. $2xy + y^2$

5. $e^{x^2+y^2}$ 6. $\sin(x^2 + y)$

7. $\cos(x^3 + xy)$ 8. $\arctan(x^2 - 2xy)$

9. e^{x+y} 10. $\sin(x + y)$.

Find $D_1D_2D_3f$ and $D_3D_2D_1f$ in the following cases.

11. xyz 12. x^2yz

13. e^{xyz} 14. $\sin(xyz)$

15. $\cos(x + y + z)$ 16. $\sin(x + y + z)$

17. $(x^2 + y^2 + z^2)^{-1}$ 18. $x^3y^2z + 2(x + y + z)$.

19. A function of three variables $f(x, y, z)$ is said to satisfy **Laplace's equation** if

$$\frac{\partial^2 f}{\partial x^2} + \frac{\partial^2 f}{\partial y^2} + \frac{\partial^2 f}{\partial z^2} = 0.$$

Verify that the following functions satisfy Laplace's equation.

(a) $x^2 + y^2 - 2z^2$
(d) $e^{3x+4y} \cos(5z)$

20. Let f, g be two functions (of two variables) with continuous partial derivatives of order $\leqq 2$ in an open set U. Assume that

$$\frac{\partial f}{\partial x} = -\frac{\partial g}{\partial y} \quad \text{and} \quad \frac{\partial f}{\partial y} = \frac{\partial g}{\partial x}.$$

Show that

$$\frac{\partial^2 f}{\partial x^2} + \frac{\partial^2 f}{\partial y^2} = 0.$$

21. Let $f(x, y) = \arctan y/x$ for $x > 0$. Show that

$$\frac{\partial^2 f}{\partial x^2} + \frac{\partial^2 f}{\partial y^2} = 0.$$

CHAPTER IV

The Chain Rule
and the Gradient

In this chapter, we prove the chain rule for functions of several variables and give a number of applications. Among them will be several interpretations for the gradient. These form one of the central points of our theory. They show how powerful the tools we have accumulated turn out to be.

There is a section at the end of the chapter giving additional techniques in partial differentiation, whose flavor is quite different from that of the chain rule used in the other applications. This section may be omitted since these techniques will play no role in the subsequent applications (conservation law, uniqueness of potential function, value of an integral when a potential function exists, etc.). However, it is important in other contexts, especially that of partial differential equations, and it may be considered useful to have exposed students to a technique which allows them, for instance, to get the Laplace operator in polar coordinates. Special drilling is necessary for that at the present level of mathematical sophistication. The section has been kept separated from the rest in order to allow for its easy omission, or alternative ordering of the material.

§1. THE CHAIN RULE

Let f be a function defined on some open set U. Let $C(t)$ be a curve such that the values $C(t)$ are contained in U. Then we can form the composite function $f \circ C$, which is a function of t, given by

$$(f \circ C)(t) = f(C(t)).$$

Example 1. Take $f(x, y) = e^x \sin(xy)$. Let $C(t) = (t^2, t^3)$. Then

$$f(C(t)) = e^{t^2} \sin(t^5).$$

The expression on the right is obtained by substituting t^2 for x and t^3 for y in

$f(x,y)$. This is a function of t in the old sense of functions of one variable. If we interpret f as the temperature, then $f(C(t))$ is the temperature of a bug traveling along the curve $C(t)$ at time t.

The chain rule tells us how to find the derivative of this function, provided we know the gradient of f and the derivative C'. Its statement is as follows.

Chain Rule. *Let f be a function which is defined and differentiable on an open set U. Let C be a differentiable curve (defined for some interval of numbers t) such that the values $C(t)$ lie in the open set U. Then the function*

$$f(C(t))$$

is differentiable (as a function of t), and

$$\frac{df(C(t))}{dt} = (\operatorname{grad} f(C(t))) \cdot C'(t).$$

Memorize this formula by repeating it out loud.

In the notation dC/dt, this also reads

$$\frac{df(C(t))}{dt} = (\operatorname{grad} f)(C(t)) \cdot \frac{dC}{dt}.$$

Proof of the Chain Rule. By definition, we must investigate the quotient

$$\frac{f(C(t+h)) - f(C(t))}{h}.$$

Let

$$K = K(t, h) = C(t + h) - C(t).$$

Then our quotient can be rewritten in the form

$$\frac{f(C(t) + K) - f(C(t))}{h}.$$

Using the definition of differentiability for f, we have

$$f(X + K) - f(X) = (\operatorname{grad} f)(X) \cdot K + \|K\| g(K)$$

and

$$\lim_{\|K\| \to 0} g(K) = 0.$$

Replacing K by what it stands for, namely $C(t + h) - C(t)$, and dividing by h, we obtain:

$$\frac{f(C(t+h)) - f(C(t))}{h} = (\operatorname{grad} f)(C(t)) \cdot \frac{C(t+h) - C(t)}{h}$$

$$\pm \left\| \frac{C(t+h) - C(t)}{h} \right\| g(K).$$

As h approaches 0, the first term of the sum approaches what we want, namely

$$(\text{grad } f)(C(t)) \cdot C'(t).$$

The second term approaches

$$\pm \|C'(t)\| \lim_{h \to 0} g(K),$$

and when h approaches 0, so does $K = C(t + h) - C(t)$. Hence the second term of the sum approaches 0. This proves our chain rule.

To use the chain rule for certain computations, it is convenient to reformulate it in terms of components, and in terms of the two notations we have used for partial derivatives

$$\frac{\partial f}{\partial x} = D_1 f(x, y), \qquad \frac{\partial f}{\partial y} = D_2 f(x, y)$$

when the variables are x, y.

Suppose $C(t)$ is given in terms of coordinates by

$$C(t) = (x_1(t), \ldots, x_n(t)),$$

then

$$\frac{d(f(C(t)))}{dt} = \frac{\partial f}{\partial x_1} \frac{dx_1}{dt} + \cdots + \frac{\partial f}{\partial x_n} \frac{dx_n}{dt}.$$

If f is a function of two variables (x, y) then

$$\frac{df(C(t))}{dt} = \frac{\partial f}{\partial x} \frac{dx}{dt} + \frac{\partial f}{\partial y} \frac{dy}{dt}.$$

In the D_1, D_2 notation, we can write this formula in the form

$$\frac{d}{dt}(f(x(t), y(t))) = (D_1 f)(x, y)\frac{dx}{dt} + (D_2 f)(x, y)\frac{dy}{dt},$$

and similarly for several variables. For simplicity we usually omit the parentheses around $D_1 f$ and $D_2 f$. Also on the right-hand side we have abbreviated $x(t), y(t)$ to x, y, respectively. Without any abbreviation, the formula reads:

$$\frac{d}{dt}(f(x(t), y(t))) = D_1 f(x(t), y(t))\frac{dx}{dt} + D_2 f(x(t), y(t))\frac{dy}{dt}.$$

Example 2. Let $C(t) = (e^t, t, t^2)$ and let $f(x, y, z) = x^2 yz$. Then putting

$$x = e^t, \qquad y = t, \qquad z = t^2$$

we get:

$$\frac{d}{dt}f(C(t)) = \frac{\partial f}{\partial x}\frac{dx}{dt} + \frac{\partial f}{\partial y}\frac{dy}{dt} + \frac{\partial f}{\partial z}\frac{dz}{dt}$$

$$= 2xyze^t + x^2z + x^2y2t.$$

If we want this function entirely in terms of t, we substitute back the values for x, y, z in terms of t, and get

$$\frac{d}{dt}f(C(t)) = 2e^t t t^2 e^t + e^{2t}t^2 + e^{2t}t2t$$

$$= 2t^3 e^{2t} + t^2 e^{2t} + 2t^2 e^{2t}.$$

In some cases, as in the next example, one does not use the chain rule in several variables, just the old one from one-variable calculus.

Example 3. Let

$$f(x, y, z) = \sin(x^2 - 3zy + xz).$$

Then keeping y and z constant, and differentiating with respect to x, we find

$$\frac{\partial f}{\partial x} = \cos(x^2 - 3zy + xz) \cdot (2x + z).$$

More generally, let

$$f(x, y, z) = g(x^2 - 3zy + xz),$$

where g is a differentiable function of one variable. [In the special case above, we have $g(u) = \sin u$.] Then the chain rule gives

$$\frac{\partial f}{\partial x} = g'(x^2 - 3zy + xz)(2x + z).$$

We denote the derivative of g by g' as usual. We do *not* write it as dg/dx, because x is a letter which is already occupied for other purposes. We could let

$$u = x^2 - 3zy + xz,$$

in which case it would be all right to write

$$\frac{\partial f}{\partial x} = \frac{dg}{du}\frac{\partial u}{\partial x},$$

and we would get the same answer as above.

EXERCISES

1. Let P, A be constant vectors. If $g(t) = f(P + tA)$, show that

$$g'(t) = (\text{grad } f)(P + tA) \cdot A.$$

2. Suppose that f is a function such that

$$\text{grad } f(1, 1, 1) = (5, 2, 1).$$

Let $C(t) = (t^2, t^{-3}, t)$. Find

$$\frac{d}{dt}(f(C(t))) \qquad \text{at} \qquad t = 1.$$

3. Let $f(x, y) = e^{9x+2y}$ and $g(x, y) = \sin(4x + y)$. Let C be a curve such that $C(0) = (0, 0)$. Given:

$$\frac{d}{dt}f(C(t))\Big|_{t=0} = 2 \qquad \text{and} \qquad \frac{d}{dt}g(C(t))\Big|_{t=0} = 1,$$

Find $C'(0)$.

4. (a) Let P be a constant vector. Let $g(t) = f(tP)$, where f is some differentiable function. What is $g'(t)$?
   (b) Let f be a differentiable function defined on all of space. Assume that $f(tP) = tf(P)$ for all numbers t and all points P. Show that for all P we have

$$f(P) = \text{grad } f(O) \cdot P.$$

5. Let f be a differentiable function of two variables and assume that there is an integer $m \geq 1$ such that

$$f(tx, ty) = t^m f(x, y)$$

for all numbers t and all x, y. Prove **Euler's relation**

$$x\frac{\partial f}{\partial x} + y\frac{\partial f}{\partial y} = mf(x, y).$$

[*Hint:* Let $C(t) = (tx, ty)$. Differentiate both sides of the given equation with respect to t, keeping x and y **constant**. Then put $t = 1$.]

6. Generalize Exercise 5 to n variables, namely let f be a differentiable function of n variables and assume that there exists an integer $m \geq 1$ such that $f(tX) = t^m f(X)$ for all numbers t and all points X in $\mathbf{R}^n$. Show that

$$x_1\frac{\partial f}{\partial x_1} + \cdots + x_n\frac{\partial f}{\partial x_n} = mf(X),$$

which can also be written $X \cdot \text{grad } f(X) = mf(X)$.

7. (a) Let $f(x, y) = (x^2 + y^2)^{1/2}$. Find $\partial f/\partial x$ and $\partial f/\partial y$.
   (b) Let $f(x, y, z) = (x^2 + y^2 + z^2)^{1/2}$. Find $\partial f/\partial x$, $\partial f/\partial y$, $\partial f/\partial z$.

8. Let $r = (x_1^2 + \cdots + x_n^2)^{1/2}$. What is $\partial r/\partial x_i$?

9. Find the derivatives with respect to x and y of the following functions.
   (a) $\sin(x^3y + 2x^2)$ (b) $\cos(3x^2y - 4x)$
   (c) $\log(x^2y + 5y)$ (d) $(x^2y + 4x)^{1/2}$

§2. TANGENT PLANE

We begin by an example analyzing a function along a curve where the values of the function are constant. This gives rise to a very important principle of perpendicularity.

三維空間

Example 1. Let f be a function on $\mathbf{R}^3$. Let us interpret f as giving the temperature, so that at any point X in $\mathbf{R}^3$, the value of the function $f(X)$ is the temperature at X. Suppose that a bug moves in space along a differentiable curve, which we may denote in parametric form by

$$B(t).$$

Thus $B(t) = (x(t), y(t), z(t))$ is the position of the bug at time t. Let us assume that the bug starts from a point where he feels that the temperature is comfortable, and therefore that the temperature is constant along the path on which he moves. In other words, f is constant along the curve $B(t)$. This means that for all values of t, we have $B(t) = (x(t), y(t), z(t))$ is the position vector

$$f(B(t)) = k,$$ has a constant value

where k is constant. Differentiating with respect to t, and using the chain rule, we find that

$$\operatorname{grad} f(B(t)) \cdot B'(t) = 0.$$

This means that the gradient of f is perpendicular to the velocity vector at every point of the curve.

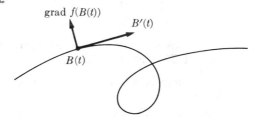

Figure 1

Let f be a differentiable function defined on an open set U in 3-space, and let k be a number. The set of points X such that

$$f(X) = k \text{ and } \operatorname{grad} f(X) \neq O$$

is called a **surface**.

Let $C(t)$ be a differentiable curve. We shall say that the curve **lies on the surface** if, for all t, we have

$$f(C(t)) = k.$$

This simply means that all the points of the curve satisfy the equation of the surface. If we differentiate this relation, we get from the chain rule:

$$\operatorname{grad} f(C(t)) \cdot C'(t) = 0.$$

Let P be a point of the surface, and let $C(t)$ be a curve on the surface passing through P. This means that there is a number t_0 such that $C(t_0) = P$. For this value t_0, we obtain

$$\operatorname{grad} f(P) \cdot C'(t_0) = 0.$$

Thus the gradient of f at P is perpendicular to the tangent vector of the curve at P. [We assume that $C'(t_0) \neq O$.] This is true for **every** differentiable curve on the surface passing through P. It is therefore very reasonable to **define** the **plane tangent** to the surface at P to be the plane passing through P and perpendicular to the vector $\operatorname{grad} f(P)$. (We know from Chapter I how to find such planes.) This definition applies only when $\operatorname{grad} f(P) \neq O$. If

$$\operatorname{grad} f(P) = O,$$

then we do not define the notion of tangent plane.

The fact that $\operatorname{grad} f(P)$ is perpendicular to every curve passing through P on the surface also gives us an interpretation of the gradient as being perpendicular to the surface

$$f(X) = k.$$

which is one of the level surfaces for the function f (Fig. 2).

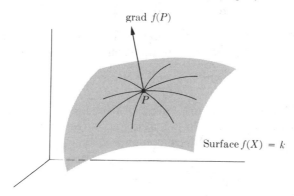

$\operatorname{grad} f(P)$

Surface $f(X) = k$

Figure 2

Example 2. Find the tangent plane to the surface

$$x^2 + y^2 + z^2 = 3$$

at the point $(1, 1, 1)$.

Let $f(X) = x^2 + y^2 + z^2$. Then at the point $P = (1, 1, 1)$,

$$\text{grad } f(P) = (2, 2, 2).$$

The equation of a plane passing through P and perpendicular to a vector N is

$$X \cdot N = P \cdot N.$$

In the present case, this yields

$$2x + 2y + 2z = 2 + 2 + 2 = 6.$$

Observe that our arguments also give us a means of finding a vector perpendicular to a curve in 2-space at a given point, simply by applying the preceding discussion to the plane instead of 3-space. A curve is defined by an equation $f(x, y) = c$, and in this case, grad $f(x_0, y_0)$ is perpendicular to the curve at the point (x_0, y_0) on the curve.

Example 3. Find the tangent line to the curve

$$x^2y + y^3 = 10$$

at the point $P = (1, 2)$, and find a vector perpendicular to the curve at that point.

Let $f(x, y) = x^2y + y^3$. Then

$$\text{grad } f(x, y) = (2xy, x^2 + 3y^2),$$

and so

$$\text{grad } f(P) = \text{grad } f(1, 2) = (4, 13).$$

This is a vector N perpendicular to the curve at the given point. The tangent line is given by $X \cdot N = P \cdot N$, and thus its equation is

$$4x + 13y = 4 + 26 = 30.$$

Example 4. A surface may also be given in the form $z = g(x, y)$ where g is some function of two variables. In this case, the tangent plane is determined by viewing the surface as expressed by the equation

$$g(x, y) - z = 0.$$

For instance, suppose the surface is given by $z = x^2 + y^2$. We wish to determine the tangent plane at $(1, 2, 5)$. Let $f(x, y, z) = x^2 + y^2 - z$. Then

$$\text{grad } f(x, y, z) = (2x, 2y, -1) \quad \text{and} \quad \text{grad } f(1, 2, 5) = (2, 4, -1).$$

The equation of the tangent plane at $P = (1, 2, 5)$ perpendicular to

$$N = (2, 4, -1)$$

is

$$2x + 4y - z = P \cdot N = 5.$$

This is the desired equation.

Example 5. Find a parametric equation for the tangent line to the curve of intersection of the two surfaces

$$x^2 + y^2 + z^2 = 6 \quad \text{and} \quad x^3 - y^2 + z = 2,$$

at the point $P = (1, 1, 2)$.

The tangent line to this curve is the line in common with the tangent planes of the two surfaces at the point P. We know how to find these tangent planes, and in Chapter I, we learned how to find the parametric representation of the line common to two planes, so we know how to do this problem. We carry out the numerical computation in full.

The first surface is defined by the equation $f(x, y, z) = 6$. A vector N_1 perpendicular to this first surface at P is given by

$$N_1 = \text{grad } f(P), \quad \text{where} \quad \text{grad } f(x, y, z) = (2x, 2y, 2z).$$

Thus for $P = (1, 1, 2)$ we find

$$N_1 = (2, 2, 4).$$

The second surface is given by the equation $g(x, y, z) = 2$, and

$$\text{grad } g(x, y, z) = (3x^2, -2y, 1).$$

Thus a vector N_2 perpendicular to the second surface at P is

$$N_2 = \text{grad } g(1, 1, 2) = (3, -2, 1).$$

A vector $A = (a, b, c)$ in the direction of the line of intersection is perpendicular to both N_1 and N_2. To find A, we therefore have to solve the equations

$$A \cdot N_1 = 0 \quad \text{and} \quad A \cdot N_2 = 0.$$

This amounts to

$$2a + 2b + 4c = 0$$
$$3a - 2b + c = 0.$$

Let, for instance, $a = 1$. Solving for b and c yields

$$a = 1, \quad b = 1, \quad c = -1.$$

Thus $A = (1, 1, -1)$. Finally, the parametric representation of the desired line is

$$P + tA = (1, 1, 2) + t(1, 1, -1).$$

EXERCISES

1. Find the equation of the tangent plane and normal line to each of the following surfaces at the specific point.
   (a) $x^2 + y^2 + z^2 = 49$ at $(6, 2, 3)$
   (b) $xy + yz + zx - 1 = 0$ at $(1, 1, 0)$
   (c) $x^2 + xy^2 + y^3 + z + 1 = 0$ at $(2, -3, 4)$
   (d) $2y - z^3 - 3xz = 0$ at $(1, 7, 2)$
   (e) $x^2y^2 + xz - 2y^3 = 10$ at $(2, 1, 4)$
   (f) $\sin xy + \sin yz + \sin xz = 1$ at $(1, \pi/2, 0)$

2. Let $f(x, y, z) = z - e^x \sin y$, and $P = (\log 3, 3\pi/2, -3)$. Find:
   (a) grad $f(P)$,
   (b) the normal line at P to the level surface for f which passes through P,
   (c) the tangent plane to this surface at P.

3. Find the parametric equation of the tangent line to the curve of intersection of the following surfaces at the indicated point.
   (a) $x^2 + y^2 + z^2 = 49$ and $x^2 + y^2 = 13$ at $(3, 2, -6)$
   (b) $xy + z = 0$ and $x^2 + y^2 + z^2 = 9$ at $(2, 1, -2)$
   (c) $x^2 - y^2 - z^2 = 1$ and $x^2 - y^2 + z^2 = 9$ at $(3, 2, 2)$
   [*Note*: The tangent line above may be defined to be the line of intersection of the tangent planes of the given point.]

4. Let $f(X) = 0$ be a differentiable surface. Let Q be a point which does not lie on the surface. Given a differentiable curve $C(t)$ on the surface, defined on an open interval, give the formula for the distance between Q and a point $C(t)$. Assume that this distance reaches a minimum for $t = t_0$. Let $P = C(t_0)$. Show that the line joining Q to P is perpendicular to the curve at P.

5. Find the equation of the tangent plane to the surface $z = f(x, y)$ at the given point P when f is the following function:
   (a) $f(x, y) = x^2 + y^2$, $P = (3, 4, 25)$
   (b) $f(x, y) = x/(x^2 + y^2)^{1/2}$, $P = (3, -4, \frac{3}{5})$
   (c) $f(x, y) = \sin(xy)$ at $P = (1, \pi, 0)$

6. Find the equation of the tangent plane to the surface $x = e^{2y-z}$ at $(1, 1, 2)$.

7. Let $f(x, y, z) = xy + yz + zx$. (a) Write down the equation of the level surface for f through the point $P = (1, 1, 0)$. (b) Find the equation of the tangent plane to this surface at P.

8. Find the equation of the tangent plane to the surface

$$3x^2 - 2y + z^3 = 9$$

at the point $(1, 1, 2)$

9. Find the equation of the tangent plane to the surface

$$z = \sin(x + y)$$

at the point where $x = 1$ and $y = 2$.

10. Find the tangent plane to the surface $x^2 + y^2 - z^2 = 18$ at the point $(3, 5, -4)$.

11. (a) Find a unit vector perpendicular to the surface

$$x^3y + xz = 1$$

at the point $(1, 2, -1)$. (b) Find the equation of the tangent plane at that point.

12. Find the cosine of the angle between the surfaces

$$x^2 + y^2 + z^2 = 3 \qquad \text{and} \qquad x - z^2 - y^2 = -3$$

at the point $(-1, 1, -1)$ (This angle is the angle between the normal vectors at the point.)

§3 DIRECTIONAL DERIVATIVE

Let f be defined on an open set and assume that f is differentiable. Let P be a point of the open set, and let A be a **unit vector** (i.e. $\|A\| = 1$). Then $P + tA$ is the parametric representation of a straight line in the direction of A and passing through P. We observe that

$$\frac{d(P + tA)}{dt} = A.$$

For instance, if $n = 2$ and $P = (p, q)$, $A = (a, b)$, then

$$P + tA = (p + ta, q + tb),$$

or in terms of coordinates,

$$x = p + ta, \qquad y = q + tb.$$

Hence

$$\frac{dx}{dt} = a \qquad \text{and} \qquad \frac{dy}{dt} = b$$

so that

$$\frac{d(P + tA)}{dt} = (a, b) = A.$$

The same argument works in higher dimensions.

We wish to consider the rate of change of f in the direction of A. It is natural to consider the values of f on the line $P + tA$, that is to consider the values

$$f(P + tA).$$

The rate of change of f along this line will then be given by taking the derivative of this expression, which we know how to do. We illustrate the line $P + tA$ in the figure.

(handwritten marginal notes:)

let $f(x, y, z) = x \ln y - e^{xz^3}$ calculate in the direction of the vector $(v) = i - j + 3k$

Evaluate the directive at $P = (-5, 1, -2)$

$$u = \frac{1}{\sqrt{11}} i - \frac{1}{\sqrt{11}} j + \frac{3}{\sqrt{11}} k$$

$$\nabla f = (\ln y - z^3 e^{xz^3}) i + (\frac{x}{y}) j - 3(x z^2 e^{xz^3}) k$$

$$\nabla f(x) \cdot u = \ln y - z^3 e^{xz^3} + \frac{x}{y} - 3xz^2 e^{xz^3}$$

$$\text{at } (-5, 1, -2) = \frac{5 + 188 e^{40\sqrt{11}}}{\sqrt{11}}$$

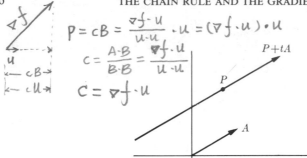

Handwritten annotations (left margin and top):

$B = u$

$A = \nabla f$

$u \cdot u = 1$

$P = cB = \dfrac{\nabla f \cdot u}{u \cdot u} \cdot u = (\nabla f \cdot u) \cdot u$

$c = \dfrac{A \cdot B}{B \cdot B} = \dfrac{\nabla f \cdot u}{u \cdot u}$

$c = \nabla f \cdot u$

Figure 3

If f represents a temperature at the point P, we look at the variation of temperature in the direction of A, starting from the point P. The value $f(P + tA)$ gives the temperature at the point $P + tA$. This is a function of t, say

$$g(t) = f(P + tA).$$

The rate of change of this temperature function is $g'(t)$, the derivative with respect to t, and $g'(0)$ is the rate of change at time $t = 0$, i.e. the rate of change of f at the point P, in the direction of A.

By the chain rule, if we take the derivative of the function

$$g(t) = f(P + tA),$$

which is defined for small values of t, we obtain

$$\frac{df(P + tA)}{dt} = \operatorname{grad} f(P + tA) \cdot A.$$

When t is equal to 0, this derivative is equal to

$$\boxed{\operatorname{grad} f(P) \cdot A}$$

For obvious geometrical reasons, we call it the **directional derivative of f in the direction of A.** We interpret it as the rate of change of f along the straight line in the direction of A, at the point P. Thus if we agree on the notation $D_A f(P)$ for the directional derivative of f at P in the direction of the unit vector A, then we have

scalar

$$\boxed{D_A f(P) = \frac{df(P + tA)}{dt}\bigg|_{t=0} = \operatorname{grad} f(P) \cdot A.}$$

In using this formula, the reader should remember that A is taken to be a **unit vector.** When a direction is given in terms of a vector whose norm is not 1, then one must first divide this vector by its norm before applying the formula.

Example 1. Let $f(x, y) = x^2 + y^3$ and let $B = (1, 2)$. Find the directional derivative of f in the direction of B, at the point $(-1, 3)$.

We note that B is not a unit vector. Its norm is $\sqrt{5}$. Let

$$A = \frac{1}{\sqrt{5}} B.$$

Then A is a unit vector having the same direction as B. Let $P = (-1, 3)$. Then $\operatorname{grad} f(P) = (-2, 27)$. Hence by our formula, the directional derivative is equal to:

$$\operatorname{grad} f(P) \cdot A = \frac{1}{\sqrt{5}}(-2 + 54) = \frac{52}{\sqrt{5}}.$$

Consider again a differentiable function f on an open set U.

Let P be a point of U. *Let us assume that* $\operatorname{grad} f(P) \neq O$, and let A be a unit vector. We know that

$$D_A f(P) = \operatorname{grad} f(P) \cdot A = \|\operatorname{grad} f(P)\| \, \|A\| \cos \theta$$

where θ is the angle between $\operatorname{grad} f(P)$ and A. Since $\|A\| = 1$, we see that the directional derivative is equal to

$$\boxed{D_A f(P) = \|\operatorname{grad} f(P)\| \cos \theta.}$$

We remind the reader that this formula holds only when A is a unit vector.

The value of $\cos \theta$ varies between -1 and $+1$ when we select all possible unit vectors A.

The maximal value of $\cos \theta$ is obtained when we select A such that $\theta = 0$, i.e. when we select A to have the same direction as $\operatorname{grad} f(P)$. In that case, the directional derivative is equal to the norm of the gradient.

Thus we have obtained another interpretation for the gradient:

> *The direction of the gradient is that of maximal increase of the function.*
> *The norm of the gradient is the rate of increase of the function in that direction (i.e. in the direction of maximal increase).*

The directional derivative in the direction of A is at a minimum when $\cos \theta = -1$. This is the case when we select A to have opposite direction to $\operatorname{grad} f(P)$. That direction is therefore the direction of maximal decrease of the function.

For example, f might represent a temperature distribution in space. At any point P, a particle which feels cold and wants to become warmer fastest should move in the direction of grad $f(P)$. Another particle which is warm and wants to cool down fastest should move in the direction of $-\operatorname{grad} f(P)$.

Example 2. Let $f(x, y) = x^2 + y^3$ again, and let $P = (-1, 3)$. Find the directional derivative of f at P, in the direction of maximal increase of f.

We had found previously that grad $f(P) = (-2, 27)$. The directional derivative of f in the direction of maximal increase is precisely the norm of the gradient, and so is equal to

$$\|\operatorname{grad} f(P)\| = \|(-2, 27)\| = \sqrt{4 + 27^2} = \sqrt{733} .$$

EXERCISES

1. Let $f(x, y, z) = z - e^x \sin y$, and $P = (\log 3, 3\pi/2, -3)$. Find:
   (a) the directional derivative of f at P in the direction of $(1, 2, 2)$,
   (b) the maximum and minimum values for the directional derivative of f at P.

2. Find the directional derivatives of the following functions at the specified points in the specified directions.
   (a) $\log(x^2 + y^2)^{1/2}$ at $(1, 1)$, direction $(2, 1)$
   (b) $xy + yz + zx$ at $(-1, 1, 7)$, direction $(3, 4, -12)$
   (c) $4x^2 + 9y^2$ at $(2, 1)$ in the direction of maximum directional derivative

3. A temperature distribution in space is given by the function

   $$f(x, y) = 10 + 6 \cos x \cos y + 3 \cos 2x + 4 \cos 3y.$$

   At the point $(\pi/3, \pi/3)$, find the direction of greatest increase of temperature, and the direction of greatest decrease of temperature.

4. In what direction are the following functions of X increasing most rapidly at the given point?
   (a) $x/\|X\|^{3/2}$ at $(1, -1, 2)$ $(X = (x, y, z))$
   (b) $\|X\|^5$ at $(1, 2, -1, 1)$ $(X = (x, y, z, w))$

5. (a) Find the directional derivative of the function

   $$f(x, y) = 4xy + 3y^2$$

   in the direction of $(2, -1)$, at the point $(1, 1)$.
   (b) Find the directional derivative in the direction of maximal increase of the function.

6. Let $f(x, y, z) = (x + y)^2 + (y + z)^2 + (z + x)^2$. What is the direction of greatest increase of the function at the point $(2, -1, 2)$? What is the directional derivative of f in this direction at that point?

7. Let $f(x, y) = x^2 + xy + y^2$. What is the direction in which f is increasing most rapidly at the point $(-1, 1)$? Find the directional derivative of f in this direction.

8. Suppose the temperature in (x, y, z)-space is given by

$$f(x, y, z) = x^2y + yz - e^{xy}.$$

Compute the rate of change of temperature at the point $(1, 1, 1)$ in the direction pointing toward the origin.

9. Let f be a differentiable function defined on an open set U. Suppose that P is a point of U such that $f(P)$ is a maximum, i.e. suppose we have

$$f(P) \geqq f(X) \qquad \text{for all } X \text{ in } U.$$

Show that grad $f(P) = O$.

10. Let f be a function on an open set U in 3-space. Let g be another function, and let S be the surface consisting of all points X such that

$$g(X) = 0 \text{ but grad } g(X) \neq O.$$

(All functions are assumed continuously differentiable.) Suppose that P is a point of the surface S such that $f(P)$ is a maximum for f on S, that is

$$f(P) \geqq f(X) \text{ for all } X \text{ on } S.$$

Prove that there exists a number λ such that

$$\text{grad } f(P) = \lambda \text{ grad } g(P).$$

[*Hint:* Recall how the tangent plane was motivated.]

§4. FUNCTIONS DEPENDING ONLY ON THE DISTANCE FROM THE ORIGIN

The first such function which comes to mind is the distance function. In 2-space, it is given by

$$r = \sqrt{x^2 + y^2} \ .$$

In 3-space, it is given by

$$r = \sqrt{x^2 + y^2 + z^2} \ .$$

In n-space, it is given by

$$r = \sqrt{x_1^2 + x_2^2 + \cdots + x_n^2} \ .$$

Let us find its gradient. For instance, in 2-space,

$$\frac{\partial r}{\partial x} = \frac{1}{2}(x^2 + y^2)^{-1/2}2x$$

$$= \frac{x}{\sqrt{x^2 + y^2}} = \frac{x}{r}.$$

Differentiating with respect to y instead of x you will find

$$\frac{\partial r}{\partial y} = \frac{y}{r}.$$

Hence

$$\text{grad } r = \left(\frac{x}{r}, \frac{y}{r} \right).$$

This can also be written

$$\text{grad } r = \frac{X}{r}.$$

Thus the gradient of r is the unit vector in the direction of the position vector. It points outward from the origin.

If we are dealing with functions on 3-space, so

$$r = \sqrt{x^2 + y^2 + z^2}$$

then the chain rule again gives

$$\frac{\partial r}{\partial x} = \frac{x}{r}, \qquad \frac{\partial r}{\partial y} = \frac{y}{r}, \qquad \text{and} \qquad \frac{\partial r}{\partial z} = \frac{z}{r}$$

so again

$$\text{grad } r = \frac{X}{r}.$$

Warning: Do **not** write $\partial r/\partial X$. This suggests dividing by a vector X and is therefore bad notation. The notation $\partial r/\partial x$ was correct and good notation since we differentiate only with respect to the single variable x. Information coming from differentiating with respect to all the variables is correctly expressed by the formula grad $r = X/r$ in the box.

In n-space, let

$$r = \sqrt{x_1^2 + \cdots + x_n^2}.$$

Then

$$\frac{\partial r}{\partial x_i} = \frac{1}{2} \left(x_1^2 + \cdots + x_n^2 \right)^{-1/2} 2x_i$$

so

$$\frac{\partial r}{\partial x_i} = \frac{x_i}{r}.$$

By definition of the gradient, it follows that

$$\text{grad } r = \frac{X}{r}.$$

We now come to other functions depending on the distance. Such functions arise frequently. For instance, a temperature function may be inversely proportional to the distance from the source of heat. A potential function may be inversely proportional to the square of the distance from a certain point. The gradient of such functions has special properties which we discuss further.

Example 1. Let

$$f(x, y) = \sin r = \sin \sqrt{x^2 + y^2} .$$

Then $f(x, y)$ depends only on the distance r of (x, y) from the origin. By the chain rule,

$$\frac{\partial f}{\partial x} = \frac{d\sin r}{dr} \cdot \frac{\partial r}{\partial x}$$

$$= (\cos r)\frac{1}{2}(x^2 + y^2)^{-1/2}2x$$

$$= (\cos r)\frac{x}{r} .$$

Similarly, $\partial f/\partial y = (\cos r)y/r$. Consequently

$$\text{grad} f(x,y) = \left((\cos r)\frac{x}{r}, (\cos r)\frac{y}{r}\right)$$

$$= \frac{\cos r}{r}(x,y)$$

$$= \frac{\cos r}{r} X.$$

The same use of the chain rule as in the special case

$$f(x, y) = \sin r$$

which we worked out in Example 1 shows:

Let g be a differentiable function of one variable, and let $f(X) = g(r)$. Then

$$\text{grad } f(X) = \frac{g'(r)}{r} X.$$

Work out all the examples given in Exercise 2. You should memorize and keep in mind this simple expression for the gradient of a function which depends only on the distance. Such dependence is expressed by the function g.

Exercises 9 and 10 give important information concerning functions which depend only on the distance from the origin, and should be seen as essential complements of this section. They will prove the following result.

A differentiable function $f(X)$ depends only on the distance of X from the origin if and only if grad $f(X)$ is parallel to X or O.

In this situation, the gradient grad $f(X)$ may point towards the origin, or away from the origin, depending on whether the function is decreasing or increasing as the point moves away from the origin.

Example 2. Suppose a heater is located at the origin, and the temperature at a point decreases as a function of the distance from the origin, say is inversely proportional to the square of the distance from the origin. Then temperature is given as

$$h(X) = g(r) = k/r^2$$

for some constant $k > 0$. Then the gradient of temperature is

$$\text{grad } h(X) = -2k\frac{1}{r^3}\frac{X}{r} = -\frac{2k}{r^4}X.$$

The factor $2k/r^4$ is positive, and we see that grad $h(X)$ points in the direction of $-X$. Each circle centered at the origin is a level curve for temperature. Thus the gradient may be drawn as on the following figure. The gradient is parallel to X but in opposite direction. A bug traveling along the circle will stay at constant temperature. If he wants to get warmer fastest, he must move toward the origin.

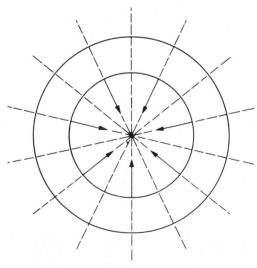

Figure 4

The dotted lines indicate the path of the bug when moving in the direction of maximal increase of the function. These lines are perpendicular to the circles of constant temperature.

Example 3. Instead of having one source of heat we may have several. For instance, suppose we have two sources of heat located at points Q_1 and Q_2, and these sources give off the same amount of heat. For *each* source the function describing temperature at a point X is then given by

$$f_1(X) = \frac{k}{\|X - Q_1\|^2} \qquad \text{and} \qquad f_2(X) = \frac{k}{\|X - Q_2\|^2}.$$

Hence the temperature at X due to the two sources is the sum

$$f(X) = f_1(X) + f_2(X).$$

The level curves are called **isothermal curves**, or curves of constant temperature. They look as they do on Fig. 5.

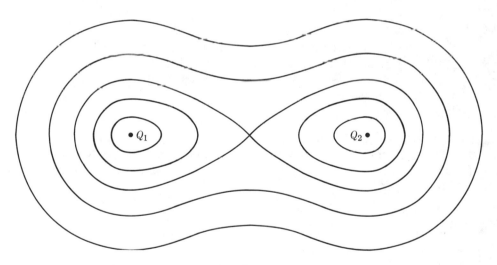

Figure 5

There is a similar picture if instead of heat we think of two magnets placed at Q_1 and Q_2, or two sources of electricity. The dotted curves represent the curves having the direction of maximal increase of the function. The non-dotted curves represent the level curves where the force of attraction between the two magnets is constant. One can verify experimentally that these two sets of curves intersect orthogonally, i.e. that the level curves are perpendicular to the curves of maximal increase, thus confirming the theoretical proof of this fact.

Sometimes we want to take a repeated derivative of a function depending only on r. It is then useful for brevity of notation *not* to expand r in terms of its definition as the square root of sum of squares.

Example 4. Let $r = \sqrt{x^2 + y^2}$ and let $f(x, y) = 1/r^3$. We wish to find

$$D_1 D_2 f(x, y) = \frac{\partial^2 f}{\partial x\, \partial y}.$$

First we find

$$D_2 f(x, y) = \frac{d}{dr}(1/r^3) \frac{\partial r}{\partial y} = -3r^{-4} \frac{y}{r} = -3 \frac{y}{r^5}.$$

[You should know from the chain rule that $\partial r / \partial y = y/r$.]

Next we take $D_1 = \partial / \partial x$ of this last expression, using the chain rule again. Then:

$$D_1 D_2 f(x, y) = \frac{\partial}{\partial x}\left(-3 \frac{y}{r^5}\right)$$

$$= -3y \frac{\partial}{\partial x}\left(\frac{1}{r^5}\right)$$

$$= -3y \frac{d}{dr}\left(\frac{1}{r^5}\right) \cdot \frac{x}{r}$$

$$= \frac{15xy}{r^7}.$$

Suppose we deal with a function of two variables $f(x, y)$. It comes up frequently in physics and mathematics and many other fields to consider the function

$$\frac{\partial^2 f}{\partial x^2} + \frac{\partial^2 f}{\partial y^2} = D_1^2 f(x, y) + D_2^2 f(x, y).$$

Without writing the variables explicitly, we may just write the function in the form

$$D_1^2 f + D_2^2 f.$$

Functions of two variables which satisfy the condition

$$D_1^2 f + D_2^2 f = 0$$

are called **harmonic**. There is, of course, a similar definition for harmonic functions of three variables $f(x, y, z)$, namely, those satisfying

$$D_1^2 f + D_2^2 f + D_3^2 f = 0.$$

This is called **Laplace's equation**, and we view $D_1^2 + D_2^2$ in 2-space, or

$$D_1^2 + D_2^2 + D_3^2$$

as an operator in 3-space on functions, called the **Laplace operator**. Examples of harmonic functions are given in Exercise 11. If (x, y, z) are the three variables, then Laplace's equation can also be written

$$\frac{\partial^2 f}{\partial x^2} + \frac{\partial^2 f}{\partial y^2} + \frac{\partial^2 f}{\partial z^2} = 0.$$

In Exercise 12 you will express this condition more simply for a function which depends only on r.

EXERCISES

1. Let g be a function of r, let $r = \|X\|$, and $X = (x, y, z)$. Let $f(X) = g(r)$. Show that

$$\left(\frac{dg}{dr}\right)^2 = \left(\frac{\partial f}{\partial x}\right)^2 + \left(\frac{\partial f}{\partial y}\right)^2 + \left(\frac{\partial f}{\partial z}\right)^2.$$

2. Let g be a function of r, and $r = \|X\|$. Let $f(X) = g(r)$. Find grad $f(X)$ for the following functions.

   (a) $g(r) = 1/r$ (b) $g(r) = r^2$ (c) $g(r) = 1/r^3$
   (d) $g(r) = e^{-r^2}$ (e) $g(r) = \log 1/r$ (f) $g(r) = 4/r^m$
   (g) $g(r) = \cos r$

   You may either work out each exercise separately, writing $r = \sqrt{x_1^2 + \cdots + x_n^2}$, and use the chain rule, finding $\partial f/\partial x_i$ in each case, or you may apply the general formula obtained in Example 1, that if $f(X) = g(r)$, we have

$$\text{grad } f(X) = \frac{g'(r)}{r} X.$$

   Probably you should do both for a while to get used to the various notations and situations which may arise.

   The next five exercises concern certain parametrizations, and some of the results from them will be used in Exercise 9.

3. Let A, B be two unit vectors such that $A \cdot B = 0$. Let $F(t) = (\cos t)A + (\sin t)B$. Show that $F(t)$ lies on the sphere of radius 1 centered at the origin, for each value of t. [*Hint*: What is $F(t) \cdot F(t)$?]

4. Let P, Q be two points on the sphere of radius 1, centered at the origin. Let $L(t) = P + t(Q - P)$, with $0 \le t \le 1$. If there exists a value of t in $[0, 1]$ such that $L(t) = O$, show that $t = \frac{1}{2}$, and that $P = -Q$.

5. Let P, Q be two points on the sphere of radius 1. Assume that $P \ne -Q$. Show that there exists a curve joining P and Q on the sphere of radius 1, centered at the origin. [*Hint*: Divide $L(t)$ in Exercise 4 by its norm.]

6. If P, Q are two unit vectors such that $P = -Q$, show that there exists a differentiable curve joining P and Q on the sphere of radius 1, centered at the origin. You may assume that there exists a unit vector A which is perpendicular to P. Then use Exercise 3.

7. Parametrize the ellipse $(x^2/a^2) + (y^2/b^2) = 1$ by a differentiable curve.

8. Let f be a differentiable function (in two variables) such that grad $f(X) = cX$ for some constant c and all X in 2-space. Show that f is constant on any circle of radius $a > 0$, centered at the origin. [*Hint*: Put $x = a \cos t$ and $y = a \sin t$ and find df/dt.]

Exercise 8 is a special case of a general phenomenon, stated in Exercise 9.

9. Let f be a differentiable function in n variables, and assume that there exists a function h such that grad $f(X) = h(X)X$. Show that f is constant on the sphere of radius $a > 0$ centered at the origin.

[That f is constant on the sphere of radius a means that given any two points P, Q on this sphere, we must have $f(P) = f(Q)$. To prove this, use the fact proved in Exercises 5 and 6 that given two such points, there exists a curve $C(t)$ joining the two points, i.e. $C(t_1) = P$, $C(t_2) = Q$, and $C(t)$ lies on the sphere for all t in the interval of definition, so

$$C(t) \cdot C(t) = a^2.$$

The hypothesis that grad $f(X)$ can be written in the form $h(X)X$ for some function h means that grad $f(X)$ is *parallel* to X (or O). Indeed, we know that grad $f(X)$ parallel to X means that grad $f(X)$ is equal to a scalar multiple of X, and this scalar may depend on X, so we have to write it as a function $h(X)$.]

10. Let $r = \|X\|$. Let g be a differentiable function of one variable whose derivative is never equal to 0. Let $f(X) = g(r)$. Show that grad $f(X)$ is parallel to X for $X \neq O$.

[This statement is the converse of Exercise 9. The proof is quite easy, cf. Example 1. The function $h(X)$ of Exercise 9 is then seen to be equal to $g'(r)/r$.]

11. Verify that the following functions are harmonic.

(a) $\log \sqrt{x^2 + y^2} = \log r$ (in two variables!)

(b) $\dfrac{1}{\sqrt{x^2 + y^2 + z^2}} = \dfrac{1}{r}$ (in three variables!)

12. (a) Let $f(x, y) = g(r)$ where $r = \sqrt{x^2 + y^2}$. Show that

$$\frac{\partial^2 f}{\partial x^2} + \frac{\partial^2 f}{\partial y^2} = \frac{d^2 g}{dr^2} + \frac{1}{r}\frac{dg}{dr} .$$

(b) If $f(x, y) = e^{-r^2}$, show that

$$\frac{\partial^2 f}{\partial x^2} + \frac{\partial^2 f}{\partial y^2} = 4f(x, y)(r^2 - 1).$$

13. Let $f(x, y, z) = g(r)$, where $r = \sqrt{x^2 + y^2 + z^2}$. Show that

$$\frac{\partial^2 f}{\partial x^2} + \frac{\partial^2 f}{\partial y^2} + \frac{\partial^2 f}{\partial z^2} = \frac{d^2 g}{dr^2} + \frac{2}{r}\frac{dg}{dr} .$$

Note. The right-hand side gives the left-hand side in terms of the single coordinate r. When we consider functions depending only on the distance from the origin, we see that the right-hand side involves only ordinary differentiation with respect to one variable, namely the distance r, whereas the left-hand side involves the three partial derivatives as shown, which is more complicated. We have seen that a function f such that the left-hand side in the relation of the exercise is equal to 0 is called harmonic. When the function depends only the distance, as arises frequently in physics, then the condition for the function to be harmonic can be expressed in terms of ordinary differentiation instead of partial differentiations, thus leading to ordinary differential equations rather than partial differential equations. The same principle occurs in many other contexts, when it is possible to get rid of some of the variables.

§5. FURTHER TECHNIQUE IN PARTIAL DIFFERENTIATION

The techniques developed in this section will not be used in the next applications and can be omitted. They have their own flavor, and have importance in other contexts, especially what is known as partial differential equations. They are included here to provide the opportunity to learn them if this is deemed important in the context of the particular given course.

The chain rule as stated in §1 can be applied to the seemingly more general situation when x, y are functions of more than one variable. Let $f(x, y)$ be a function of two variables. Suppose that

$$x = \varphi(t, u) \qquad \text{and} \qquad y = \psi(t, u)$$

are differentiable functions of two variables. Let

$$g(t, u) = f(\varphi(t, u), \psi(t, u)).$$

If we keep u fixed and take the partial derivative of g with respect to t, then we can apply our chain rule, and obtain

$$\frac{\partial g}{\partial t} = \frac{\partial f}{\partial x}\frac{\partial x}{\partial t} + \frac{\partial f}{\partial y}\frac{\partial y}{\partial t}.$$

In the D_1, D_2 notation, this also reads

$$D_1 g(t, u) = D_1 f(x, y) D_1 \varphi(t, u) + D_2 f(x, y) D_1 \psi(t, u)$$

or also

$$\boxed{D_1 g(t, u) = D_1 f(x, y)\frac{\partial x}{\partial t} + D_2 f(x, y)\frac{\partial y}{\partial t}.}$$

Experience will show you which is the most convenient notation.

Example 1. Let $f(x, y) = x^2 + 2xy$. Let $x = r \cos \theta$ and $y = r \sin \theta$. Let $g(r, \theta) = f(r \cos \theta, r \sin \theta)$ be the composite function. Find $\partial g/\partial \theta$.

We have

$$\frac{\partial f}{\partial x} = 2x + 2y, \qquad \frac{\partial f}{\partial y} = 2x,$$

$$\frac{\partial x}{\partial \theta} = -r \sin \theta \qquad \text{and} \qquad \frac{\partial y}{\partial \theta} = r \cos \theta.$$

Hence

(*) $$\frac{\partial g}{\partial \theta} = (2x + 2y)(-r \sin \theta) + 2x(r \cos \theta).$$

If you want the answer completely in terms of r, θ, you can substitute $r \cos \theta$ and $r \sin \theta$ for x and y respectively in this expression. Written in full, the answer reads:

$$D_2 g(r, \theta) = (D_1 f)(r \cos \theta, r \sin \theta)(-r \sin \theta) + (D_2 f)(r \cos \theta, r \sin \theta) r \cos \theta$$
$$= (2r \cos \theta + 2r \sin \theta)(-r \sin \theta) + 2(r \cos \theta)(r \cos \theta).$$

Such an expression is clumsy to write, and that is why we leave it in abbreviated form as in (*).

Example 2. Sometimes the letters x and y are occupied to denote variables which are not the first and second variables of the function f. In this case, **other letters must be used if we wish to replace $D_1 f$ and $D_2 f$ by partial derivatives with respect to these variables.** For example, let

$$u = f(x^2 - y, xy).$$

To find $\partial u / \partial x$, we let

$$s = x^2 - y \qquad \text{and} \qquad t = xy.$$

Then

$$\frac{\partial u}{\partial x} = \frac{\partial f}{\partial s}\frac{\partial s}{\partial x} + \frac{\partial f}{\partial t}\frac{\partial t}{\partial x}$$
$$= \frac{\partial f}{\partial s}2x + \frac{\partial f}{\partial t}y = D_1 f(s, t)2x + D_2 f(s, t)y.$$

The function u depends on x, y and we may write

$$u = g(x, y) = f(x^2 - y, xy).$$

Then we write

(1) $$\frac{\partial g}{\partial x} = \frac{\partial f}{\partial s}2x + \frac{\partial f}{\partial t}y.$$

Similarly,

(2)
$$\frac{\partial g}{\partial y} = \frac{\partial f}{\partial s}(-1) + \frac{\partial f}{\partial t}x.$$

The advantage of the $D_1 f$, $D_2 f$ notation is that it does not depend on a choice of letters, and makes it clear that we take the partial derivatives of f with respect to the first and second variables. In that notation, (1) and (2) read:

(1*) $D_1 g(x, y) = D_1 f(x^2 - y, xy)2x + D_2 f(x^2 - y, xy)y$

(2*) $D_2 g(x, y) = D_1 f(x^2 - y, xy)(-1) + D_2 f(x^2 - y, xy)x.$

When written in that form, which is the *only* correct form, the formula has the property that it is invariant under permutations of the alphabet. We can change x, y to any other two letters and the formula remains valid (provided the two letters are different from f and g, and D, of course). Thus we would have:

(1**) $D_1 g(v, w) = D_1 f(v^2 - w, vw)2v + D_2 f(v^2 - w, vw)w$

(2**) $D_2 g(v, w) = D_1 f(v^2 - w, vw)(-1) + D_2 f(v^2 - w, vw)v.$

We avoided the letter u also because at the beginning of the discussion we let $u = f(x^2 - y, xy)$, so for purposes of the discussion, the letter u was already occupied. On the other hand, it is slightly more clumsy to write $D_1 f(s, t)$ rather than $\partial f/\partial s$. Thus the second notation, **when used with an appropriate choice of variables**, is shorter and a little more mechanical. We emphasize, however, that it can only be used when the letters denoting the variables have been fixed properly.

Example 3. Let $g(t, x, y) = f(t^2 x, ty)$. Then

$$\frac{\partial g}{\partial t} = D_1 f(t^2 x, ty)2tx + D_2 f(t^2 x, ty)y.$$

Here again, since the letter x is occupied, we cannot write $\partial f/\partial x$ for $D_1 f$. In this example, we view x, y as fixed, and $g(t, x, y)$ as the function of t alone. If we put

$$C(t) = (t^2 x, ty),$$

then

$$C'(t) = (2tx, y).$$

We see that $\partial g/\partial t$ has the form

$$\frac{\partial g}{\partial t} = \operatorname{grad} f(C(t)) \cdot C'(t).$$

Evaluating at special numbers then gives:

$$D_1 g(1, x, y) = D_1 f(x, y)2x + D_2 f(x, y)y$$
$$D_1 g(0, x, y) = D_2 f(0, 0)y$$
$$D_1 g(1, x, 1) = D_1 f(x, 1)2x + D_2 f(x, 1)$$

and so forth.

Example 4. Keeping the same functions as in Example 3, we now find the repeated derivative $\partial^2 g/\partial t^2$. We apply the same principle as before, but to the two functions $D_1 f$ and $D_2 f$. Also we have to use the rule for the derivative of a product, because $D_1 f(t^2 x, ty)2tx$ is a product of two functions of t. We then find

$$\frac{\partial^2 g}{\partial t^2} = D_1 f(t^2 x, ty)2x + \left[D_1 D_1 f(t^2 x, ty)2tx + D_2 D_1 f(t^2 x, ty)y \right]2tx$$

$$+ D_1 D_2 f(t^2 x, ty)2txy + D_2 D_2 f(t^2 x, ty)yy.$$

Of course, we may replace $D_1 D_1 f$ by $D_1^2 f$ and $D_2 D_2 f$ by $D_2^2 f$.

EXERCISES

(All functions are assumed to be differentiable as needed.)

1. If $x = u(r, s, t)$ and $y = v(r, s, t)$ and $z = f(x, y)$, write out the formula for

$$\frac{\partial z}{\partial r} \quad \text{and} \quad \frac{\partial z}{\partial t}.$$

2. Find the partial derivatives with respect to x, y, s, and t for the following functions.
   (a) $f(x, y, z) = x^3 + 3xyz - y^2 z, \ x = 2t + s, y = -t - s, z = t^2 + s^2$
   (b) $f(x, y) = (x + y)/(1 - xy), \ x = \sin 2t, y = \cos(3t - s)$

3. Let f be a differentiable function on $\mathbf{R}^3$ and suppose that

$$D_1 f(0, 0, 0) = 2, \qquad D_2 f(0, 0, 0) = D_3 f(0, 0, 0) = 3.$$

   Let $g(u, v) = f(u - v, u^2 - 1, 3v - 3)$. Find $D_1 g(1, 1)$.

4. Assume that f is a function satisfying

$$f(tx, ty) = t^m f(x, y).$$

   for all numbers x, y, and t. Show that

$$x^2 \frac{\partial^2 f}{\partial x^2} + 2xy \frac{\partial^2 f}{\partial x \, \partial y} + y^2 \frac{\partial^2 f}{\partial y^2} = m(m - 1)f(x, y).$$

   [*Hint*: Differentiate twice with respect to t. Then put $t = 1$.]

5. If $u = f(x - y, y - x)$, show that

$$\frac{\partial u}{\partial x} + \frac{\partial u}{\partial y} = 0.$$

6. (a) Let $g(x, y) = f(x + y, x - y)$, where f is a differentiable function of two variables, say $f = f(u, v)$. Show that

$$\frac{\partial g}{\partial x} \frac{\partial g}{\partial y} = \left(\frac{\partial f}{\partial u}\right)^2 - \left(\frac{\partial f}{\partial v}\right)^2.$$

(b) Let $g(x, y) = f(2x + 7y)$, where f is a differentiable function of one variable. Show that

$$2\frac{\partial g}{\partial y} = 7\frac{\partial g}{\partial x}.$$

(c) Let $g(x, y) = f(2x^3 + 3y^2)$. Show that

$$y\frac{\partial g}{\partial x} = x^2\frac{\partial g}{\partial y}.$$

7. Let $x = u \cos \theta - v \sin \theta$, and $y = u \sin \theta + v \cos \theta$, with θ equal to a constant. Let $f(x, y) = g(u, v)$. Show that

$$\left(\frac{\partial g}{\partial u}\right)^2 + \left(\frac{\partial g}{\partial v}\right)^2 = \left(\frac{\partial f}{\partial x}\right)^2 + \left(\frac{\partial f}{\partial y}\right)^2.$$

8. (a) Let $x = r \cos \theta$ and $y = r \sin \theta$. Let $z = f(x, y)$. Show that

$$\frac{\partial z}{\partial r} = \frac{\partial f}{\partial x} \cos \theta + \frac{\partial f}{\partial y} \sin \theta, \qquad \frac{1}{r}\frac{\partial z}{\partial \theta} = -\frac{\partial f}{\partial x} \sin \theta + \frac{\partial f}{\partial y} \cos \theta.$$

(b) If we let $z = g(r, \theta) = f(r \cos \theta, r \sin \theta)$, show that

$$\left(\frac{\partial g}{\partial r}\right)^2 + \frac{1}{r^2}\left(\frac{\partial g}{\partial \theta}\right)^2 = \left(\frac{\partial f}{\partial x}\right)^2 + \left(\frac{\partial f}{\partial y}\right)^2.$$

9. Let c be a constant, and let $z = \sin(x + ct) + \cos(2x + 2ct)$. Show that

$$\frac{\partial^2 z}{\partial t^2} = c^2 \frac{\partial^2 z}{\partial x^2}.$$

10. Let c be a constant and let $z = f(x + ct) + g(x - ct)$. Let

$$u = x + ct \quad \text{and} \quad v = x - ct.$$

Show that

$$\frac{\partial^2 z}{\partial t^2} = c^2\frac{\partial^2 z}{\partial x^2} = c^2(f''(u) + g''(v)).$$

11. Let $z = f(u, v)$ and $u = x + y$, $v = x - y$. Show that

$$\frac{\partial^2 z}{\partial x \partial y} = \frac{\partial^2 z}{\partial u^2} - \frac{\partial^2 z}{\partial v^2}.$$

12. Let $z = f(x + y) - g(x - y)$. Let $u = x + y$ and $v = x - y$. Show that

$$\frac{\partial^2 z}{\partial x^2} = \frac{\partial^2 z}{\partial y^2} = f''(u) - g''(v).$$

13. Let $x = r \cos \theta$, $y = r \sin \theta$ be the formulas for the polar coordinates. Let

$$f(x, y) = f(r \cos \theta, r \sin \theta) = g(r, \theta).$$

Show that

$$\boxed{\frac{\partial^2 g}{\partial r^2} + \frac{1}{r} \frac{\partial g}{\partial r} + \frac{1}{r^2} \frac{\partial^2 g}{\partial \theta^2} = \frac{\partial^2 f}{\partial x^2} + \frac{\partial^2 f}{\partial y^2}.}$$

Note. This exercise gives the Laplace operator in polar coordinates. It is important because it shows you how the right-hand side can be expressed in terms of polar coordinates on the left-hand side. The right-hand side occurs frequently in the theory of wave motions.

For the proof, start with the formulas of Exercise 8(a), namely,

$$\frac{\partial g}{\partial r} = \frac{\partial f}{\partial x} \cos \theta + \frac{\partial f}{\partial y} \sin \theta \qquad \frac{\partial g}{\partial \theta} = -\frac{\partial f}{\partial x} r \sin \theta + \frac{\partial f}{\partial y} r \cos \theta,$$

and take further derivatives with respect to r and with respect to θ, using the rule for derivative of a product, together with the chain rule. Then add the expression you obtain to form the left-hand side of the relation you are supposed to prove. There should be enough cancellation on the right-hand side to prove the desired relation.

14. Let n be a positive integer. For each of the following functions $g(r, \theta)$ show that

$$\frac{\partial^2 g}{\partial r^2} + \frac{1}{r} \frac{\partial g}{\partial r} + \frac{1}{r^2} \frac{\partial^2 g}{\partial \theta^2} = 0.$$

(a) $g(r, \theta) = r^n \cos n\theta$ (b) $g(r, \theta) = r^n \sin n\theta$.

Note. A function $f(x, y) = g(r, \theta)$ which satisfies the condition of this exercise is called **harmonic**, and is important in the theory of wave motions. This exercise gives the basic example of harmonic functions.

CHAPTER V

Potential Functions

§1. CONSERVATION LAW

Let U be an open set. By a **vector field** on U we mean an association which to every point of U associates a vector of the same dimension.

If F is a vector field on U, and X a point of U, then we denote by $F(X)$ the vector associated to X by F and call it the **value of** F **at** X, as usual.

Example 1. Let $F(x, y) = (x^2y, \sin xy)$. Then F is a vector field which to the point (x, y) associates $(x^2y, \sin xy)$, having the same number of coordinates, namely two of them in this case.

A vector field in physics is often interpreted as a field of forces. A vector field may be visualized as a field of arrows, which to each point associates an arrow as shown on the figure.

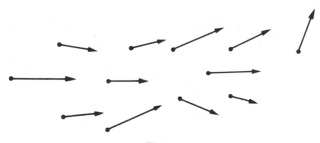

Figure 1

Each arrow points in the direction of the force, and the length of the arrow represents the magnitude of the force.

If f is a differentiable function on U, then we observe that grad f is a vector field, which associates the vector grad $f(P)$ to the point P of U.

If F is a vector field, and if there exists a differentiable function f such that $F = \operatorname{grad} f$, then the vector field is called **conservative**. Since

$$-\operatorname{grad} f = \operatorname{grad}(-f)$$

it does not matter whether we use f or $-f$ in the definition of conservative.

Let us assume that F is a conservative field on U, and let ψ be a differentiable function such that for all points X in U we have

$$F(X) = -\operatorname{grad} \psi.$$

In physics, one interprets ψ as the **potential energy**. Suppose that a particle of mass m moves along a differentiable curve $C(t)$ in U, and let us assume that this particle obeys Newton's law:

$$F(C(t)) = mC''(t)$$

for all t where $C(t)$ is defined. (Force equals mass times acceleration.)

Physicists define the **kinetic energy** to be

$$\tfrac{1}{2}mC'(t)^2 = \tfrac{1}{2}mv(t)^2.$$

Conservation Law. *The sum of the potential energy and kinetic energy is constant.*

Proof. We have to prove that

$$\psi(C(t)) + \tfrac{1}{2}mC'(t)^2$$

is constant. To see this, we differentiate this sum. By the chain rule, we see that the derivative is equal to

$$\operatorname{grad} \psi(C(t)) \cdot C'(t) + mC'(t) \cdot C''(t).$$

By Newton's law, $mC''(t) = F(C(t)) = -\operatorname{grad} \psi(C(t))$. Hence this derivative is equal to

$$\operatorname{grad} \psi(C(t)) \cdot C'(t) - \operatorname{grad} \psi(C(t)) \cdot C'(t) = 0.$$

This proves what we wanted.

It is not true that all vector fields are conservative. We shall discuss the problem of determining which ones are conservative in the next chapter.

The fields of classical physics are for the most part conservative.

Example 2. Consider a force $F(X)$ which is inversely proportional to the square of the distance from the point X to the origin, and in the direction of X. Then there is a constant k such that for $X \neq O$ we have

$$F(X) = k \frac{1}{\|X\|^2} \frac{X}{\|X\|},$$

because $X/\|X\|$ is the unit vector in the direction of X. Thus

$$F(X) = k \frac{1}{r^3} X,$$

where $r = \|X\|$. A potential energy for F is given by

$$\psi(X) = \frac{k}{r}.$$

This is immediately verified by taking the partial derivatives of this function.

If there exists a function $\varphi(X)$ such that

$$F(X) = (\text{grad } \varphi)(X), \qquad \text{that is} \quad F = \text{grad } \varphi,$$

then we shall call such a function φ a **potential function** for F. Our conventions are such that a potential function is equal to *minus* the potential energy.

EXERCISES

1. Find a potential function for a force field $F(X)$ that is inversely proportional to the distance from the point X to the origin and is in the direction of X.

2. Same question, replacing "distance" with "cube of the distance."

3. Let k be an integer ≥ 1. Find a potential function for the vector field F given by

$$F(X) = \frac{1}{r^k} X, \qquad \text{where} \quad r = \|X\|.$$

[*Hint*: Recall the formula that if $\varphi(X) = g(r)$, then

$$\text{grad } \varphi(X) = \frac{g'(r)}{r} X.$$

Set $F(X)$ equal to the right-hand side and solve for g.]

homework

Review of notions which we have had so far.

We have met three types of associations, which we list systematically.

Functions, which associated numbers to numbers or numbers to points for functions of several variables. For instance,

$$f(x, y) = \sin xy - x^3 y$$

is a function of two variables, and its values are numbers.

Curves, which associate points in space to numbers. For instance,

$$C(t) = (2t, t^2, t^3)$$

is a curve in 3-space. Here t is a number, but $C(t)$ is in $\mathbf{R}^3$.

Vector fields, which associate n-tuples to n-tuples (the same n). For instance,

$$F(x, y) = (x^2 y, \sin xy)$$

is a vector field on $\mathbf{R}^2$. Furthermore,

$$F(x, y, z) = (xz, y + z, e^{xyz})$$

defines a vector field on $\mathbf{R}^3$.

Do not confuse these various notions.

§2. POTENTIAL FUNCTIONS

We are going to deal systematically with the possibility of finding a potential function for a vector field. We begin with the case of two variables, which is typical. You should then be able to work out the case of three variables as an exercise (the answer to which will actually be carried out in the back of the book).

Let F be a vector field on an open set U. If φ is a differentiable function on U such that $F = \operatorname{grad} \varphi$, then we say that φ is a **potential function** for F.

One can raise two questions about potential functions. Are they unique, and do they exist?

We consider the first question, and we shall be able to give a satisfactory answer to it. The problem is analogous to determining an integral for a function of one variable, up to a constant, and we shall formulate and prove the analogous statement in the present situation.

We recall that even in the case of functions of one variable, it is *not* true that whenever two functions f, g are such that

$$\frac{df}{dx} = \frac{dg}{dx},$$

then f and g differ by a constant, unless we assume that f, g are defined on some interval. As we emphasized in the *First Course*, we could for instance take

$$f(x) = \begin{cases} \dfrac{1}{x} + 5 & \text{if } x < 0, \\[2mm] \dfrac{1}{x} - \pi & \text{if } x > 0, \end{cases}$$

$$g(x) = \frac{1}{x} \quad \text{if } x \neq 0.$$

Then f, g have the same derivative, but there is no constant C such that for all $x \neq 0$ we have $f(x) = g(x) + C$.

In the case of functions of several variables, we shall have to make a similar restriction on the domain of definition of the functions.

Let U be an open set and let P, Q be two points of U. We shall say that P, Q can be **joined by a differentiable curve** if there exists a differentiable curve $C(t)$ (with t ranging over some interval of numbers) which is contained in U, and two values of t, say t_1 and t_2 in that interval, such that

$$C(t_1) = P \quad \text{and} \quad C(t_2) = Q.$$

For example, if U is the entire plane, then any two points can be joined by a straight line. In fact, if P, Q are two points, then we take

$$C(t) = P + t(Q - P), \quad \text{with } 0 \leq t \leq 1.$$

When $t = 0$, then $C(0) = P$. When $t = 1$, then $C(1) = Q$.

It is not always the case that two points of an open set can be joined by a straight line. We have drawn a picture of two points P, Q in an open set U which cannot be so joined (Fig. 2). Part of the segment lies outside U.

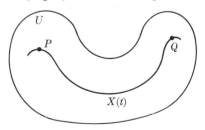

Figure 2

An open set U will be said to be **connected** if, given two points P, Q in U, there exists a differentiable curve in U which joins the two points. We are now in a position to state the theorem we had in mind.

Theorem 1. *Let U be a connected open set. Let f, g be two differentiable functions on U. If* grad $f(X) = $ grad $g(X)$ *for every point of U, then there exists a constant k such that*

$$f(X) = g(X) + k$$

for all points X of U.

Proof. We note that grad $(f - g) = $ grad $f - $ grad $g = O$, and we must prove that $f - g$ is constant. Letting $\varphi = f - g$, we see that it suffices to prove: If grad $\varphi(X) = O$ for every point X of U, then φ is constant.

Let P be a fixed point of U and let Q be any other point. Let $X(t)$ be a differentiable curve joining P to Q, which is contained in U, and defined over an interval. The derivative of the function $\varphi(X(t))$ is, by the chain rule,

$$\frac{d\varphi(X(t))}{dt} = \text{grad } \varphi(X(t)) \cdot X'(t).$$

But $X(t)$ is a point of U for all values of t in the interval. Hence by our assumption, grad $\varphi(X(t)) = O$, and so the derivative of $\varphi(X(t))$ is 0 for all t in the interval. Hence there is a constant k such that

$$\varphi(X(t)) = k$$

for all t in the interval. In other words, the function φ is constant on the curve. Hence $\varphi(P) = \varphi(Q)$.

This result is true for any point Q of U. Hence φ is constant on U, as was to be shown.

Our theorem proves the uniqueness of potential functions (within the restrictions placed by our extra hypothesis on the open set U).

We still have the problem of determining when a vector field F admits a potential function.

We first make some remarks in the case of functions of two variables.

Let F be a vector field (in 2-space), so that we can write

$$F(x, y) = (f(x, y), g(x, y))$$

with functions f and g, defined over a suitable open set. We want to know when there exists a function $\varphi(x, y)$ such that

$$\frac{\partial \varphi}{\partial x} = f \quad \text{and} \quad \frac{\partial \varphi}{\partial y} = g.$$

Such a function would be a potential function for F, by definition. (We assume throughout that all hypotheses of differentiability are satisfied as needed.)

Suppose that such a function φ exists. Then

$$\frac{\partial f}{\partial y} = \frac{\partial}{\partial y}\left(\frac{\partial \varphi}{\partial x}\right) \quad \text{and} \quad \frac{\partial g}{\partial x} = \frac{\partial}{\partial x}\left(\frac{\partial \varphi}{\partial y}\right).$$

By Theorem 3 of Chapter III, §4, the two partial derivatives on the right are equal. This means that if there exists a potential function for F, then

$$\frac{\partial f}{\partial y} = \frac{\partial g}{\partial x}, \quad \text{that is} \quad D_2 f = D_1 g.$$

This gives us a simple test in practice to tell whether a potential function may exist.

Theorem 2. *Let* f, g *be differentiable functions having continuous partial derivatives on an open set U in 2-space. If*

$$\frac{\partial f}{\partial y} \neq \frac{\partial g}{\partial x}, \quad \text{that is if} \quad D_2 f \neq D_1 g,$$

then the vector field given by $F(x, y) = (f(x, y), g(x, y))$ *does not have a potential function.*

Example. Consider the vector field given by

$$F(x, y) = (x^2 y, \sin xy).$$

Then we let $f(x, y) = x^2 y$ and $g(x, y) = \sin xy$. We have:

$$\frac{\partial f}{\partial y} = x^2 \quad \text{and} \quad \frac{\partial g}{\partial x} = y \cos xy.$$

Since $\partial f / \partial y \neq \partial g / \partial x$, it follows that the vector field does not have a potential function.

We shall prove in §3 and §6 that the converse of Theorem 2 is true in some very important cases.

EXERCISES

Determine which of the following vector fields have potential functions. The vector fields are described by the functions $(f(x, y), g(x, y))$.

1. $(1/x, xe^{xy})$

2. $(\sin(xy), \cos(xy))$

3. (e^{xy}, e^{x+y})

4. $(3x^4 y^2, x^3 y)$

5. $(5x^4 y, x \cos(xy))$

6. $\left(\dfrac{x}{\sqrt{x^2 + y^2}}, 3xy^2\right)$

§3. LOCAL EXISTENCE OF POTENTIAL FUNCTIONS

We shall state a theorem which will give us conditions under which the converse of Theorem 2 is true.

Theorem 3. (**In dimension 2.**) *Let f, g be differentiable functions on an open set of the plane. If this open set is the entire plane, or a rectangle, if the partial derivatives of f, g exist and are continuous, and if*

$$\frac{\partial f}{\partial y} = \frac{\partial g}{\partial x}, \qquad \text{that is} \qquad D_2 f = D_1 g,$$

then the vector field F(x, y) = (f(x, y), g(x, y)) has a potential function.

We shall indicate how a proof of Theorem 3 goes after we have discussed some examples.

Example 1. Determine whether the vector field F given by

$$F(x, y) = (e^{xy}, e^{x+y})$$

has a potential function.

Here, $f(x, y) = e^{xy}$ and $g(x, y) = e^{x+y}$. We have:

$$\frac{\partial f}{\partial y} = xe^{xy} \qquad \text{and} \qquad \frac{\partial g}{\partial x} = e^{x+y}.$$

Since these are not equal, we know that there cannot be a potential function.

If the partial derivatives $\partial f/\partial y$ and $\partial g/\partial x$ turn out to be equal, then one can try to find a potential function by integrating with respect to one of the variables. Thus we try to find

$$\int f(x, y) \, dx,$$

keeping y constant, and taking the ordinary integral of functions of one variable. If we can find such an integral, it will be a function $\psi(x, y)$, whose partial with respect to x will be equal to $f(x, y)$ (by definition). Adding a function of y, we can then adjust it so that its partial with respect to y is equal to $g(x, y)$.

Example 2. Let $F(x, y) = (2xy, x^2 + 3y^2)$. Determine whether this vector field has a potential function, and if it does, find it.

By definition, we have

$$f(x, y) = 2xy \qquad \text{and} \qquad g(x, y) = x^2 + 3y^2.$$

We find at once that $D_2 f = D_1 g$, so a potential function exists and we want to find it. We thus want to find $\varphi(x, y)$ such that

$$\frac{\partial \varphi}{\partial x} = 2xy, \qquad \text{and} \qquad \frac{\partial \varphi}{\partial y} = x^2 + 3y^2.$$

We first solve the problem with respect to x, and thus it is natural to use the integral

$$\int 2xy \, dx = x^2 y.$$

However, we may add to this integral any function of y alone, because y behaves like a constant with respect to x. Thus it is natural to let

$$\varphi(x, y) = \int 2xy \, dx + u(y) = x^2 y + u(y),$$

with some function $u(y)$ which is unspecified for the moment. Then certainly

$$\frac{\partial \varphi}{\partial x} = 2xy \qquad \text{because} \qquad \frac{\partial u(y)}{\partial x} = 0.$$

So half of our problem is solved. There remains to check $\dfrac{\partial \varphi}{\partial y}$. We have

$$\frac{\partial \varphi}{\partial y} = x^2 + \frac{\partial u}{\partial y},$$

and we require that $\partial \varphi / \partial y = x^2 + 3y^2$. For this it suffices that

$$\frac{\partial u}{\partial y} = 3y^2,$$

and therefore it suffices that

$$u(y) = \int 3y^2 \, dy = y^3,$$

so our final solution is

$$\varphi(x, y) = x^2 y + y^3$$

which is a potential function for F.

The analogue of Theorem 3 is also true in arbitrary dimension. We state it in dimension 3.

Theorem 4. *Let $F = (f_1, f_2, f_3)$ be a vector field on a rectangular box in 3-space, such that the functions f_1, f_2, f_3 have continuous partial derivatives. Assume that $D_i f_j = D_j f_i$ for all pairs of indices i, j. This means*

$$D_1 f_2 = D_2 f_1, \qquad D_1 f_3 = D_3 f_1, \qquad D_2 f_3 = D_3 f_2.$$

Then F has a potential function.

The same statement is valid replacing 3 by n.

Warning. It is **very important** that the domain of definition of the vector field in Theorems 3 and 4 be a rectangle (or conceivably a quite special type of open set, as discussed in the proof in §6). We shall see later that for more general types of open sets, even if $D_i f_j = D_j f_i$ for all pairs of indices i, j we **cannot** necessarily conclude that there exists a potential function.

In practice, suppose we want to find a potential function explicitly when Theorems 3 and 4 are applicable, i.e. when the vector field is defined over a rectangular box. We first integrate $f_1(x, y, z)$ with respect to x, and then the desired potential function φ will be of the form

$$\varphi(x, y, z) = \int f_1(x, y, z) \, dx + \psi(y, z)$$

where $\psi(y, z)$ is independent of x. *Note that we cannot write*

$$\psi(y, z) = u(y) + v(z)$$

as a sum of a function of y alone plus a function of z alone. It might turn out that $\psi(y, z)$ might be $y^2 z^3$ for instance, which cannot be written as such a sum.

Example 3. Find a potential function for the vector field

$$F(x, y, z) = (y \cos(xy), \, x \cos(xy) + 2yz^3, \, 3y^2 z^2).$$

We first find

$$\int y \cos(xy) \, dx = \sin xy.$$

The potential function will have the form

$$\varphi(x, y, z) = \sin xy + \psi(y, z).$$

We note that

$$\frac{\partial}{\partial y} \sin xy = x \cos(xy).$$

Hence to satisfy the condition $D_2 \varphi(x, y, z) = x \cos(xy) + 2yz^3$ we need only that

$$\frac{\partial \psi}{\partial y} = 2yz^3.$$

Integrating with respect to y yields

$$\psi(y, z) = \int 2yz^3 \, dy = y^2 z^3 + u(z)$$

where $u(z)$ is the "constant of integration" with respect to y, so

$$\varphi(x, y, z) = \sin(xy) + y^2 z^3 + u(z),$$

where $u(z)$ depends only on z. However we now see that

$$\frac{\partial(y^2 z^3)}{\partial z} = 3y^2 z^2,$$

so we can take $u(z) = 0$, and the desired potential function is

$$\varphi(x, y, z) = \sin(xy) + y^2 z^3.$$

The hypothesis $D_i f_j = D_j f_i$ *guarantees* that the above procedure can be carried out to the end to yield the desired potential function. The *proof of this*, i.e. the proof of Theorem 3 will be given in §6.

In some cases, we can tell the existence of a potential function from another principle than that of Theorems 3 and 4.

Example 4. Let $r = \sqrt{x^2 + y^2}$ and let

$$F(x, y) = \left(\frac{e^r}{r} x, \frac{e^r}{r} y \right).$$

Then F has a potential function, because we recall from Chapter IV, §4 that if $f(X) = g(r)$, then

$$\text{grad } f(X) = \frac{g'(r)}{r} X.$$

We wish to solve

$$\frac{e^r}{r} = \frac{g'(r)}{r}.$$

This amounts to solving $g'(r) = e^r$, so $g(r) = e^r$. Then

$$f(x, y) = e^r$$

is the potential function.

Of course this is also compatible with the method of Exercise 3, because $\partial r / \partial x = x/r$ and so

$$\int \frac{e^r}{r} x \, dx = \int e^r \, dr = e^r.$$

EXERCISES

Determine which of the following vector fields admit potential functions.

1. $(e^x, \sin xy)$
2. $(2x^2 y, y^3)$
3. $(2xy, y^2)$
4. $(y^2 x^2, x + y^4)$

Find potential functions for the following vector fields. We let $r = \|X\|$ and $X \neq O$.

5. (a) $F(X) = \dfrac{1}{r} X$ (b) $F(X) = \dfrac{1}{r^2} X$

(c) $F(X) = r^n X$ (if n is an integer).

6. $(4xy, 2x^2)$ 7. $(xy \cos xy + \sin xy, x^2 \cos xy)$

8. $(3x^2y^2, 2x^3y)$ 9. $(2x, 4y^3)$

10. (a) (ye^{xy}, xe^{xy}) (b) $(y \cos xy, x \cos xy)$

(c) $(2xy \cos x^2y, x^2 \cos x^2y)$

11. Let $r = \|X\|$. Let g be a differentiable function of one variable. Show that the vector field defined by

$$F(X) = \frac{g'(r)}{r} X$$

in the domain $X \neq O$ always admits a potential function. What is this potential function?

12. Find a potential function $\varphi(x, y)$ for the vector field

$$F(x, y) = (3x^2y + 2y^2, x^3 + 4xy - 1),$$

with the property that $\varphi(1, 1) = 4$.

13. Find a potential function φ for the following vector fields $F(x, y, z)$:

(a) $(2x, 3y, 4z)$ (b) $(y + z, x + z, x + y)$

(c) $(e^{y+2z}, xe^{y+2z}, 2xe^{y+2z})$ (d) $(y \sin z, x \sin z, xy \cos z)$

(e) $(yz, xz + z^3, xy + 3yz^2)$ (f) $(e^{yz}, xze^{yz}, xye^{yz})$

(g) $(z^2, 2y, 2xz)$ (h) $(yz \cos xy, xz \cos xy, \sin xy)$

(i) $(y^3z + y, 3xy^2z + x + z, xy^3 + y)$

14. What is the gradient of $\varphi(x, y) = \arctan(y/x)$, defined over any rectangle not containing the line $x = 0$?

15. Let F be a vector field on an open set in 3-space, so that F is given by three coordinate functions, say $F = (f_1, f_2, f_3)$. Define the **curl** of F to be the vector field given by

$$(\text{curl } F)(x_1, x_2, x_3) = \left(\frac{\partial f_3}{\partial x_2} - \frac{\partial f_2}{\partial x_3}, \frac{\partial f_1}{\partial x_3} - \frac{\partial f_3}{\partial x_1}, \frac{\partial f_2}{\partial x_1} - \frac{\partial f_1}{\partial x_2} \right).$$

Define the **divergence** of F to be the function $g = \text{div } F$ given by

$$g(x, y, z) = \frac{\partial f_1}{\partial x} + \frac{\partial f_2}{\partial y} + \frac{\partial f_3}{\partial z}.$$

In terms of the D_i notation, we can also write

$$\text{curl } F = (D_2 f_3 - D_3 f_2, D_3 f_1 - D_1 f_3, D_1 f_2 - D_2 f_1)$$

and

$$\text{div } F = D_1 f_1 + D_2 f_2 + D_3 f_3.$$

(a) Prove that div curl $F = 0$.

(b) Prove that curl grad $\varphi = O$, for any function φ.

Remark 1. The condition on the vector field F expressed in Theorem 4 (for three variables) is equivalent to the condition

$$\text{curl } F = O.$$

Indeed, curl $F = O$ if and only if its three coordinate functions are 0, and this is exactly equivalent with

$$D_i f_j = D_j f_i \qquad \text{for} \qquad i, j = 1, 2, 3.$$

Remark 2. The divergence was defined purely algebraically above. It has a very interesting physical interpretation, but we need more machinery to be able to derive this interpretation. See the chapter on Green's theorem and the divergence theorem.

§4. AN IMPORTANT SPECIAL VECTOR FIELD

Consider the vector field

$$G(x, y) = \left(\frac{-y}{x^2 + y^2}, \frac{x}{x^2 + y^2} \right).$$

It can be drawn pictorially as follows. Suppose that we look at its value on a circle of fixed radius r, and vary θ. Substituting

$$x = r \cos \theta \qquad \text{and} \qquad y = r \sin \theta,$$

we find that

$$G(x, y) = \left(\frac{-\sin \theta}{r}, \frac{\cos \theta}{r} \right) = \frac{1}{r}(-\sin \theta, \cos \theta).$$

On the other hand, let us parametrize the circle of fixed radius r by the usual coordinates

$$C(\theta) = (r \cos \theta, r \sin \theta),$$

so that

$$C'(\theta) = (-r \sin \theta, r \cos \theta).$$

Then we see that $C'(\theta)$ and $G(x, y)$ have the same direction, which is tangent to the circle, counterclockwise. Thus the vector field consists of forces which rotate around the circle, and has been drawn in Fig. 3.

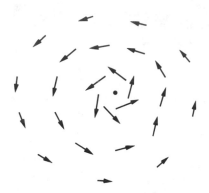

Figure 3

Note that

$$\|G(x, y)\| = \frac{1}{r} \quad \text{because} \quad \|(-\sin \theta, \cos \theta)\| = 1.$$

When $r = \sqrt{x^2 + y^2}$ is very small, then $\|G(x, y)\|$ is very large. The vector field may be viewed as representing the rotation of a fluid in a sink. The fluid rotates much more rapidly near the point where the water flows out, and rotates more slowly further away from that point.

Observe that this vector field is not defined at the origin. Indeed, the vectors (arrow) have arbitrarily large norms as we get closer to the origin. The domain of definition is the plane from which the origin has been deleted.

On the other hand, this vector field can be easily verified to satisfy the condition

$$D_2 f = D_1 g.$$

Hence by Theorem 3, if R is a rectangle **which does not contain the origin**, then G has a potential function on R. It is easy to find this potential function. We begin by trying the integral

$$\int \frac{-y}{x^2 + y^2} \, dx.$$

Since $-y$ behaves like a constant when integrating with respect to x, this amounts to finding

$$\int \frac{1}{x^2 + y^2} \, dx.$$

You should know how to do this from the first course in calculus, and by a change of variables, you should know that this integral leads to an arctangent.

In any case, we are led to the function

$$\varphi(x, y) = \arctan\frac{y}{x},$$

defined at first over any rectangle which does not meet the line $x = 0$. Taking the partial derivatives, we find:

$$\frac{\partial\varphi}{\partial x} = \frac{1}{1 + (y/x)^2}\frac{-y}{x^2} = \frac{-y}{x^2 + y^2}$$

and

$$\frac{\partial\varphi}{\partial y} = \frac{1}{1 + (y/x)^2}\frac{1}{x} = \frac{x}{x^2 + y^2}.$$

Thus $\varphi(x, y)$ is a potential function for $G(x, y)$.

We emphasize that this potential function has been defined so far by the above formula only on rectangles which do not meet the line $x = 0$. However, we can do better than that, for this special vector field.

We recognize $y/x = \tan\theta$, where θ is the usual angle as shown on the figure (Fig. 4).

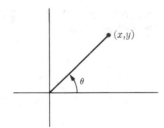

Figure 4

Let us delete a thin sector from the plane as shown on Figure 5.

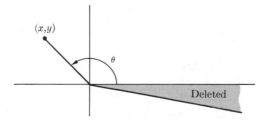

Figure 5

Let us define the function

$$\varphi(x, y) = \theta$$

where θ is not allowed to range over the deleted part, so we can describe the allowable range of values of θ by an inequality

$$0 \leq \theta \leq 2\pi - c,$$

where c is some small fixed number > 0. *Then $\varphi(x, y)$ is a potential function for $F(x, y)$.* For the values of x, y such that $x > 0, y \geq 0$ we can use the formula already given, namely

$$\theta = \arctan y/x.$$

On the line $x = 0$ we have, for instance,

$$\varphi(0, y) = \pi/2 \qquad \text{if} \quad y > 0$$
$$\varphi(0, y) = 3\pi/2 \qquad \text{if} \quad y < 0,$$

and we also have the value

$$\varphi(x, 0) = \pi \qquad \text{if} \quad x < 0.$$

It can then be easily verified that this function $\varphi(x, y) = \theta$ is a potential function for the vector field F on the plane from which the shaded region has been deleted. On the half plane to the left of the vertical line $x = 0$, this function θ differs by a constant of integration from the function

$$\arctan y/x.$$

When taking partial derivatives, this constant of integration vanishes.

There is also a formula which will give the potential function θ in the whole half-plane with $y \geq 0$, namely

$$\psi(x, y) = \arccos \frac{x}{\sqrt{x^2 + y^2}} = \arccos \frac{x}{r}.$$

A direct differentiation with respect to x, and then with respect to y, will give the first and second component of the vector field, that is

$$\text{grad } \psi(x, y) = \left(\frac{-y}{x^2 + y^2}, \frac{x}{x^2 + y^2} \right).$$

Do it as an exercise. With this formula no constant of integration is needed to get the potential function θ with $0 \leq \theta \leq \pi$.

Our construction of the potential function has been adapted especially to the special vector field of this section, which has its own peculiar behavior.

The impossibility of finding a potential function for this vector field **over the whole plane from which the origin has been deleted** should already be intuitively apparent, and will be proved in the next chapter by considering integrals along curves. See Example 3, §3 of the next chapter. Thus there is no coherent way of defining a potential function on the whole domain of definition of the vector field.

EXERCISES

1. Verify that the vector field discussed in this section satisfies the condition

$$D_2 f = D_1 g.$$

2. Verify that the function $\psi(x, y) = -\arctan x/y$ is a potential function of the vector field of this section on any rectangle not intersecting the line $y = 0$.

3. Verify that the function $\psi(x, y) = \arccos x/r$ is a potential function for this vector field in the upper half of the plane, where it is defined.

§5. DIFFERENTIATING UNDER THE INTEGRAL

As already stated, this section gives some background for the proof of Theorems 3 and 4.

Let f be a continuous function on a rectangle $a \leqq x \leqq b$ and $c \leqq y \leqq d$. We can then form a function of y by taking

$$\psi(y) = \int_a^b f(x, y)\, dx.$$

Example 1. We can determine explicitly the function ψ if we let $f(x, y) = \sin(xy)$, namely:

$$\psi(y) = \int_0^\pi \sin(xy)\, dx = -\left.\frac{\cos(xy)}{y}\right|_{x=0}^{x=\pi} = -\frac{\cos(\pi y) - 1}{y}.$$

Integrating $\sin xy$ with respect to x between definite numbers 0 and π has eliminated the variable x and left us with a function of y only.

We are interested in finding the derivative of ψ. The next theorem allows us to do this in certain cases, by differentiating with respect to y under the integral sign.

Theorem 5. *Assume that f is continuous on the preceding rectangle, and that $D_2 f$ exists and is continuous. Let*

$$\psi(y) = \int_a^b f(x, y)\, dx.$$

Then ψ is differentiable, and

$$\frac{d\psi}{dy} = D\psi(y) = \int_a^b D_2 f(x, y)\, dx = \int_a^b \frac{\partial f(x, y)}{\partial y}\, dx.$$

Proof. By definition, we have to investigate the Newton quotient for ψ. We have

$$\frac{\psi(y + h) - \psi(y)}{h} = \int_a^b \left[\frac{f(x, y + h) - f(x, y)}{h} \right] dx.$$

We then have to find

$$\lim_{h \to 0} \int_a^b \frac{f(x, y + h) - f(x, y)}{h} dx.$$

It can be shown (but we omit the proof) that we can take the limit under the integral sign, so we get

$$\int_a^b \lim_{h \to 0} \frac{f(x, y + h) - f(x, y)}{h} dx = \int_a^b D_2 f(x, y) dx,$$

thus proving our theorem.

Example 2. Letting $f(x, y) = \sin(xy)$ as before, we find that

$$D_2 f(x, y) = x \cos(xy).$$

If we let

$$\psi(y) = \int_0^\pi f(x, y) dx,$$

then

$$D\psi(y) = \int_0^\pi D_2 f(x, y) dx = \int_0^\pi x \cos(xy) dx.$$

By evaluating this last integral, or by differentiating the expression found for ψ at the beginning of the section, the reader will find the same value, namely

$$D\psi(y) = -\left[\frac{\pi y \sin(\pi y) - \cos(\pi y)}{y^2} + \frac{1}{y^2} \right].$$

We can apply the previous theorem using any x as upper limit of the integration. Thus we may let

$$\psi(x, y) = \int_a^x f(t, y) dt,$$

in which case the theorem reads

$$\frac{\partial \psi}{\partial y} = D_2 \psi(x, y) = \int_a^x D_2 f(t, y) dt = \int_a^x \frac{\partial f(t, y)}{\partial y} dt.$$

We use t as a variable of integration to distinguish it from the x which is now used as an end point of the interval $[a, x]$ instead of $[a, b]$.

The preceding way of determining the derivative of ψ with respect to y is called **differentiating under the integral sign.** *Note that it is completely different from the differentiation in the fundamental theorem of calculus.* In this case, we have an integral

$$g(x) = \int_a^x f(t)\, dt,$$

and

$$\frac{dg}{dx} = Dg(x) = f(x).$$

Thus when $f(x,y)$ is a function of two variables, and $\psi\,(x, y)$ is defined as above, the fundamental theorem of calculus states that

$$\frac{\partial \psi}{\partial x} = D_1\psi(x, y) = f(x, y).$$

For example, if we let

$$\psi(x, y) = \int_0^x \sin(ty)\, dt,$$

then

$$D_1\psi(x, y) = \sin(xy),$$

but by Theorem 3,

$$D_2\psi(x, y) = \int_0^x \cos(ty)t\, dt.$$

EXERCISES

In each of the following cases, find $D_1\psi(x, y)$ and $D_2\psi(x, y)$, by evaluating the integrals.

1. $\psi(x, y) = \int_1^x e^{ty}\, dt$

2. $\psi(x, y) = \int_0^x \cos(ty)\, dt$

3. $\psi(x, y) = \int_1^x (y + t)^2\, dt$

4. $\psi(x, y) = \int_1^x e^{y+t}\, dt$

5. $\psi(x, y) = \int_1^x e^{y-t}\, dt$

6. $\psi(x, y) = \int_0^x t^2y^3\, dt$

7. $\psi(x, y) = \int_1^x \frac{\log(ty)}{t}\, dt$

8. $\psi(x, y) = \int_1^x \sin(3ty)\, dt$

§6. PROOF OF THE LOCAL EXISTENCE THEOREM

In this section, we prove Theorem 3.

We suppose that the vector field F is defined on a rectangle R and we select any point (x_0, y_0) in the rectangle. We let $F = (f, g)$ and assume $D_2 f = D_1 g$. We wish to find a potential function $\varphi(x, y)$.

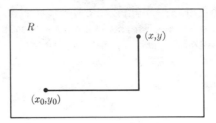

Figure 6

We first integrate $f(x, y)$ with respect to x, and add an arbitrary function of y, so we let

$$\varphi(x, y) = \int_{x_0}^{x} f(t, y) \, dt + u(y).$$

By the fundamental theorem of calculus, we find

$$\frac{\partial \varphi}{\partial x} = \frac{\partial}{\partial x} \int_{x_0}^{x} f(t, y) \, dt + \frac{\partial u(y)}{\partial x}$$

$$= f(x, y)$$

because $\partial u(y)/\partial x = 0$. So

$$D_1 \varphi(x, y) = f(x, y)$$

as wanted. We now have to check $D_2 \varphi(x, y)$. Using Theorem 5, and differentiating with respect to y, we get:

$$D_2 \varphi(x, y) = \int_{x_0}^{x} D_2 f(t, y) \, dt + \frac{\partial u}{\partial y}$$

$$= \int_{x_0}^{x} D_1 g(t, y) \, dt + \frac{\partial u}{\partial y} \qquad \text{(because } D_2 f = D_1 g\text{)}$$

$$= g(t, y)\Big|_{x_0}^{x} + \frac{\partial u}{\partial y}$$

$$= g(x, y) - g(x_0, y) + \frac{\partial u}{\partial y}.$$

Since we want $D_2 \varphi = g$ it suffices that $-g(x_0, y) + \dfrac{\partial u}{\partial y} = 0$, that is:

$$\frac{\partial u}{\partial y} = g(x_0, y).$$

Thus we let

$$u(y) = \int g(x_0, y) \, dy,$$

to conclude the proof.

Observe that the additional function $u(y)$ is also obtained as an integral, so we may write at once our function $\varphi(x, y)$ in the form

$$\varphi(x, y) = \int_{x_0}^{x} f(t, y) \, dt + \int_{y_0}^{y} g(x_0, t) \, dt.$$

Warning: Suppose that the vector field F is defined on an arbitrary open set U, and that $D_2 f = D_1 g$. Then we do *not* have a theorem asserting the existence of the potential function, in general. It was essential in the previous theorem to make additional assumptions on U, because we needed to integrate over intervals when we took for instance

$$\int_{x_0}^{x} D_2 f(t, y) \, dt.$$

In a more general open set U, the corresponding interval may not be contained in U, as illustrated on the next picture (Fig. 7).

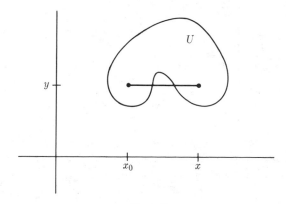

Figure 7

In such a case, the proof cannot apply. In the next section, we shall investigate the situation in more general open sets.

If the open set is a disc, then the same proof does apply, and the corresponding picture is as follows (Fig. 8).

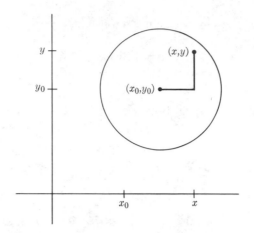

Figure 8

The proof would apply equally to any open set such that the analogous line segments were contained in U, as drawn on the next figure (Fig. 9).

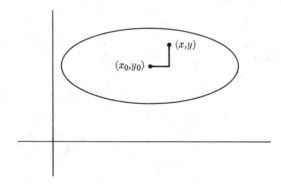

Figure 9

The proof of Theorem 4 for functions of three variables proceeds along entirely similar lines. Suppose $F = (f_1, f_2, f_3)$, and the three variables are x, y, z. Let (x_0, y_0, z_0) be some fixed point in the rectangular box, and define

$$\varphi(x, y, z) = \int_{x_0}^{x} f_1(t, y, z)\, dt + \int_{y_0}^{y} f_2(x_0, t, z)\, dt + \int_{z_0}^{z} f_3(x_0, y_0, t)\, dt.$$

Then

$$D_1\varphi(x, y, z) = f_1(x, y, z)$$

by the fundamental theorem of calculus applied to a function of one variable x. We leave it to you as an exercise to verify that $D_2\varphi = f_2$ and $D_3\varphi = f_3$. The complete proof will be given in the answers to the exercises, but it is more profitable for you to try to work it out first without looking it up.

EXERCISE

Complete the proof of Theorem 4.

Let C be a line segment from A(0,0) to B(1,1)
and let $w = x + y^2$ Evaluate $\int_C w\,ds$ for C

$x = t$
$y = t$ for $0 \le t \le t$

$ds^2 = dx^2 + dy^2$
$dx = 1\,dt$ $ds^2 = (1+1)dt^2$
$dy = 1dt$
$ds = \sqrt{2}\,dt$

$\int_0^1 (t + t^2)\sqrt{2}\,dt$

$\sqrt{2}\left[\dfrac{t^2}{2} + \dfrac{t^3}{3}\right]\Big|_0^1 = \dfrac{5\sqrt{2}}{6}$

disk
shell

$V = \dfrac{2}{3}\pi r^3 = \dfrac{2}{3}\pi 8 = \dfrac{16\pi}{3}$
$r = 2$

$y = x^2$

$y = x^2$
$\nabla V = \pi r^2 h$
$= \pi y^2 \Delta x$
$= \pi x^4 \Delta x$

$\int_0^2 \pi x^4 dx = \dfrac{32\pi}{5}$

$x = 2$

$\nabla V = 2\pi y (2-x)\Delta y$
$= 2\pi y (2 - \sqrt{y})\Delta y$

$\int_0^4 2\pi y (2 - \sqrt{y})\,dy$

area

$\int_{-2}^2 \dfrac{1}{2}\pi r^2$

$= \int_{-2}^2 \dfrac{1}{2}\pi (4-x^2)\,dx$

$= \dfrac{1}{2}\pi \left[4x - \dfrac{x^3}{3}\right]_{-2}^2$

$= \dfrac{32\pi}{6} = \dfrac{16\pi}{3}$

$\nabla V = \pi r^2 h$ $x^2 + y^2 = 4^2$

$\int_0^2 \pi x^2 dy$

shell $x^2 + y^2 = 4$

$V = \int_0^2 2\pi r f(x)\,dx$

$V = \int_0^2 2\pi x \sqrt{4-x^2}\,dx$

$V = \dfrac{2\pi}{-2}\int \sqrt{4-x^2}(-2x)\,dx$

$= \dfrac{16\pi}{3}$

$= \int_0^2 \pi (4 - y^2)\,dy$

$= \int_0^2 (4\pi - \pi y^2)\,dy$

$= 4\pi y - \dfrac{y^3 \pi}{3}\Big|_0^2 = 8\pi - \dfrac{8\pi}{3}$

CHAPTER VI

Curve Integrals

Exam   Nov 12

Let F be a vector field on an open set U in the plane, as shown on the figure. We interpret F as a field of forces.

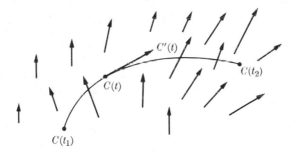

Figure 1

Suppose we move a particle along a curve $C(t)$ in U. It is natural to ask for the work done when moving the particle from a point $C(t_1)$ to a point $C(t_2)$ along the curve. For instance the force field may represent the wind, and the particle may be an airplane flying in the wind's path. The wind may be blowing in an entirely different direction, thereby hindering the plane. 阻碍

To find the work done against this force field along the curve, we shall first take the component of the force along the curve. This is given by a dot product, which becomes a function of time t. We then integrate this function along the curve, and interpret the result as the work. We now discuss this systematically.

§1. DEFINITION AND EVALUATION OF CURVE INTEGRALS

Let U be an open set in n-space. As usual, the important cases will be when $n = 2$ or 3 but to cover these two cases, we must leave n unspecified. Much of what we say will be true in general.

127

Let F be a vector field on U. We can represent F by components. When $n = 2$, we usually write

$$F(X) = (f(x, y), g(x, y)).$$

When $n = 3$ we write

$$F(X) = (f_1(X), f_2(X), f_3(X)),$$

each f_i being a function, the i-th coordinate function. If each function $f_1(X), \dots, f_n(X)$ is continuous, then we shall say that F is a **continuous** vector field. If each function $f_1(X), \dots, f_n(X)$ is differentiable, then we shall say that F is a **differentiable** vector field.

We shall also deal with curves. Rather than use the letter X to denote a curve, we shall use another letter, for instance C, to avoid certain confusions which might arise in the present context. Furthermore, it is now convenient to assume that our curve C is defined on a **closed** interval $I = [a, b]$, with $a < b$. For each number t in I, the value $C(t)$ is a point in space. We shall say that the curve C **lies in** U if $C(t)$ is a point of U for all t in I. We say that C is **continuously differentiable** if its derivative $C'(t) = dC/dt$ exists and is continuous. We abbreviate the expression "continuously differentiable" by saying that the curve is a C^1-curve, or of class C^1.

From now on, all vector fields will be assumed as differentiable as needed wherever they are defined, and similarly all curves will be assumed of class C^1 or as differentiable as needed. This will not be repeated to simplify statements of theorems.

Let F be a vector field on U, and let C be a curve in U. The dot product

$$F(C(t)) \cdot \frac{dC}{dt}$$

is a **function** of t.

Example 1. Let $F(x, y) = (e^{xy}, y^2)$, and $C(t) = (t, \sin t)$. Then

$$C'(t) = (1, \cos t)$$

and

$$F(C(t)) = (e^{t \sin t}, \sin^2 t).$$

Hence

$$F(C(t)) \cdot C'(t) = e^{t \sin t} + (\cos t)(\sin^2 t).$$

Suppose that C is defined on the interval $[a, b]$. We define the **integral of F along C** to be

$$\int_C F = \int_a^b F(C(t)) \cdot \frac{dC}{dt} \, dt.$$

This integral is a direct generalization of the familiar notion of the integral of functions of one variable. If we are given a function $f(u)$, and u is a function of t, then

$$\int_{u(a)}^{u(b)} f(u)\, du = \int_a^b f(u(t)) \frac{du}{dt}\, dt.$$

(This is the formula describing the substitution method for evaluating integrals.)

In n-space, $C(a)$ and $C(b)$ are points, and our curve passes through these two points. Thus the integral we have written down can be interpreted as an integral of the vector field, along the curve, between the two points. It will also be convenient to write the integral in the form

$$\boxed{\int_{P,\, C}^{Q} F = \int_{C(a)}^{C(b)} F(C) \cdot dC}$$

to denote the integral along the curve C, from P to Q.

Warning. Do not confuse the *numbers* a, b which are the ends of the *interval* over which the curve is defined, and the *points*

$$P = C(a) \qquad \text{and} \qquad Q = C(b),$$

which are the beginning point and end point of the curve itself.

The integral along the curve C from P to Q may depend on this curve, and so it is essential to use the symbol for this curve in the notation of the integral

$$\int_{P,\, C}^{Q} F, \qquad \text{or} \qquad \int_C F(C) \cdot dC.$$

Example 2. Let $F(x, y) = (x^2 y, y^3)$. Find the integral of F along the straight line from the origin to the point $(1, 1)$.

We can parametrize the line segment by

$$C(t) = (t, t), \qquad \text{with} \quad 0 \le t \le 1.$$

Thus

$$F(C(t)) = (t^3, t^3).$$

Furthermore,

$$C'(t) = \frac{dC}{dt} = (1, 1).$$

Hence

$$F(C(t)) \cdot \frac{dC}{dt} = 2t^3.$$

The integral we must find is therefore equal to:

$$\int_C F = \int_0^1 2t^3 \, dt = \frac{2t^4}{4} \Big|_0^1 = \frac{1}{2}.$$

It is also convenient to introduce still another symbolic notation for the integral of F over the curve C. In 2-space, suppose

$$F = (f, g),$$

so f, g are the coordinate functions of F. We write

$$\boxed{\int_C F = \int_C f dx + g dy.}$$

Symbolically, the expression on the right is the dot product

$$(f, g) \cdot (dx, dy).$$

The meaning of the symbolic notation for the integral is of course the expression obtained by inserting the dt, namely,

$$\int_a^b \left[f(x(t), y(t)) \frac{dx}{dt} + g(x(t), y(t)) \frac{dy}{dt} \right] dt,$$

which is none other than

$$\int_a^b F(C(t)) \cdot \frac{dC}{dt} \, dt.$$

Remark 1. Our integral of a vector field along a curve is defined for **parametrized curves**. In practice, a curve is sometimes given in a non-parametrized way. For instance, we may want to integrate over the curve defined by $y = x^2$. Then we select some parametrization which is usually the most natural, in this case

$$x = t, \qquad y = t^2.$$

In general, if a curve is defined by a function $y = g(x)$, we select the parametrization

$$\boxed{x = t, \qquad y = g(t).}$$

For a **circle of radius** r centered at the origin, we select the parametrization

$$x = r \cos t, \qquad y = r \sin t, \qquad 0 \leq t \leq 2\pi.$$

whenever we wish to integrate counterclockwise.

For a **straight line segment** between two points P and Q, we take the parametrization C given by

$$C(t) = P + t(Q - P), \qquad 0 \leqq t \leqq 1.$$

The context should always make it clear which parametrization is intended. It can be shown that the integral is independent of the choice of parametrization.

Example 3. Let us find the integral of the vector field

$$F(x, y) = (x^2, xy)$$

over the parabola $x = y^2$ between $(1, -1)$ and $(1, 1)$.
    We take the parametrization

$$y = t \quad \text{and} \quad x = t^2, \quad \text{with} \quad -1 \leqq t \leqq 1$$

as illustrated on the figure (Fig. 2).

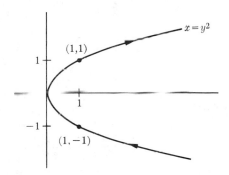

Figure 2

Then $dx = 2t\, dt$ and $dy = dt$, so that:

$$\int_C F = \int_C x^2\, dx + xy\, dy$$

$$= \int_{-1}^{1} t^4 2t\, dt + t^3\, dt$$

$$= \frac{2t^6}{6} + \frac{t^4}{4} \bigg|_{-1}^{1} = 0.$$

Example 4. Find the integral of the vector field

$$G(x, y) = \left(\frac{-y}{x^2 + y^2}, \frac{x}{x^2 + y^2} \right)$$

around the circle of radius 3 counterclockwise from the point (3, 0) to the point

$$\left(\frac{3\sqrt{3}}{2}, \frac{3}{2}\right).$$

We parametrize the circle by

$$x = 3 \cos \theta \quad . \quad \text{and} \quad y = 3 \sin \theta,$$

and the desired arc is given by the values of θ such that

$$0 \leqq \theta \leqq \pi/6.$$

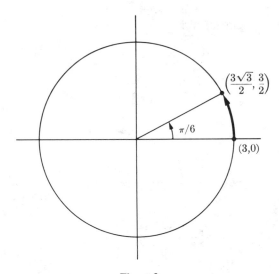

Figure 3

We know that θ ranges from 0 to $\pi/6$ because

$$\frac{3/2}{3\sqrt{3}/2} = \frac{1}{\sqrt{3}} = \tan \pi/6.$$

We then have $dx = -3 \sin \theta \, d\theta$, $dy = 3 \cos \theta \, d\theta$, so that

$$\int_C G = \int_C \frac{-y}{x^2 + y^2} \, dx + \frac{x}{x^2 + y^2} \, dy$$

$$= \int_0^{\pi/6} \frac{-3 \sin \theta}{9} (-3 \sin \theta) \, d\theta + \frac{3 \cos \theta}{9} (3 \cos \theta) \, d\theta$$

$$= \int_0^{\pi/6} d\theta = \pi/6.$$

The vector field of this example is very important, cf. Example 3 of §3. Write $x = r\cos\theta$ and $y = r\sin\theta$. Also write

$$dx = \frac{\partial x}{\partial r} dr + \frac{\partial x}{\partial \theta} d\theta \quad \text{and} \quad dy = \frac{\partial y}{\partial r} dr + \frac{\partial y}{\partial \theta} d\theta.$$

Verify that

$$\boxed{\frac{-y}{x^2 + y^2} dx + \frac{x}{x^2 + y^2} dy = d\theta.}$$

It is worth keeping this relation in mind when working with integrals of this vector field.

Remark 2. We may be given a finite number of curves forming a path as indicated in the following figure (Fig. 4):

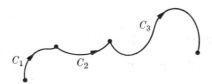

Figure 4

Thus formally, we define a path C to be a finite sequence $\{C_1, \ldots, C_m\}$, where each C_i is a curve, defined on an interval $[a_i, b_i]$, such that the end point of C_i is the beginning point of C_{i+1}. Thus if $P_i = C_i(a_i)$ and $Q_i = C_i(b_i)$, then

$$Q_i = P_{i+1}.$$

We define the integral of F along such a path C to be the sum

$$\int_C F = \int_{C_1} F + \int_{C_2} F + \cdots + \int_{C_m} F.$$

We say that the path C is a **closed path** if the end point of C_m is the beginning point of C_1.

In Fig. 5, we have drawn a closed path such that the beginning point of C_1, namely P_1, is the end point of the path C_4, which joins P_4 to P_1.

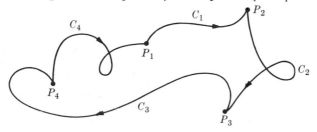

Figure 5

Example 5. Let $F(x, y) = (x^2, xy)$ and let the path consist of the segment of the parabola $y = x^2$ between $(0, 0)$ and $(1, 1)$, and the line segment from $(1, 1)$ and $(0, 0)$. (Cf. Fig. 6.)

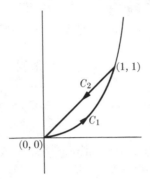

Figure 6

The segment of parabola can be parametrized by

$$C_1(t) = (t, t^2) \qquad \text{with} \quad 0 \le t \le 1.$$

Thus

$$x = t \quad \text{and} \quad y = t^2.$$

Then $dx = dt$, $dy = 2t\, dt$, and so

$$\int_{C_1} F = \int_{C_1} x^2\, dx + xy\, dy = \int_0^1 t^2\, dt + t^3 2t\, dt$$

$$= \frac{t^3}{3} + 2\frac{t^5}{5}\Big|_0^1$$

$$= \frac{1}{3} + \frac{2}{5}.$$

The line segment can be parametrized by

$$C_2(t) = (1 - t, 1 - t) \qquad \text{with} \quad 0 \le t \le 1.$$

Thus

$$x = 1 - t \quad \text{and} \quad y = 1 - t.$$

Then

$$\int_{C_2} F = \int_{C_2} x^2\, dx + xy\, dy = \int_0^1 (1 - 2t + t^2)(-1)\, dt + (1 - 2t + t^2)(-1)\, dt$$

$$= -\frac{2}{3}.$$

We let the path $C = \{C_1, C_2\}$. Then

$$\int_C F = \int_{C_1} F + \int_{C_2} F = \frac{1}{3} + \frac{2}{5} - \frac{2}{3} = -\frac{1}{3} + \frac{2}{5}.$$

Observe how we integrated F around a closed path, and we found a value for the integral $\neq 0$.

EXERCISES

Compute the curve integrals of the vector field over the indicated curves.

1. $F(x, y) = (x^2 - 2xy, y^2 - 2xy)$ along the parabola $y = x^2$ from $(-2, 4)$ to $(1, 1)$.

2. $(x, y, xz - y)$ over the line segment from $(0, 0, 0)$ to $(1, 2, 4)$.

3. Let $r = (x^2 + y^2)^{1/2}$. Let $F(X) = r^{-1}X$. Find the integral of F over the circle of radius 2, taken in counterclockwise direction.

4. Let C be a circle of radius 20 with center at the origin. Let F be a vector field such that $F(X)$ has the same direction as X. What is the integral of F around C?

5. What is the work done by the force $F(x, y) = (x^2 - y^2, 2xy)$ moving a particle of mass m along the square bounded by the coordinate axes and the lines $x = 3, y = 3$ in counterclockwise direction?

6. Let $F(x, y) = (cxy, x^6y^2)$, where c is a positive constant. Let a, b be numbers > 0. Find a value of a in terms of c such that the curve integral of F along the curve $y = ax^b$ from $(0, 0)$ to the line $x = 1$ is independent of b.

Find the values of the indicated integrals of vector fields along the given curves in Exercises 7 through 12.

7. $(y^2, -x)$ along the parabola $x = y^2/4$ from $(0, 0)$ to $(1, 2)$.

8. $(x^2 - y^2, x)$ along the arc in the first quadrant of the circle $x^2 + y^2 = 4$ from $(0, 2)$ to $(2, 0)$.

9. (x^2y^2, xy^2) along the closed path formed by parts of the line $x = 1$ and the parabola $y^2 = x$, counterclockwise.

10. $(x^2 - y^2, x)$ counterclockwise around the circle $x^2 + y^2 = 4$.

11. (a) The vector field

$$G(x, y) = \left(\frac{-y}{x^2 + y^2}, \frac{x}{x^2 + y^2} \right)$$

counterclockwise along the circle $x^2 + y^2 = 2$ from $(1, 1)$ to $(-\sqrt{2}, 0)$.
(b) The same vector field counterclockwise around the whole circle.
(c) Around the circle $x^2 + y^2 = 1$.
(d) Around the circle $x^2 + y^2 = r^2$.

(e) Verify that for this vector field, we have $\partial f/\partial y = \partial g/\partial x$. For a continuation of this train of thought, see Green's theorem.

12. Find the integral of the vector field $F(x, y) = (xy, x)$ along the parabola $x = 2y^2$ from the point $(2, -1)$ to the point $(8, 2)$.

§2. THE REVERSE PATH

Let $C(t)$ be a curve defined over an interval $a \leq t \leq b$. We think of a bug traveling along the curve in the indicated direction. The bug may wish to retrace its steps, and go backward along the curve. Thus if C is a curve joining a point P to a point Q, the bug may wish to travel backward from Q to P. How shall we parametrize its path? Pictorially, this is clear, but we want to give the backward curve a parametrization over some interval, possibly the same interval as for the curve itself.

For this purpose we define the **opposite curve** C^-, or the **reverse curve**, by letting

$$C^-(t) = C(a + b - t).$$

Thus when $t = b$ we find that $C^-(b) = C(a)$, and when $t = a$ we find that $C^-(a) = C(b)$. As t increases from a to b, we see that $a + b - t$ decreases from b to a and thus we visualize C^- as going from $C(b)$ to $C(a)$ in reverse direction from C (Fig. 7).

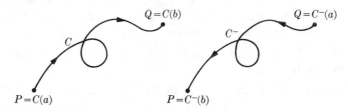

Figure 7

Lemma. *Let F be a vector field on the open set U, and let C be a curve in U, defined on the interval $[a, b]$. Then*

$$\int_{C^-} F = -\int_C F.$$

Proof. This is a simple application of the change of variables formula. Let $u = a + b - t$. Then $du/dt = -1$. By definition and the chain rule, we get:

$$\int_{C^-} F = \int_a^b F(C^-(t)) \cdot \frac{dC^-}{dt}\, dt$$

$$= \int_a^b F(C(a + b - t)) \cdot C'(a + b - t)(-1)\, dt.$$

We now change variables, with $du = -dt$. When $t = a$ then $u = b$, and when $t = b$ then $u = a$. Thus our integral is equal to

$$\int_b^a F(C(u)) \cdot C'(u) \, du = -\int_a^b F(C(u)) \cdot C'(u) \, du,$$

thereby proving the lemma.

The lemma expresses the expected result, that if we integrate the vector field along the opposite direction, then the value of the integral is the negative of the value obtained by integrating F along the curve itself. Therefore, if the curve C is defined on the interval $[a, b]$, the integral of F over the reverse curve C^- will often be written directly as

$$\boxed{\int_{C^-} F = \int_b^a F(C(t)) \cdot C'(t) \, dt = -\int_a^b F(C(t)) \cdot C'(t) \, dt.}$$

For integration over line segments, it is particularly convenient to use the reverse path, as shown in the following example.

Example. Integrate the vector field $F(x, y) = (x^2, xy)$ from the point $(1, 1)$ to the origin $(0, 0)$, along the line segment.

Note that this is precisely one of the integrals considered in Example 5 of the preceding section. Instead of parametrizing the segment as we did in that section, we parametrize the reverse segment, the easy one, namely we let

$$C(t) = (t, t) \qquad \text{with} \quad 0 \leq t \leq 1.$$

Then this segment, with its orientation, looks as on the figure (Fig. 8). In terms of the variables, we have

$$x = t \qquad \text{and} \qquad y = t.$$

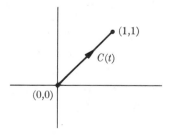

Figure 8

The desired integral is that of F over C^-. Consequently:

$$\int_{C^-} F = -\int_C F = -\int_0^1 t^2 \, dt + t^2 \, dt$$

$$= -\int_0^1 2t^2 \, dt = -\frac{2}{3}.$$

Observe that the algebra here is much easier than the algebra in Example 5 of §1.

If a path C consists of curves $\{C_1, \ldots, C_m\}$, then the **reverse path** consists of the reverse curves in **opposite order**:

$$C^- = \{C_m^-, \ldots, C_1^-\}.$$

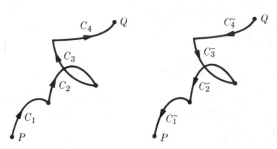

Figure 9

On Fig. 9 when coming back from Q to P, we start with the reverse curve C_4^- and end with the reverse curve C_1^-.

EXERCISES

1. Find the integral of the vector field
$$F(x, y) = (2xy, -3xy)$$
clockwise around the square bounded by the lines $x = 3$, $x = 5$, $y = 1$, $y = 3$.

2. What is the work done by the force $F(x, y) = (x^2 - y^2, 2xy)$ moving a particle of mass m along the square bounded by the coordinate axes and the lines $x = 3$, $y = 3$, in counterclockwise direction? [This is the same exercise as No. 5 in the preceding section. Do it now with the new possibility of parametrizing segments in reverse direction.]

§3. CURVE INTEGRALS WHEN THE VECTOR FIELD HAS A POTENTIAL FUNCTION

When the vector field F admits a potential function φ, then the integral of F along a curve has a simple expression in terms of φ.

Theorem 1. *Let F be a vector field on the open set U and assume that F = grad φ for some function φ on U. Let C be a curve in U, joining the points P and Q. Then*

$$\int_{P,\,C}^{Q} F = \varphi(Q) - \varphi(P).$$

In particular, the integral of F is independent of the curve C joining P and Q.

Proof. Let C be defined on the interval $[a, b]$, so that $C(a) = P$ and $C(b) = Q$. By definition, we have

$$\int_{P,\,C}^{Q} F = \int_{a}^{b} F(C(t)) \cdot C'(t)\,dt = \int_{a}^{b} \mathrm{grad}\ \varphi(C(t)) \cdot C'(t)\,dt.$$

But the expression inside the integral is nothing but the derivative with respect to t of the function g given by $g(t) = \varphi(C(t))$, because of the chain rule. Thus our integral is equal to

$$\int_{a}^{b} g'(t)\,dt = g(b) - g(a) = \varphi(C(b)) - \varphi(C(a)).$$

This proves our theorem.

This theorem is easily extended to paths. See Exercise 1.

In physical terms, the theorem expresses the fact that when a potential function exists, the work done by moving a particle along a curve between points P, Q, is equal to the difference of the potential function at Q and P.

Corollary. *Let F be a vector field on an open set U. If F has a potential function, then the integral of F along every closed path in U is equal to 0. If there exists a closed path C in U such that*

$$\int_{C} F \neq 0$$

then F does not have a potential function.

Proof. Let C be a closed path whose beginning point and end point is the same point P. If φ is a potential function, then

$$\int_{C} F = \varphi(P) - \varphi(P) = 0.$$

Therefore, if the integral around C is $\neq 0$ then there cannot exist a potential function.

For an example, see Example 3 below.

Example 1. Let

$$F(x, y, z) = (2xy^3z, 3x^2y^2z, x^2y^3).$$

Then F has a potential function φ, namely,

$$\varphi(x, y, z) = x^2 y^3 z.$$

You can check this easily by taking the three partial derivatives $\dfrac{\partial \varphi}{\partial x}, \dfrac{\partial \varphi}{\partial y}, \dfrac{\partial \varphi}{\partial z}$ and finding the coordinate functions of F. Let

$$P = (1, -1, 2) \quad \text{and} \quad Q = (-3, 2, 5).$$

Then

$$\int_P^Q F = \varphi(Q) - \varphi(P) = 360 - (-2) = 362.$$

Evaluating the integral of the vector field by means of the potential function (when it exists) avoids the hassle of parametrizing the curve, taking the dot product and going through the process of evaluating the integral in terms of the parameter t. Thus one gets the answer much faster.

Example 2. Let $F(X) = kX/r^3$, where $r = \|X\|$, and k is a constant. This is the vector field inversely proportional to the square of the distance from the origin, used so often in physics. Then F has a potential function, namely the function φ such that $\varphi(X) = -k/r$. Let $P = (1, 1, 1)$ and $Q = (1, 2, -1)$. Then

$$\int_P^Q F = \varphi(Q) - \varphi(P) = -k\left(\frac{1}{\|Q\|} - \frac{1}{\|P\|}\right) = -k\left(\frac{1}{\sqrt{6}} - \frac{1}{\sqrt{3}}\right).$$

On the other hand, if P_1, Q_1 are two points at the *same* distance from the origin (i.e. lying on the same circle, centered at the origin), then

$$\int_{P_1}^{Q_1} F = \varphi(Q_1) - \varphi(P_1) = -k\left(\frac{1}{\|Q_1\|} - \frac{1}{\|P_1\|}\right) = 0.$$

Example 3. Let C be a closed curve, whose end point is equal to the beginning point P. In Theorem 1 when the vector field F admits a potential function φ, it follows that the integral of F over the closed curve is then equal to 0, because it is equal to

$$\varphi(P) - \varphi(P) = 0.$$

This allows us to give an example for a situation when a vector field $F = (f, g)$ satisfies the condition

$$\frac{\partial f}{\partial y} = \frac{\partial g}{\partial x} \quad (\text{i.e. } D_2 f = D_1 g)$$

but F does *not* have a potential function. Let

$$G(x, y) = \left(\frac{-y}{x^2 + y^2}, \frac{x}{x^2 + y^2} \right).$$

A simple computation, left as an exercise, shows that it satisfies the above condition. Compute the integral of G over the closed circle of radius 1, centered at the origin. You will find a value $\neq 0$. This does *not* contradict Theorem 3 in the preceding chapter, because the vector field is defined on the open set obtained from the plane by deleting the origin, i.e. the vector field is not defined at $(0, 0)$. The open set has a "hole" in it (a pinhole, in fact).

You will see the above vector field come up quite frequently. It is typical of vector fields $F = (f, g)$ such that $D_2 f = D_1 g$ but for which no potential function exists. In fact, there is a very good reason why you essentially won't see any other example, because the following result is true.

Let U be the plane from which the origin has been deleted. Let F be a vector field on U such that $D_2 f = D_1 g$. Let

$$G(x, y) = \left(\frac{-y}{x^2 + y^2}, \frac{x}{x^2 + y^2} \right).$$

Then there exists a constant k and a function φ such that

$$F = kG + \text{grad } \varphi,$$

or in terms of (x, y),

$$F(x, y) = kG(x, y) + \text{grad } \varphi(x, y)$$

for all (x, y) in U.

The proof will be given in the next section.

Warning. Just because a vector field is *not* defined at the origin does not **necessarily** mean this vector has no potential function. See Case 4 of Table 1.

We summarize the story on potential functions in a table. We are given a vector field F on a connected open set U, and

$$F = (f, g).$$

Case 1. If $D_2 f \neq D_1 g$, then there is no potential function.

Case 2. If $D_2 f = D_1 g$ and U is a rectangle, then a potential function exists. It can be found by integrating one variable at a time as in the proof of Theorem 3, Chapter V, §3.

Case 3. If $D_2 f = D_1 g$ but U is *not* a rectangle, then a potential function may exist or may not exist.

(a) If there exists *some* closed curve C in U such that

$$\int_C F \neq 0,$$

then a potential function does *not* exist by Theorem 1 and its corollary.

Example. $G(x,y) = (\dfrac{-y}{x^2 + y^2}, \dfrac{x}{x^2 + y^2})$, U is the plane from which the origin is deleted, integral around the circle is 2π.

(b) If the integral of F around *every* closed curve in U is 0, then there *exists* a potential function by Theorem 3 below. [This is not a useful test for us since it involves infinitely many possible closed curves, and we do not apply it.]

Case 4. There may be a vector field on an open set U which is *not* a rectangle, with $D_2 f = D_1 g$, for which a potential function exists.

Example $F(x, y) = \dfrac{g'(r)}{r} (x, y)$, where g is a function of one variable. The potential function is $\varphi(X) = g(r)$. The *proof* that this is a potential function is obtained by taking the gradient directly, and seeing by the chain rule that it gives $F(x, y)$, see Chapter IV, §4, Example 1. The test $D_2 f = D_1 g$ is *not* applicable since the domain of definition of F is the whole plane from which the origin is deleted, not a rectangle.

Example 4. Let $G(x, y)$ be the same vector field as discussed above. Find the integral of G along the path shown on Fig. 10, between the points (1, 0) and (0, 1).

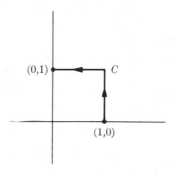

Figure 10

Let C be that path. We know from Chapter V, §4 that the vector field has a potential function on an open set U containing the path, and that this potential function is

$$\varphi(x, y) = \theta.$$

Consequently, on this particular path, the integral is independent of the path, and we have

$$\int_C G = \varphi(0, 1) - \varphi(1, 0) = \frac{\pi}{2} - 0 = \frac{\pi}{2}.$$

EXERCISES

1. Let $C = (C_1, \ldots, C_m)$ be a path in an open set U. Let F be a vector field on U, admitting a potential function φ. Let P be the beginning point of the path and Q its end point. Show that

$$\int_{P, C}^{Q} F = \varphi(Q) - \varphi(P).$$

[*Hint*: Apply Theorem 1 to the beginning point P_i and end point P_{i+1} for each curve C_i.]

2. Find the integral of the vector field $F(x, y, z) = (2x, 3y, 4z)$ along the straight line $C(t) = (t, t, t)$ between the points (0, 0, 0) and (1, 1, 1).

3. Find the integral of the vector field $F(x, y, z) = (y + z, x + z, x + y)$ along the straight line $C(t) = (t, t, t)$ between (0, 0, 0) and (1, 1, 1).

4. Find the integral of the vector field given in Exercises 2 and 3 between the given points along the curve $C(t) = (t, t^2, t^4)$. Compare your answers with those previously found. Is there a general reason why they came out as they did?

5. Let $F(x, y, z) = (y, x, 0)$. Find the integral of F along the straight line from $(1, 1, 1)$ to $(3, 3, 3)$.

6. Let P, Q be points of 3-space. Show that the integral of the vector field given by

$$F(x, y, z) = (z^2, 2y, 2xz)$$

from P to Q is independent of the curve selected between P and Q.

7. Let $F(x, y) = (x/r^3, y/r^3)$ where $r = (x^2 + y^2)^{1/2}$. Find the integral of F along the curve $C(t) = (e^t \cos t, e^t \sin t)$ from the point $(1, 0)$ to the point $(e^{2\pi}, 0)$.

8. Let $F(x, y, z) = (z^3y, z^3x, 3z^2xy)$. Show that the integral of F between two points is independent of the curve between the points.

9. Let $F(x, y) = (x^2y, xy^2)$.
   (a) Does this vector field admit a potential function?
   (b) Compute the integral of this vector field from O to the point P indicated on the figure, along the line segment from $(0, 0)$ to $(1/\sqrt{2}, 1/\sqrt{2})$.

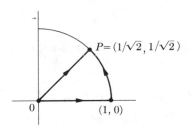

Figure 11

   (c) Compute the integral of this vector field from O to P along the path which consists of the segment from $(0, 0)$ to $(1, 0)$, and the arc of circle from $(1, 0)$ to P. Compare with the value found in (b).

10. Let

$$F(x, y) = \left(\frac{x \cos r}{r}, \frac{y \cos r}{r} \right),$$

where $r = \sqrt{x^2 + y^2}$. Find the value of the integral of this vector field:
   (a) Counterclockwise along the circle of radius 1, from $(1, 0)$ to $(0, 1)$.
   (b) Counterclockwise along the entire circle.
   (c) Does this vector field admit a potential function? Why?

11. Let

$$F(x, y) = \left(\frac{x - y}{x^2 + y^2}, \frac{x + y}{x^2 + y^2} \right).$$

   (a) Find the integral of this vector field around the circle of radius 1 centered at the origin, counterclockwise.

(b) Does this vector field admit a potential function on the plane, from which the origin has been deleted?

12. Let

$$F(x, y) = \left(\frac{-y + 3x}{x^2 + y^2}, \frac{x + 3y}{x^2 + y^2} \right).$$

(a) Does this vector field admit a potential function inside the square

$$1 \leq x \leq 2 \quad \text{and} \quad 1 \leq y \leq 2?$$

Why?

(b) Find the integral of this vector field around the circle of radius 1 centered at the origin, counterclockwise.

(c) Does this vector field admit a potential function on the plane from which the origin has been deleted? Why?

13. Let

$$F(x, y) = \left(\frac{xe^r}{r}, \frac{ye^r}{r} \right)$$

where $r = \sqrt{x^2 + y^2}$. Find the value of the integral of this vector field:

(a) Counterclockwise along the circle of radius 1 centered at the origin.

(b) Counterclockwise along the circle of radius 5 centered at the point $(14, -17)$.

(c) Does this vector field admit a potential function? Why?

14. Let again

$$F(x, y) = \left(\frac{xe^r}{r}, \frac{ye^r}{r} \right).$$

Find the value of the integral of this vector field:

(a) From $(2, 1)$ to $(-3, 4)$ along any path not passing through the origin.

(b) From $(2, 0)$ to $(0, 2)$ along the circle of radius 2.

(c) From $(2, 0)$ to $(\sqrt{2}, \sqrt{2})$ along the circle of radius 2.

(d) All the way around the circle of radius 2.

15. Find the integral of the vector field

$$G(x, y) = \left(\frac{-y}{x^2 + y^2}, \frac{x}{x^2 + y^2} \right):$$

(a) Along the line $x + y = 1$ from $(0, 1)$ to $(1, 0)$.

(b) From the point $(2, 0)$ to the point $(-1, \sqrt{3})$ along the path shown on the figure.

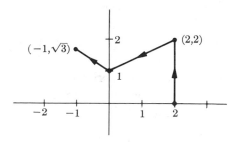

Figure 12

16. Find the integral of the vector field $(x, y^2, 4z^3)$ along the path shown on the figure, from the point $(0, 0, 0)$ to the point $(1, 1, 2)$.

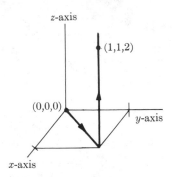

Figure 13

§4. DEPENDENCE OF THE INTEGRAL ON THE PATH

By a path from now on, we mean a piecewise C^1-path, and all vector fields are assumed continuous.

Given two points P, Q in some open set U, and a vector field F on U, it may be that the integral of F along two paths from P to Q depends on the path. We are going to prove the converse of Theorem 1.

Theorem 2. *Let U be a connected open set and let F be a vector field on U. Assume that given two points P, Q in U, the integral*

$$\int_{P,\,C}^{Q} F$$

is independent of the path C in U joining P and Q. Then there exists a potential function for F on U.

Proof. We select some fixed point P_0 in U, and for an arbitrary point X in U, we define

$$\varphi(X) = \int_{P_0}^{X} F,$$

where the integral is taken along any path from P_0 to X. By assumption, this integral does not depend on the path, so we don't need to specify the path in the notation. We must show that the partial derivatives $D_i\varphi(X)$ exist for all P in U, and if the vector field F has coordinate functions

$$F = (f_1, \ldots, f_n),$$

then $D_i\varphi(X) = f_i(X)$.

To do this, let E_i be the unit vector with 1 in the i-th component and 0 in the other components. Then we shall use the obvious relation

$$F(X) \cdot E_i = f_i(X).$$

To determine $D_i \varphi(X)$ we must consider the Newton quotient

$$\frac{\varphi(X + hE_i) - \varphi(X)}{h} = \frac{1}{h} \left[\int_{P_0}^{X+hE_i} F - \int_{P_0}^{X} F \right]$$

and show that its limit as $h \to 0$ is $f_i(X)$. The integral from P_0 to $X + hE_i$ can be taken along a path going first from P_0 to X and then from X to $X + hE_i$ (Fig. 14).

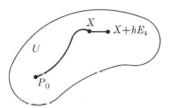

Figure 14

We can then cancel the integrals from P_0 to X and obtain

$$\frac{\varphi(X + hE_i) - \varphi(X)}{h} = \frac{\int_{X}^{X+hE_i} F(C) \cdot dC}{h},$$

taking the integral along any curve C between X and $X + hE_i$. In fact, we take C to be the parametrized straight line segment given by

$$C(t) = X + thE_i \qquad \text{with} \quad 0 \leq t \leq 1.$$

[This is the standard way of parametrizing a line segment between two points P, Q, namely $P + t(Q - P)$.] Then

$$C'(t) = hE_i$$

and

$$F(C(t)) \cdot C'(t) = f_i(X + thE_i)h,$$

so

$$\frac{\varphi(X + hE_i) - \varphi(X)}{h} = \frac{1}{h} \int_0^1 f_i(X + thE_i)h \, dt.$$

Change variables. Let $u = th$ and $du = h\,dt$. Then

$$\frac{\varphi(X + hE_i) - \varphi(X)}{h} = \frac{1}{h}\int_0^h f_i(X + uE_i)\,du.$$

Let $g(u) = f_i(X + uE_i)$. This is an ordinary function of one variable u, and our last expression for the Newton quotient of φ has the form

$$\frac{1}{h}\int_0^h g(u)\,du.$$

By the fundamental theorem of calculus, for any continuous function g we have (cf. *Remark* after the proof):

$$\lim_{h \to 0} \frac{1}{h}\int_0^h g(u)\,du = g(0).$$

Applying this to $g(u) = f_i(X + uE_i)$ we note that $g(0) = f_i(X)$, and therefore we obtain the limit

$$\lim_{h \to 0} \frac{\varphi(X + hE_i) - \varphi(X)}{h} = f_i(X).$$

This proves what we wanted.

Remark. The use of the fundamental theorem of calculus in the preceding proof should be recognized as absolutely straightforward. If G is an indefinite integral for g, then

$$\int_0^h g(t)\,dt = G(h) - G(0),$$

and hence

$$\frac{1}{h}\int_0^h g(t)\,dt = \frac{G(h) - G(0)}{h}$$

is the ordinary Newton quotient for G. The fundamental theorem of calculus asserts precisely that the limit as $h \to 0$ is equal to $G'(0) = g(0)$.

We can also formulate an equivalent condition in terms of closed paths.

Theorem 3. *Let U be an open connected set, and let F be a vector field on U. If the integral of F around every closed path in U is equal to 0, then F has a potential function on U.*

Proof. Let P, Q be points in U. Let C and D be paths from P to Q in U. Let $D = (D_1, \ldots, D_k)$ where each D_j is a C^1-curve. Then we may form the opposite path

$$D^- = (D_k^-, \ldots, D_1^-),$$

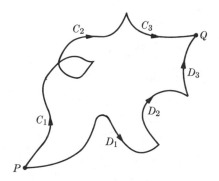

Figure 15

and by the lemma of §2

$$\int_D F = - \int_D F.$$

If $C = (C_1, \ldots, C_m)$, then the path $(C_1, \ldots, C_m, D_k^-, \ldots, D_1^-)$ is a closed path from P to P (Fig. 15). By hypothesis, the integral of F along this closed path is equal to 0. Thus

$$\int_C F + \int_{D^-} F = 0.$$

From this it follows that

$$\int_C F = \int_D F.$$

Hence the integral from P to Q is independent of the path. We can now apply Theorem 2 to conclude the proof.

Theorem 3 is not useful because the hypothesis involves *every* closed path, which amounts to infinitely many paths, so it cannot be verified in practice. In the next theorem, we find a situation where *one* closed path suffices.

Theorem 4. *Let F be a vector field defined on the plane from which the origin is deleted, and write* $F = (f, g)$. *Assume that* $D_2 f = D_1 g$. *Let C be the circle of radius* 1 *centered at the origin, oriented counterclockwise.*

Case 1. *If*

$$\int_C F = 0,$$

then F has a potential function.

Case 2. *Let*

$$k = \frac{1}{2\pi} \int_C F.$$

Then there exists a function φ such that

$$F = kG + \text{grad } \varphi,$$

where

$$G(x, y) = \left(\frac{-y}{x^2 + y^2}, \frac{x}{x^2 + y^2} \right).$$

Proof. Assume that the integral of F around C is 0. Then we shall prove that there is a function φ such that $F = \text{grad } \varphi$. Indeed, for any point $X \neq 0$, we define

$$\varphi(X) = \text{integral of } F \text{ along the path shown on the figure (Fig. 16).}$$

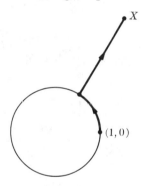

Figure 16

The assumption shows that φ is well-defined, and the same argument used in proving Theorem 2 then shows that grad $\varphi = F$.

For Case 2, let

$$k = \frac{1}{2\pi} \int_C F.$$

Then

$$\int_C F - kG = \int_C F - k \int_C G = 2\pi k - 2\pi k = 0.$$

Hence Case 1 applies, and there is a function φ such that

$$F - kG = \text{grad } \varphi.$$

This shows that $F = kG + \text{grad } \varphi$, and proves the theorem.

CHAPTER VII

Double Integrals

areas
moment (centroids)
surface areas
volume.

When studying functions of one variable, it was possible to give essentially complete proofs for the existence of an integral of a continuous function over an interval. The investigation of the integral involved lower sums and upper sums.

It is important to understand the notion of upper and lower sums in the higher dimensional context. To give complete proofs for the theory in two or more variables becomes more involved, and hence we shall omit the proofs. However, the basic theorem that an integral defined as the unique number between lower sums and upper sums can be evaluated by repeated integration with respect to the variables successively allows us to compute integrals in several variables using *only* one-variable techniques, combined with a geometric description of the domain of integration, usually in terms of inequalities. We shall therefore discuss in detail both of these aspects.

We shall also list various formulas giving double integrals in terms of polar coordinates, and we give a geometric argument to make them plausible.

合理业

§1. DOUBLE INTEGRALS

We begin by discussing the analogue of upper and lower sums associated with partitions.

Let R be a region of the plane (Fig. 1), and let f be a function defined on R. We shall say that f is **bounded** if there exists a number M such that $|f(X)| \leq M$ for all X in R.

Let a, b be two numbers with $a \leq b$, and let c, d be two numbers with $c \leq d$. We consider the closed interval $[a, b]$ on the x-axis and the closed interval $[c, d]$ on the y-axis. These determine a rectangle R in the plane, consisting of all pairs of points (x, y) with $a \leq x \leq b$ and $c \leq y \leq d$.

The rectangle R above will be denoted by $[a, b] \times [c, d]$.

151

① Find the area bounded by the parabola $y = x^2$ and the line $y = 2x + 3$

② Find the areas of the region bounded by $y = x^2$ and the line $y = x$

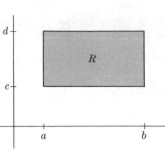

Figure 1

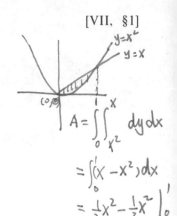

$$A = \int_0^1 \int_{x^2}^{x} dy\, dx$$

$$= \int_0^1 (x - x^2)\, dx$$

$$= \tfrac{1}{2}x^2 - \tfrac{1}{3}x^3 \Big|_0^1$$

$$= \tfrac{1}{2} - \tfrac{1}{3}$$

$$= \tfrac{3}{6} - \tfrac{2}{6} = \tfrac{1}{6}$$

Let I denote the interval $[a, b]$. By a partition P_I of I we mean a sequence of numbers

$$x_1 = a \le x_2 \le \cdots \le x_m = b$$

which we also write as $P_I = (x_1, \ldots, x_m)$. Similarly, by a partition P_J of the interval $J = [c, d]$ we mean a sequence of numbers

$$y_1 = c \le y_2 \le \cdots \le y_n = d$$

which we write as $P_J = (y_1, \ldots, y_n)$.

Each pair of small intervals $[x_i, x_{i+1}]$ and $[y_j, y_{j+1}]$ determines a rectangle

$$S_{ij} = [x_i, x_{i+1}] \times [y_j, y_{j+1}].$$

(Cf. Fig. 2(a).) We denote symbolically by $P = P_I \times P_J$ the partition of R into rectangles S_{ij} and we call such S_{ij} a **subrectangle** of the **partition** (Fig. 2(b)).

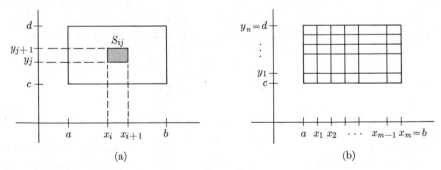

(a)   (b)

Figure 2

If R is a rectangle as above, we define its **area** to be the obvious thing, namely

$$\text{Area}(R) = (d - c)(b - a).$$

Thus the area of each subrectangle S_{ij} is $(y_{j+1} - y_j)(x_{i+1} - x_i)$.

Let A be a region in the plane, and let f be a function defined on A. As usual, we say that f is **continuous** at a point P of A if

$$\lim_{X \to P} f(X) = f(P).$$

We say that f is continuous on A if it is continuous at every point of A.

If S is a set and f a function on S which reaches a maximum on S, we let

$$\max_S f$$

denote this maximum value. It is a value $f(v)$ for some point v in S such that $f(v) \geqq f(w)$ for all w in S. Similarly, we let

$$\min_S f$$

be the minimum value of the function on S, if it exists. We recall a fact which we do not prove, that a continuous function on a closed and bounded set always takes on a maximum and minimum value. For instance, a continuous function on a closed interval $[a, b]$ always has a maximum. A continuous function on a rectangle as above also has a maximum, and a minimum.

We then form sums which are analogous to the lower and upper sums used to define the integral of functions of one variable. If P denotes the partition as above, and f is a continuous function on R, we define

$$L(P, f) = \sum_S (\min_S f)\, \text{Area}(S),$$

$$U(P, f) = \sum_S (\max_S f)\, \text{Area}(S).$$

The symbol $\sum_S$ means that we must take the sum over all subrectangles of the partition. In terms of the indices i, j, we can rewrite say the lower sum as

$$L(P, f) = \sum_{i=1}^{m-1} \sum_{j=1}^{n-1} (\min_{S_{ij}} f)(y_{j+1} - y_j)(x_{i+1} - x_i)$$

$$= \sum_i \sum_j (\min_{S_{ij}} f)\, \text{Area}(S_{ij}),$$

and similarly for the upper sum.

Let v_{ij} be a point in the small rectangle S_{ij} such that $f(v_{ij})$ is a maximum of f on this rectangle. Then the upper sum $U(P, f)$ can be written also in the form

$$U(P, f) = \sum_{i=1}^{m-1} \sum_{j=1}^{n-1} f(v_{ij})(y_{j+1} - y_j)(x_{i+1} - x_i)$$

$$= \sum_i \sum_j f(v_{ij})\, \text{Area}(S_{ij}).$$

If v_{ij} is a point in S_{ij} such that $f(v_{ij})$ is neither a maximum nor a minimum for f on S_{ij}, then the above sum lies between the upper and lower sum, and is called a **Riemann sum** for f.

Since the lower sums are defined by taking minima, and the upper sums are defined by taking maxima of f over certain rectangles, it is clear that

$$L(P, f) \leqq U(P, f),$$

and in fact every lower sum is less than or equal to every upper sum.

We define f to be **integrable** on R if there exists a unique number which is greater than or equal to every lower sum, and less than or equal to every upper sum.

If this number exists, we call it the **integral** of f, and denote it by

$$\iint_R f \quad \text{or} \quad \iint_R f(x, y) \, dy \, dx.$$

Theorem 1. *Let R be a rectangle, and let f be a function defined and continuous on R. Then f is integrable on R.*

Interpretation of the integral as volume.

We can interpret the integral as a **volume** under certain conditions. Namely, suppose that $f(x, y) \geqq 0$ for all (x, y) in R. The value $f(x, y)$ may be viewed as a height above the point (x, y), and we may consider the integral of f as the volume of the 3-dimensional region lying above the rectangle R and bounded from above by the graph of f (Fig. 3).

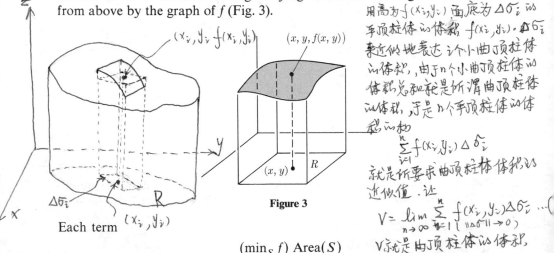

用高为 $f(x_i, y_i)$ 面底为 $\Delta \sigma_i$ 的
平顶柱体的体积 $f(x_i, y_i) \cdot \Delta \sigma_i$
来近似地表示 这个小曲顶柱体
的体积,由于 n 个小曲顶柱体的
体积总和就是所谓曲顶柱体
的体积,于是 n 个平顶柱体的体
积的和

$$\sum_{i=1}^{n} f(x_i, y_i) \Delta \sigma_i$$

就是所要求曲顶柱体体积的
近似值. 让

$$V = \lim_{n \to \infty} \sum_{i=1}^{n} f(x_i, y_i) \Delta \sigma_i \quad (\|\Delta \sigma\| \to 0)$$

V 就是曲顶柱体的体积.

Figure 3

Each term (x_i, y_i)

$$(\min_S f) \, \text{Area}(S)$$

is the volume of a rectangular box whose base is the rectangle S in the (x, y)-plane, and whose height is $\min_S f$. The volume of such a box is precisely

$(\min_S f)$ Area(S), where, as we said above, Area(S) is the area of S. This box lies below the 3-dimensional region bounded from above by the graph of f. Similarly, the term

$$(\max_S f)\ \text{Area}(S)$$

is the volume of a box whose base is S and whose height is $\max_S f$. This box lies above the above region. This makes our interpretation of the integral as volume clear.

Interpretation of the integral as mass.

Also, as in one variable, a positive function on a region may be viewed as a density, and thus if $f \geq 0$ on R, then we also interpret

$$\iint_R f(x, y)\, dy\, dx$$

as the **mass** of R.

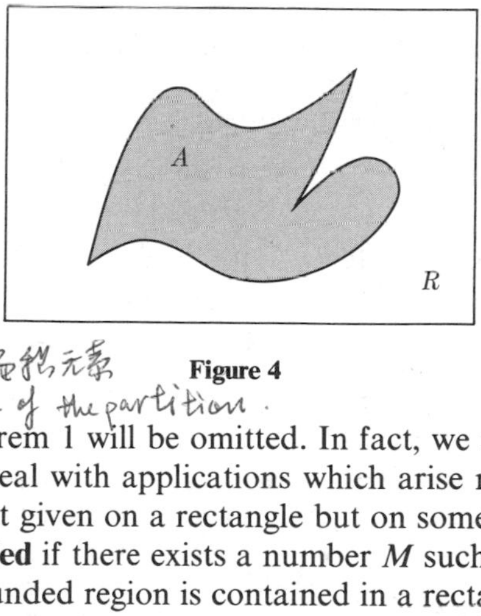

Figure 4

The proof of Theorem 1 will be omitted. In fact, we need a somewhat more general discussion to deal with applications which arise naturally in practice. A function f is usually not given on a rectangle but on some region A in the plane. We say that A is **bounded** if there exists a number M such that $\|X\| \leq M$ for all points X in A. Any bounded region is contained in a rectangle, as shown on Fig. 4.

The set of boundary points of the region A will be called the **boundary** of A. We shall say that the boundary is **smooth** if it consists of a finite number of smooth curves. A smooth curve means a C^1 curve, i.e. a curve parametrized such that the coordinate functions have continuous derivatives, as studied in Chapter II. The boundary of A in Fig. 4 consists of three such curves. We draw a finite number of C^1 curves in the next figure (Fig. 5).

Suppose the function f is defined on a region A as in Fig. 4, so that A is bounded and has a boundary which is smooth. If we want to integrate f over the

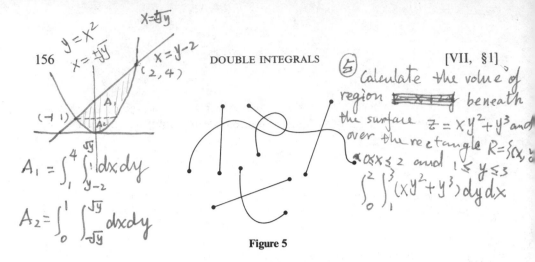

(handwritten, top left figure)
$y = x^2$
$x = \sqrt{y}$
$x = \sqrt{y}$
$x = y - 2$
$(2, 4)$
$(-1, 1)$
A_1
A_2

$$A_1 = \int_1^4 \int_{y-2}^{\sqrt{y}} dx\,dy$$

$$A_2 = \int_0^1 \int_{-\sqrt{y}}^{\sqrt{y}} dx\,dy$$

(handwritten, top right)
⑤ Calculate the volume of region ~~$z = xy$~~ beneath the surface $z = xy^2 + y^3$ and over the rectangle $R = \{(x, y)\}$ $0 \le x \le 2$ and $1 \le y \le 3$

$$\int_0^2 \int_1^3 (xy^2 + y^3)\,dy\,dx$$

Figure 5

(handwritten left margin) $z = 1$

region A, then it is natural to extend the definition of f to the whole rectangle R, by letting

$$f(v) = 0$$

for every point v in R such that v does not lie in A. Then even if we assume that f is continuous on A, we see that f is not continuous on R. The points of discontinuity are precisely the points of the boundary of A. Therefore we cannot apply Theorem 3 directly, and we need a minor adjustment of our definitions to deal with this case, which we now discuss.

Suppose that instead of being continuous on R the function f is merely bounded, and so has a least upper bound and a greatest lower bound. Let P be a partition of R, and let S be a subrectangle of the partition. By

$$\text{lub}_S f = \text{lub}_{v \text{ in } S} f(v)$$

we mean the least upper bound of all values $f(v)$ for v in S. We take it as a known property of the real numbers that any bounded set of numbers has a least upper bound, and also a greatest lower bound. Similarly, we denote by

$$\text{glb}_S f = \text{glb}_{v \text{ in } S} f(v)$$

the greatest lower bound of all values of f on S. We may then form upper and lower sums with the least upper bound and greatest lower bound, respectively, that is:

$$U(P, f) = \sum_S (\text{lub}_S f)\, \text{Area}(S)$$

and

$$L(P, f) = \sum_S (\text{glb}_S f)\, \text{Area}(S)$$

(handwritten right margin) 特点定理：如果函数 $f(x, y)$ 在闭区间 R 上连续，则 $\iint_R f(x) $ 的极限速存在；也就是在 R 上 f 为=定积分 定存在。换言之，连续函数 在有界闭区域上是可积的。

Theorem 2. *Let R be a rectangle and let f be a function defined on R, bounded, and continuous except possibly at the points lying on a finite number of smooth curves. Then f is integrable on R.*

Again, we shall not prove Theorem 2, nor the following routine properties.

Theorem 3. *Assume that* f, g *are functions on the rectangle R, and are integrable. Then $f + g$ is integrable. If k is a number, then kf is integrable. We have:*

$$\iint_R (f + g) = \iint_R f + \iint_R g \quad \text{and} \quad \iint_R (kf) = k \iint_R f.$$

Theorem 4. *If f, g are integrable on R, and $f \leqq g$, then*

$$\iint_R f \leqq \iint_R g.$$

Let A be a region in the plane, contained in a rectangle R (Fig. 4). Let f be a function defined on A. We denote by f_A the function which has the same values as f at points of A, and such that $f_A(Q) = 0$ if Q is a point not in A. Then f_A is defined on the rectangle R, and we define

$$\iint_A f = \iint_R f_A$$

provided that f_A is integrable. By Theorem 2, we note that if the boundary of A is smooth, and if f is continuous on A, then f_A is continuous except at all points lying on the boundary of A, and hence f_A is integrable.

We now have one more property of the integral which is convenient to integrate a function over several regions.

Theorem 5. *Let A be a bounded region in the plane, expressed as a union of two regions A_1 and A_2 having no points in common except possibly a finite number of curves. If f is a function defined on A and continuous except at a finite number of smooth curves, then*

$$\iint_A f = \iint_{A_1} f + \iint_{A_2} f.$$

Furthermore, if A is itself some smooth curve, contained in a rectangle R, and if f is a bounded function on R which has the value 0 except possibly for points of A, then

$$\iint_A f = 0.$$

We shall not give the proof of Theorem 4, which anyhow is intuitively clear. In Fig. 6(a) we have drawn a smooth curve in R where f may not be 0, and such that $f(v) = 0$ if v lies in R but v is not a point of A. Then

ind the volume of the solid whose base is a triangle lying in the plane, bounded by the x-axis, the line $y = x$ and the line $x = 1$ and whose top lies in the plane $z = x + y + 1$ (1)

This is reasonable because the 2-dimensional area of a curve is 0. In Fig. 6(b) we have drawn three regions A_1, A_2, A_3 which have only smooth curves in common. The integral of a function f over the three regions is then the sum of the integrals of f over each region separately.

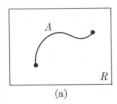

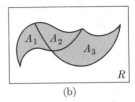

(a)   (b)

Figure 6

§2. REPEATED INTEGRALS

To compute the integral we shall investigate repeated integrals.

Let f be a function defined on our rectangle. For each x in the interval $[a, b]$ we have a function f_x of y given by $f_x(y) = f(x, y)$, and this function f_x is defined on the interval $[c, d]$. Assume that for each x the function f_x is integrable over this interval (in the old sense of the word, for functions of one variable). We may then form the integral

$$\int_c^d f_x(y)\, dy = \int_c^d f(x, y)\, dy.$$

The expression we obtain depends on the particular value of x chosen in the interval $[a, b]$, and is thus a function of x. Assume that this function is integrable over the interval $[a, b]$. We can then take the integral

$$\int_a^b \left[\int_c^d f(x, y)\, dy \right] dx, \quad \text{also written} \quad \int_a^b \int_c^d f(x, y)\, dy\, dx,$$

which is called the **repeated integral** of f.

Example 1. Let $f(x, y) = x^2 y$. Find the repeated integral of f over the rectangle determined by the intervals $[1, 2]$ on the x-axis and $[-3, 4]$ on the y-axis.

We must find the repeated integral

$$\int_1^2 \int_{-3}^4 f(x, y)\, dy\, dx.$$

To do this, we first compute the integral with respect to y, namely

$$\int_{-3}^4 x^2 y\, dy.$$

For a fixed value of x, we can take x^2 out of the integral, and hence this inner integral is equal to

$$x^2 \int_{-3}^{4} y \, dy = x^2 \frac{y^2}{2} \Big|_{-3}^{4}$$

$$= \frac{7x^2}{2}.$$

We then integrate with respect to x, namely

$$\int_{1}^{2} \frac{7x^2}{2} \, dx = \frac{49}{6}.$$

Thus

$$\int_{1}^{2} \int_{-3}^{4} x^2 y \, dy \, dx = \frac{49}{6}.$$

The repeated integral is useful in computing a double integral because of the following theorem, whose proof will also be omitted.

Theorem 6. *Let R be a rectangle $[a, b] \times [c, d]$, and let f be integrable on R. Assume that for each x in $[a, b]$ the function f_x given by*

$$f_x(y) = f(x, y)$$

is integrable on $[c, d]$. Then the function of x given by

$$\int_{c}^{d} f(x, y) \, dy$$

is integrable on $[a, b]$, and

$$\iint_{R} f = \int_{a}^{b} \left[\int_{c}^{d} f(x, y) \, dy \right] dx.$$

In Example 1, we may now write

$$\iint_{R} x^2 y \, dy \, dx = \int_{1}^{2} \left[\int_{-3}^{4} x^2 y \, dy \right] dx = \frac{49}{6}.$$

Geometrically speaking, the inner integral for a fixed value of x gives the area of a cross section as indicated in the following figure. Then integrating such areas yields the volume of the 3-dimensional figure bounded below by the rectangle R, and above by the graph of f.

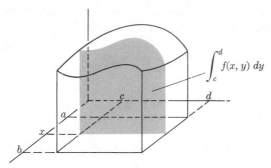

Figure 7

The following situation will arise frequently in practice.

Let g_1, g_2 be two smooth functions on a closed interval $[a, b]$ $(a \leq b)$ such that $g_1(x) \leq g_2(x)$ for all x in that interval. Let c, d be numbers such that

$$c < g_1(x) \leq g_2(x) < d$$

for all x in the interval $[a, b]$. Then g_1, g_2 determine a region A lying between $x = a$, $x = b$, and the two curves $y = g_1(x)$ and $y = g_2(x)$. (Cf. Fig. 8.)

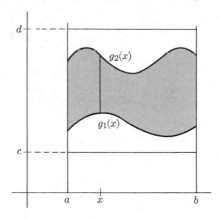

Figure 8

Let f be a function which is continuous on the region A, and define f on the rectangle $[a, b] \times [c, d]$ to be equal to 0 at any point of the rectangle not lying in the region A. For any value x in the interval $[a, b]$ the integral

$$\int_c^d f(x, y) \, dy$$

can be written as a sum:

$$\int_c^{g_1(x)} f(x, y) \, dy + \int_{g_1(x)}^{g_2(x)} f(x, y) \, dy + \int_{g_2(x)}^d f(x, y) \, dy.$$

Since $f(x, y) = 0$ whenever $c \leq y < g_1(x)$ and $g_2(x) < y \leq d$, it follows that the two extreme integrals are equal to 0. Thus the repeated integral of f over the rectangle is in fact equal to the repeated integral

$$\int_a^b \left[\int_{g_1(x)}^{g_2(x)} f(x, y) \, dy \right] dx.$$

Regions of the type described by two functions g_1, g_2 as above are the most common type of regions with which we deal.

From Theorem 6 and the preceding discussion, we obtain:

Corollary. *Let g_1, g_2 be two smooth functions defined on a closed interval $[a, b]$ ($a \leq b$) such that $g_1(x) \leq g_2(x)$ for all x in that interval. Let f be a continuous function on the region A lying between $x = a$, $x = b$, and the two curves $y = g_1(x)$ and $y = g_2(x)$. Then*

$$\iint_A f = \int_a^b \left[\int_{g_1(x)}^{g_2(x)} f(x, y) \, dy \right] dx;$$

in other words, the double integral is equal to the repeated integral.

We give examples showing how to apply Theorem 6, or rather its corollary.

Example 2. Let $f(x, y) = x^2 + y^2$. Find the integral of f over the region A bounded by the straight line $y = x$ and the parabola $y = x^2$ (Fig. 9).

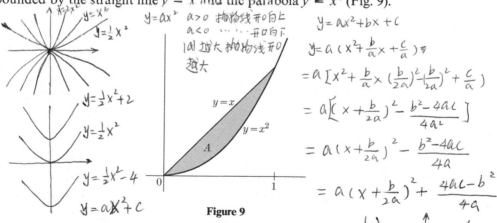

Figure 9

In this case, the region A consists of all points (x, y) such that

$$0 \leq x \leq 1 \quad \text{and} \quad x^2 \leq y \leq x.$$

Thus our integral is equal to

$$\int_0^1 \left[\int_{x^2}^x (x^2 + y^2) \, dy \right] dx.$$

Now the inner integral is given by

$$\int_{x^2}^{x}(x^2 + y^2)\, dy = x^2 y + \frac{y^3}{3}\Big|_{x^2}^{x} = x^3 + \frac{x^3}{3} - x^4 - \frac{x^6}{3}.$$

Hence the repeated integral is equal to

$$\iint_A f = \int_0^1 \left(x^3 + \frac{x^3}{3} - x^4 - \frac{x^6}{3}\right) dx = \frac{x^4}{4} + \frac{x^4}{12} - \frac{x^5}{5} - \frac{x^7}{21}\Big|_0^1$$

$$= \frac{1}{4} + \frac{1}{12} - \frac{1}{5} - \frac{1}{21}.$$

(We don't need to simplify the number on the right.)

Given a region A, it is frequently possible to break it up into smaller regions having only boundary points in common, and such that each smaller region is of the type we have just described. In that case, to compute the integral of a function over A, we can apply Theorem 5.

Example 3. Let $f(x, y) = 2xy$. Find the integral of f over the triangle bounded by the lines $y = 0, y = x$, and the line $x + y = 2$.

The region is as shown in Fig. 10.

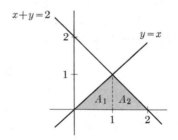

Figure 10

We break up our region into the portion from 0 to 1 and the portion from 1 to 2. These correspond to the small triangles A_1, A_2, as indicated in the picture. Then

$$\iint_{A_1} f = \int_0^1 \left[\int_0^x 2xy\, dy\right] dx \qquad \text{and} \qquad \iint_{A_2} f = \int_1^2 \left[\int_0^{2-x} 2xy\, dy\right] dx.$$

Then

$$\iint_A f = \iint_{A_1} f + \iint_{A_2} f.$$

There is no difficulty in evaluating these integrals, and we leave them to you.

On the other hand, we may also view the region A to be the set of all points (x, y) satisfying the inequalities

$$0 \leq y \leq 1 \qquad y \leq x \leq 2 - y.$$

Hence from this point of view, we do not have to split the integral over two regions A_1 and A_2, but we may evaluate it directly as

$$\int_0^1 \int_y^{2-y} 2xy \, dx \, dy.$$

The inner integral is

$$y \int_y^{2-y} 2x \, dx = y \left[x^2 \right] \Big|_y^{2-y} = y(2 - y)^2 - y^3$$

$$= 4y - 4y^2.$$

Hence the desired double integral is equal to

$$\int_0^1 (4y - 4y^2) \, dy$$

which we leave to you.

Finally, the **area** of a region A is the integral of the function 1 over A, i.e.

$$\text{Area}(A) = \iint_A 1 \, dy \, dx.$$

This is obviously true when A is a rectangle, and it follows for general regions A by using upper and lower sums.

Example 4. Find the area of the region bounded by the straight line $y = x$ and the curve $y = x^2$.

The region has been sketched in Example 2. By definition,

$$\text{Area}(A) = \int_0^1 \int_{x^2}^x dy \, dx = \int_0^1 (x - x^2) \, dx$$

$$= \frac{x^2}{2} - \frac{x^3}{3} \Big|_0^1 = \frac{1}{2} - \frac{1}{3} = \frac{1}{6}.$$

We also observe that the same arguments as before apply if we interchange the role of x and y. Thus for the rectangle R we also have

$$\iint_R f(x, y) \, dy \, dx = \iint_R f(x, y) \, dx \, dy = \int_c^d \left[\int_a^b f(x, y) \, dx \right] dy.$$

The same goes for a region consisting of all points (x, y) such that

$$c \leqq y \leqq d \quad \text{and} \quad g_1(y) \leqq x \leqq g_2(y).$$

If A is a region in the plane bounded by a finite number of smooth curves, and f is a function on A such that $f(x) \geqq 0$ for $x \in A$, then we can interpret f as a **density function**, and we also call the integral $\int\int_A f$ the **mass** of A.

Example 5. Find the integral of the function $f(x, y) = x^2 y^2$ over the region bounded by the lines $y = 1, y = 2, x = 0$, and $x = y$ (Fig. 11).

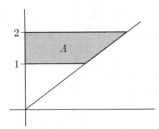

Figure 11

We have to compute the integral as prescribed, namely:

$$\int_1^2 \left[\int_0^y x^2 y^2 \, dx \right] dy = \int_1^2 y^2 \frac{x^3}{3} \Big|_0^y \, dy = \int_1^2 \frac{y^5}{3} \, dy = \frac{7}{2}.$$

We can also say that the preceding integral, namely $7/2$, is the mass of A corresponding to the density given by the function f. Of course the units of mass are those determined by the units of density.

Example 6. Find the mass of a disc of radius a if the density is proportional to the square of the distance from a point on the circumference.
We take the circle surrounding the disc to have equation

$$x^2 + y^2 = a^2,$$

and select the point on the circumference to be $(a, 0)$, as shown on Fig. 12.

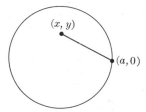

Figure 12

Then the density function is

$$f(x, y) = k\left[(x - a)^2 + y^2\right].$$

The upper half of the disc of radius a consists of all points (x, y) such that

$$-a \leq x \leq a \quad \text{and} \quad 0 \leq y \leq \sqrt{a^2 - x^2}.$$

By symmetry, the mass of the disc is twice the mass of the upper half.
The mass is therefore given by the integral

$$2 \int_{-a}^{a} \int_{0}^{\sqrt{a^2 - x^2}} k\left[(x - a)^2 + y^2\right] dy\, dx = \tfrac{3}{2} k\pi a^4.$$

The actual evaluation of the integral is done as follows. First the inner integral,

$$\int_{0}^{\sqrt{a^2 - x^2}} \left[(x - a)^2 + y^2\right] dy = \left[(x - a)^2 y + \frac{y^3}{3}\right]\Big|_{0}^{\sqrt{a^2 - x^2}}$$

$$= (x - a)^2 \sqrt{a^2 - x^2} + \frac{1}{3}(a^2 - x^2)^{3/2}.$$

Next the outer integral is

$$\int_{-a}^{a} \left[(x - a)^2 \sqrt{a^2 - x^2} + \frac{1}{3}(a^2 - x^2)^{3/2}\right] dx.$$

We evaluate this by polar coordinates, putting

$$x = a \cos \theta, \quad dx = -a \sin \theta\, d\theta.$$

Then the integral becomes

$$\int_{\pi}^{0} \left[a^2(\cos \theta - 1)^2(a \sin \theta) + \frac{1}{3} a^3 \sin^3 \theta\right](-a \sin \theta)\, d\theta$$

$$= -a^4 \int_{\pi}^{0} \left[(\cos^2 \theta - 2 \cos \theta + 1) \sin^2 \theta + \frac{1}{3} \sin^4 \theta\right] d\theta.$$

You should know how to integrate powers of the sine to find the answer as given.

Remark. The shift to polar coordinates in the middle of the proof, after setting up the problem in rectangular coordinates, creates additional steps. In addition, the equation of a circle in polar coordinates has a simpler expression for present purposes, and so you will rework this example after reading the next section somewhat more easily. Cf. Exercise 6 of the next section.

Example 7. Sketch the region defined by the inequalities 木竹求

$$-2 \leq x \leq 1 \quad \text{and} \quad 0 \leq y \leq |x|.$$

Since $0 \leq y$ the region lies above the x-axis. If $x \geq 0$, then the condition $0 \leq y \leq x$ means that the region lies below the line $y = x$. Hence for $x \geq 0$ the region looks like the piece shaded on the right of the y-axis in Fig. 13.

If $x \leq 0$, then $|x| = -x$. The inequality $0 \leq y \leq -x$ means that the region lies below the line $y = -x$ for $x \leq 0$. Hence the region looks like that shaded in the figure, to the left of the y-axis.

Example 8. Sketch the region defined by the inequalities

$$-2 \leq x \leq 0 \quad \text{and} \quad |y| \geq |x|.$$

For $y \geq 0$ and $y \geq |x|$ the point (x, y) will lie above the line $y = -x$. Furthermore, we have symmetry in the sense that if (x, y) satisfies the desired inequalities, then so does the point

$$(x, -y).$$

Hence the region is symmetric with respect to the x-axis. Hence the region looks as on Fig. 14.

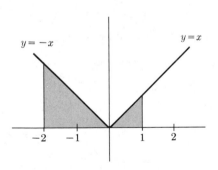

Figure 13

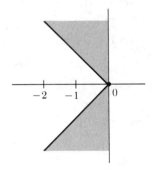

Figure 14

EXERCISES

1. Find the value of the following repeated integrals.

(a) $\int_0^2 \int_1^3 (x + y) \, dx \, dy$ (b) $\int_0^2 \int_1^2 x^2 y \, dy \, dx$ (c) $\int_0^1 \int_{y^2}^y \sqrt{x} \, dx \, dy$

(d) $\int_0^\pi \int_0^x x \sin y \, dy \, dx$ (e) $\int_1^2 \int_y^{y^2} dx \, dy$ (f) $\int_0^\pi \int_0^{\sin x} y \, dy \, dx$

Area

Limits   ① $x+y = 2$ $2x = x - 4$ $y = 0$ answer

         ② $x^2 = 4y$ $8y = x^2 + 16$ $\dfrac{18}{32}$
                                                            $\dfrac{}{3}$

② The solid bounded by the cylinders
$$x^2 + y^2 = 4 \quad y^2 + z^2 = 4 \quad (≧) \quad \frac{128}{3}$$

③ The ellipsoid $x^2 + 4y^2 + 9z^2 = 36$ 48π

(g) $\int_0^{\pi/2} \int_0^2 r^2 \cos\theta \, dr \, d\theta$

(h) $\int_0^{2\pi} \int_0^{1-\cos\theta} r^3 \cos^2\theta \, dr \, d\theta$

(i) $\int_0^{\arctan 3/2} \int_0^{2\sec\theta} r \, dr \, d\theta$

④ PP 168 #10

2. Sketch the regions described by the following inequalities.
(a) $|x| \leq 1, -1 \leq y \leq 2$ (b) $|x| \leq 3, |y| \leq 4$
(c) $x + y \leq 1, x \geq 0, y \geq 0$ (d) $0 \leq |y| \leq x, 0 \leq x \leq 5$
(e) $0 \leq x \leq y, 0 \leq y \leq 5$ (f) $|x| + |y| \leq 1$

3. Find the integral of the following functions.
(a) $x \cos(x + y)$ over the triangle whose vertices are $(0, 0)$, $(\pi, 0)$, and (π, π).
(b) e^{x+y} over the region defined by $|x| + |y| \leq 1$.
(c) $x^2 - y^2$ over the region bounded by the curve $y = \sin x$ between 0 and π.
(d) $x^2 + y$ over the triangle whose vertices are $(-\frac{1}{2}, \frac{1}{2})$, $(1, 2)$, $(1, -1)$.

4. Find the integrals of the following functions over the indicated region.
(a) $f(x, y) = x$ over the region bounded by $y = x^2$ and $y = x^3$.
(b) $f(x, y) = y$ over the same region as in (a).
(c) $f(x, y) = x^2$ over the region bounded by $y = x, y = 2x$, and $x = 2$.

5. Let a be a number > 0. Show that the area of the region consisting of all points (x, y) such that $|x| + |y| \leq a$, is $(2a)^2/2!$.

6. Find the following integrals and sketch the region of integration in each case.
(a) $\int_1^2 \int_{x^2}^{x^3} x \, dy \, dx$ (b) $\int_0^2 \int_1^3 |x - 2| \sin y \, dx \, dy$
(c) $\int_0^{\pi/2} \int_{-y}^y \sin x \, dx \, dy$ (d) $\int_{-1}^1 \int_0^{|x|} dy \, dx$
(e) $\int_0^{\pi/2} \int_0^{\cos y} x \sin y \, dx \, dy$ (f) $\int_0^1 \int_1^{e^x} (x + y) \, dy \, dx$
(g) $\int_{-3}^2 \int_0^{y^2} (x^2 + y) \, dx \, dy$

7. Find the mass of a square plate of side a if the density is proportional to the square of the distance from a vertex.

8. Integrate the function f over the indicated region.
(a) $f(x, y) = 1/(x + y)$ over the region bounded by the lines $y = x, x = 1, x = 2$, $y = 0$.
(b) $f(x, y) = x^2 - y^2$ over the region defined by the inequalities

$$0 \leq x \leq 1 \quad \text{and} \quad x^2 - y^2 \geq 0.$$

(c) $f(x, y) = x \sin xy$ over the rectangle $0 \leq x \leq \pi$ and $0 \leq y \leq 1$.
(d) $f(x, y) = x^2 - y^2$ over the triangle whose vertices are $(-1, 1)$, $(0, 0)$, $(1, 1)$.

find the volume of the given solid
9. The solid bound by the plane $x + y + z = 3$ and the three coordinate axis. $\frac{1}{2}$

(e) $f(x, y) = 1/(x + y + 1)$ over the square $0 \leqq x \leqq 1, 0 \leqq y \leqq 1$.

9. Compute the integral of the function $f(x, y) = xy$ over the region sketched below.

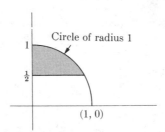

Circle of radius 1

1

$\frac{1}{2}$

$(1, 0)$

Figure 15

10. Find the volume of the region in 3-space lying above the triangle with vertices $(-1, 0)$, $(0, 1)$, $(1, 0)$ and under the graph of the function $f(x, y) = x^2 y$.

11. Find the integral of the function $f(x, y) = x - y$ over the region bounded by the curve $y = \sin x$ between $x = 0$ and $x = \pi$.

12. Find the mass of a plate bounded by one arch of the curve $y = \sin x$, and the x-axis, if the density is proportional to the distance from the x-axis.

§3. POLAR COORDINATES

We shall first review polar coordinates. A point (x, y) in the plane can be described by two other types of coordinates (r, θ), where r is the distance of the point from the origin, so

$$r = \sqrt{x^2 + y^2},$$

and θ is the angle which the ray from $(0, 0)$ through (x, y) makes with the x-axis, as shown on the figure (Fig. 16).

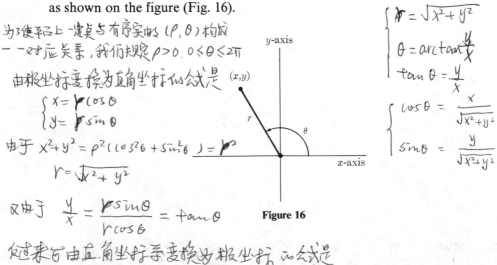

Figure 16

Then

$$x = r \cos \theta \quad \text{and} \quad y = r \sin \theta.$$

A curve may be described by giving r as a function of θ.

Example 1. Let us sketch the curve $r = 1 - \cos \theta$. We make a table of values.

θ	$\cos \theta$	$1 - \cos \theta$
0 to π	1 to -1	0 to 2
π to 2π	-1 to 1	2 to 0

Hence the curve looks like this (Fig. 17). The coordinates given are the polar coordinates (r, θ).

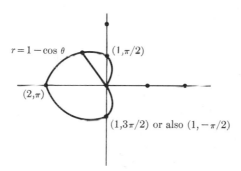

Figure 17

Note that when $\theta = \pi/2$ we have $r = 1 - \cos \pi/2 = 1$, as drawn. The curve encloses a region, which may be described by the inequalities in terms of the polar coordinates:

$$0 \leq \theta \leq 2\pi$$
$$0 \leq r \leq 1 - \cos \theta.$$

At first we shall be interested in somewhat simpler regions, namely sectors. Arbitrary regions will then be approximated by sectors.

Consider the piece S of a sector as shown on Fig. 18(b). This region S is the set of points whose polar coordinates (r, θ) satisfy the inequalities

$$a \leq \theta \leq b \quad \text{and} \quad c \leq r \leq d,$$

where a, b are numbers chosen such that $0 \leq a \leq b \leq 2\pi$, and

$$0 \leq c \leq d.$$

These inequalities also describe a rectangle in the (r, θ)-plane, and the region S in the (x, y)-plane corresponds to the rectangle $R = S^*$ in the (r, θ)-plane, under the transformation

$$x = r \cos \theta \quad \text{and} \quad y = r \sin \theta.$$

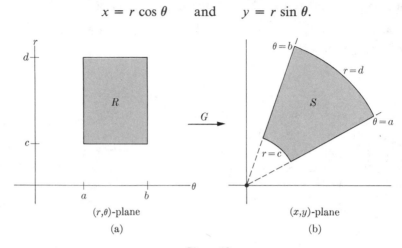

(a) (r,θ)-plane   (b) (x,y)-plane

Figure 18

Consider partitions

$$a = \theta_1 \leq \theta_2 \leq \cdots \leq \theta_n = b, \qquad c = r_1 \leq r_2 \leq \cdots \leq r_m = d$$

of the two intervals $[a, b]$ and $[c, d]$. Each pair of intervals $[\theta_i, \theta_{i+1}]$ and $[r_j, r_{j+1}]$ determines a small region as shown in the following figure (Fig. 19).

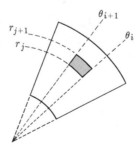

Figure 19

This region is the small sectorial piece S_{ij} consisting of all points whose polar coordinates (r, θ) satisfy the inequalities

$$\theta_i \leq \theta \leq \theta_{i+1}$$
$$r_j \leq r \leq r_{j+1}.$$

The area of such a region is equal to the difference between the area of the sector having angle $\theta_{i+1} - \theta_i$ and radius r_{j+1}, and the area of the sector having the same angle but radius r_j. The area of a sector having angle θ and radius r is equal to

$$\frac{\theta}{2\pi}\pi r^2 = \frac{\theta r^2}{2}.$$

Consequently the difference mentioned above is equal to

$$\frac{(\theta_{i+1} - \theta_i)r_{j+1}^2}{2} - \frac{(\theta_{i+1} - \theta_i)r_j^2}{2} = (\theta_{i+1} - \theta_i)\frac{(r_{j+1} + r_j)}{2}(r_{j+1} - r_j).$$

We note that

$$\frac{r_{j+1} + r_j}{2} = \bar{r}_j,$$

and therefore

$$\boxed{\text{Area of small region } S_{ij} = \bar{r}_j(r_{j+1} - r_j)(\theta_{i+1} - \theta_i).}$$

If f is a function on the (x, y)-plane, it determines a function of (r, θ) by the formula

$$\boxed{f^*(r, \theta) = f(r \cos \theta, r \sin \theta).}$$

Example 2. Let $f(x, y) = 2x^2y$. Then

$$f^*(r, \theta) = 2r^2 \cos^2 \theta \, r \sin \theta = 2r^3 \cos^2 \theta \sin \theta.$$

This is obtained by substituting $r \cos \theta$ for x and $r \sin \theta$ for y in the expression for $f(x, y)$.

We may then take the product of the value of the function $f^*(r_j, \theta_j)$ and the area of the small sectorial piece S_{ij} consisting of all points (r, θ) whose polar coordinates satisfy the inequalities

$$\theta_i \leq \theta \leq \theta_{i+1}, \quad r_j \leq r \leq r_{j+1}.$$

Taking the sum for all pairs (i, j) we see that

$$\sum_{j=1}^{m-1} \sum_{i=1}^{n-1} f^*(r_j, \theta_i)r_j(r_{j+1} - r_j)(\theta_{i+1} - \theta_i)$$

is a Riemann sum for the function $f^*(r, \theta)r$ on the rectangle $[a, b] \times [c, d]$ in the (r, θ) plane. Consequently the following theorem is now very plausible.

Theorem 7. *Let S be the piece of sector consisting of those points in the* (x, y)-*plane whose polar coordinates satisfy the inequalities as above,*

$$a \leqq \theta \leqq b \quad and \quad c \leqq r \leqq d.$$

Let $S^* = R$ *be the corresponding rectangle in the* (r, θ)-*plane. Let f be bounded and continuous on S except possibly on a finite number of smooth curves. Let* f^* *be the corresponding function of* (r, θ). *Then*

$$\iint_{S^*} f^*(r, \theta) r \, dr \, d\theta = \iint_S f(x, y) \, dy \, dx.$$

Symbolically, we write

$$dy \, dx = r \, dr \, d\theta.$$

As with rectangular coordinates, we can deal with more general regions. Let g_1, g_2 be two smooth functions defined on the interval $[a, b]$ and assume

$$0 \leqq g_1(\theta) \leqq g_2(\theta).$$

Consider the region A of the (x, y)-plane consisting of all points (x, y) whose polar coordinates (r, θ) satisfy the inequalities

$$a \leqq \theta \leqq b \quad and \quad g_1(\theta) \leqq r \leqq g_2(\theta).$$

This region is illustrated in Fig. 20.

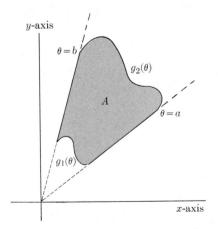

Figure 20

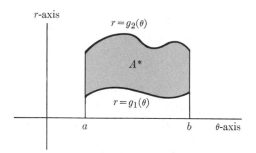

r-axis

$r = g_2(\theta)$

A^*

$r = g_1(\theta)$

a b θ-axis

Figure 21

It corresponds to the region A^* of the (r, θ)-plane described by these inequalities, and this region A^* is of a type considered in the last section, illustrated in the next figure (Fig. 21).

In Theorem 6, the integral over the rectangle is just the integral

$$\int_a^b \int_c^d f^*(r, \theta\,)r\;dr\;d\theta.$$

Suppose now that we consider a function f which is equal to 0 outside the region A, so that f^* is equal to 0 outside the region A^*. We are in the situation already described in the last section. Hence if we want to integrate f over A, we get the formula:

$$\iint_A f(x, y)\; dy\; dx = \iint_{A^*} f^*(r, \theta\,)r\;dr\;d\theta$$

which in terms of the inequalities can also be written:

$$\iint_A f(x, y)\; dy\; dx = \int_a^b \int_{g_1(\theta)}^{g_2(\theta)} f^*(r, \theta\,)r\;dr\;d\theta.$$

In dealing with polar coordinates, it is useful to remember the equation of a circle. Let $a > 0$. Then

$$r = a \cos \theta, \qquad -\pi/2 \leq \theta \leq \pi/2,$$

is the equation of a circle of radius $a/2$ and center $(a/2, 0)$. Similarly,

$$r = a \sin \theta, \qquad 0 \leq \theta \leq \pi,$$

is the equation of a circle of radius $a/2$ and center $(0, a/2)$. You can easily show this, as an exercise, using the relations

$$r = \sqrt{x^2 + y^2}, \qquad x = r \cos \theta, \qquad y = r \sin \theta.$$

改以圆心半径是 r，分别足式 (1). 圆心与原点重合 (2). 圆心左点 (p, 0) 对 (3) 圆心左点 (0, ?

对他直角坐标方程，并把它仍化为极坐标方程。

(1) 圆心直角坐标方程是

$$x^2 + y^2 = p^2$$

时 $x^2 + y^2 = r^2$ The circles have been drawn on Fig. 22.

$$p^2 = r^2$$

$$p = r$$

(2) 圆心直角坐标方程是

$$(x - p)^2 + y^2 = p^2$$

$$x = r\cos\theta$$

$$y = r\sin\theta$$

$$r^2\cos^2\theta - 2rp\cos\theta + p^2 + p^2\sin^2\theta = p^2$$

$$r^2(\cos^2\theta + \sin^2\theta) - 2rp\cos\theta = 0$$

$$r(r - 2p\cos\theta) = 0$$

$$r = 2p\cos\theta$$

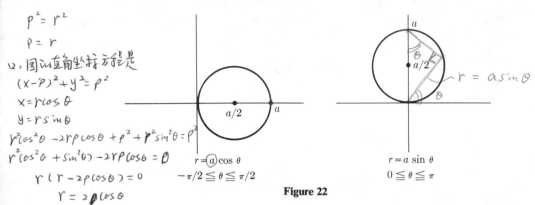

$r = (a)\cos\theta$

$-\pi/2 \leq \theta \leq \pi/2$

$r = a\sin\theta$

$0 \leq \theta \leq \pi$

Figure 22

(*Note.* The coordinates of the center above are given in rectangular coordinates.)

Example 3. Find the integral of the function $f(x, y) = x$ over the region bounded by the semicircle and the x-axis as shown on the figure (Fig. 23).

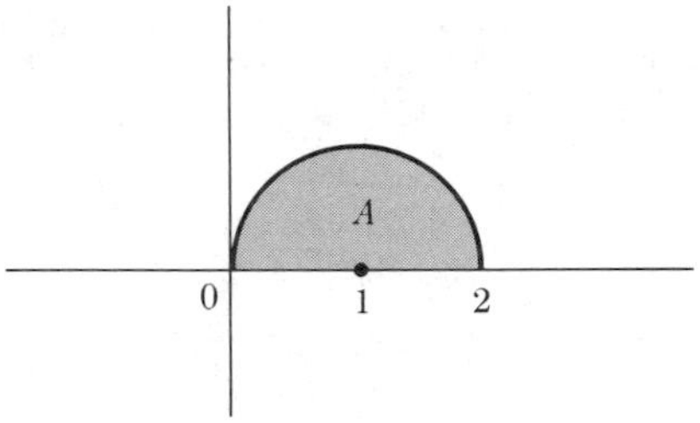

3 圆的直角坐标方程是

$$x^2 + (y - p)^2 = p^2$$

$$x^2 + y^2 = 2py$$

$$r^2 = 2py$$

$$r^2 = 2p\,r\sin\theta$$

$$r = 2p\sin\theta$$

通过此例可以看出
圆的极坐标
方程很简单
只用 $r = \rho$ 来
表示。

Figure 23

The region A consists of all points whose polar coordinates (r, θ) satisfy the inequalities

$$0 \leq \theta \leq \pi/2 \quad\text{and}\quad 0 \leq r \leq 2\cos\theta.$$

Hence

$$\iint_A x \, dy \, dx = \iint_{A^*} r \cos\theta \, r \, dr \, d\theta$$

$$= \int_0^{\pi/2} \int_0^{2\cos\theta} r^2 \cos\theta \, dr \, d\theta.$$

The inner integral is

$$\int_0^{2\cos\theta} r^2 \, dr = \frac{r^3}{3}\Big|_0^{2\cos\theta} = \frac{8}{3}\cos^3\theta.$$

Hence

$$\iint_A x \, dy \, dx = \frac{8}{3}\int_0^{\pi/2}\cos^4\theta \, d\theta,$$

which you should know how to do. One technique is to write

$$\cos^2 \theta = \frac{1 + \cos 2\theta}{2},$$

and repeat the use of this formula to lower the powers of the cosine appearing in the integral. The second time you apply this, you do it to

$$\cos^2 2\theta = \frac{1 + \cos 4\theta}{2}.$$

Of course you can also look the integral up in integral tables. The answer comes out $\pi/2$.

Example 4. Find the integral of the function $f(x, y) = x^2$ over the region enclosed by the curve given in polar coordinates by the equation

$$r = 1 - \cos \theta.$$

The function of the polar coordinates (r, θ) corresponding to f is given by

$$f^*(r, \theta) = r^2 \cos^2 \theta.$$

The region in the polar coordinate space is described by the inequalities

$$0 \leq r \leq 1 - \cos \theta \quad \text{and} \quad 0 \leq \theta \leq 2\pi.$$

This region in the (x, y)-plane looks like Fig. 24:

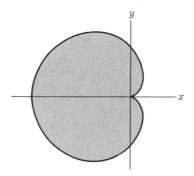

Figure 24

The desired integral is therefore the integral

$$\int_0^{2\pi} \int_0^{1-\cos \theta} r^3 \cos^2 \theta \, dr \, d\theta.$$

We integrate first with respect to r, which is easy, and see that our integral is equal to

$$\int_0^{2\pi} \tfrac{1}{4}(1 - \cos\theta)^4 \cos^2\theta \, d\theta.$$

The evaluation of this integral is done by techniques of the first course in calculus. We expand out the expression of the fourth power, and get a sum of terms involving $\cos^k \theta$ for $k = 0, \ldots, 6$. The reader should know how to integrate powers of the cosine, using repeatedly the formula

$$\cos^2\theta = \frac{1 + \cos 2\theta}{2},$$

or using the recursion formula in terms of lower powers. No matter what method the reader uses, he will find the final answer to be

$$\frac{49\pi}{32}.$$

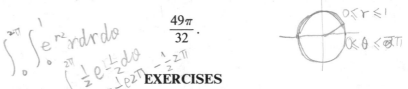

EXERCISES

1. By changing to polar coordinates, find the integral of $e^{x^2+y^2}$ over the region consisting of the points (x, y) such that $x^2 + y^2 \leq 1$.

2. Find the volume of the region lying over the disc $x^2 + (y - 1)^2 \leq 1$ and bounded from above by the function $z = x^2 + y^2$.

3. Find the integral of $e^{-(x^2+y^2)}$ over the circular disc bounded by

$$x^2 + y^2 = a^2, \qquad a > 0.$$

4. In Exercise 3, find the limit of the integral as a becomes large. This limit is interpreted as the integral

$$\int_{-\infty}^{\infty}\int_{-\infty}^{\infty} e^{-(x^2+y^2)} \, dx \, dy.$$

5. Sketch the region defined by $x \geq 0$, $x^2 + y^2 \leq 2$, and $x^2 + y^2 \geq 1$. Determine the integral of $f(x, y) = x^2$ over this region.

6. Find the mass of a circular disk of radius a if the density is proportional to the square of the distance from a point on the circumference.

7. Let A be the disc of radius 1 and center 0. Find

$$\iint_A (x^2 + y^2) \, e^{(x^2+y^2)^2} \, dy \, dx.$$

Evaluate the following integrals. Take $a > 0$.

8. $\int_{-a}^{a}\int_{-\sqrt{a^2-x^2}}^{\sqrt{a^2-x^2}} dy \, dx$

9. $\int_0^a \int_0^{\sqrt{a^2-y^2}} (x^2 + y^2) \, dx \, dy$

10. $\int_0^{a/\sqrt{2}} \int_y^{\sqrt{a^2-y^2}} x \, dx \, dy$

11. Find the area inside the curve $r = a(1 + \cos \theta)$ and outside the circle $r = a$.

12. The base of a solid is the region of Exercise 11 and the top is given by the function $f(x, y) = x$. Find the volume.

13. Find the area enclosed by the following curves.
    (a) $r^2 = \cos \theta$ (b) $r^2 = 2a^2 \cos 2\theta$

14. The base of a solid is the area of Exercise 13, and the top is bounded by the function (in terms of polar coordinates) $f(r, \theta) = \sqrt{2a^2 - r^2}$. Find the volume.

15. Find the integral of the function

$$f(x, y) = \frac{1}{(x^2 + y^2 + 1)^{3/2}}$$

over the disc of radius a centered at the origin. Letting a tend to infinity, show that

$$\int_{-\infty}^{\infty} \int_{-\infty}^{\infty} f(x, y) \, dy \, dx = 2\pi.$$

16. Answer the same question for the function

$$f(x, y) = \frac{1}{(x^2 + y^2 + 2)^2}.$$

17. Find the integral of the function

$$f(x, y) = \frac{1}{(x^2 + y^2)^3}$$

over the region between the two circles of radius 2 and radius 3, centered at the origin.

18. (a) Find the integral of the function $f(x, y) = x$ over the region bounded in polar coordinates by $r = 1 - \cos \theta$.
    (b) Let a be a number > 0. Find the integral of the function $f(x, y) = x^2$ over the region bounded in polar coordinates by $r = a(1 - \cos \theta)$.

19. Sketch the region defined by $x \geq 0$, $x^2 + y^2 \leq 2$ and $x^2 + y^2 \geq 1$. Determine the integral of the following functions over this region.
    (a) $f(x, y) = x^2$ (b) $f(x, y) = x$ (c) $f(x, y) = y$.

20. Sketch the region defined by $y \geq x$, $x^2 + y^2 \leq 2$, and $x^2 + y^2 \geq 1$. Find the

integral of the function

$$f(x,y) = \frac{xy}{x^2 + y^2}$$

over this region.

21. A cylindrical hole of radius 1 is bored through the center of a sphere of radius 2. What volume is removed?

22. Let n be a positive integer, and let $f(x,y) = 1/r^n$, where $r = \sqrt{x^2 + y^2}$.
     (a) Find the integral of this function over the region contained between two circles of radii a and b respectively, with $0 < a < b$.
     (b) For which values of n does this integral approach a limit as $a \rightarrow 0$?

A solid has a circular base of radius 4 units
Find the volume of the solid if every plane
section $\perp$ to a fixed diameters is an
equilateral triangle.

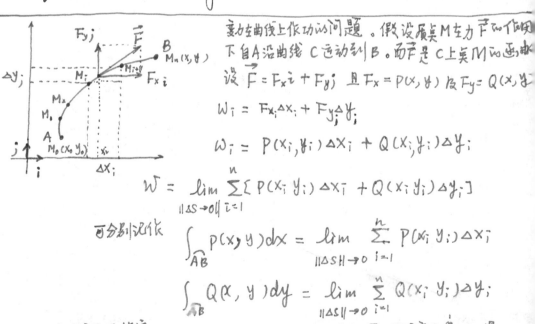

CHAPTER VIII

Green's Theorem
(a way of finding w [line Intergrals)

§1. THE STANDARD VERSION

Suppose we are given a vector field on some open set U in the plane. Then this vector field has two components, i.e. we can write

$$F(x, y) = (p(x, y), q(x, y)),$$

where p, q are functions of two variables (x, y). In everything that follows, we assume that all functions we deal with are C^1, i.e. that these functions have continuous partial derivatives, and similarly for vector fields and curves.

Let C be a curve in U, defined on an interval $[a, b]$. For the integral of F over C we use the notation

$$\int_C F = \int_a^b F(C(t)) \cdot C'(t)\, dt = \int_C p(x, y)\, dx + q(x, y)\, dy,$$

and abbreviate this as

$$\int_C p\, dx + q\, dy.$$

This is reasonable since the curve gives

$$x = x(t)$$

and

$$y = y(t)$$

as functions of t, and

$$F(C(t)) \cdot \frac{dC}{dt} = p(x, y)\frac{dx}{dt} + q(x, y)\frac{dy}{dt}.$$

179

Green's Theorem. *Let p, q be functions on a region A, which is the interior of a closed path C, parametrized counterclockwise. Then*

$$\int_C p\,dx + q\,dy = \iint_A \left(\frac{\partial q}{\partial x} - \frac{\partial p}{\partial y} \right) dy\,dx.$$

The region and its boundary may look as follows (Fig. 1):

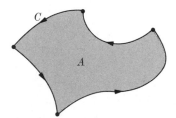

Figure 1

It is difficult to prove Green's theorem in general, partly because it is difficult to make rigorous the notion of "interior" of a path, and also the notion of counterclockwise. In practice, for any specifically given region, it is always easy, however. That it may be difficult in general is already suggested by drawing a somewhat less simple region as follows:

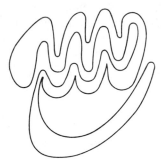

Figure 2

We shall therefore prove Green's theorem only in special cases, where we can give the region and the parametrization of its boundary explicitly.

Case 1. Suppose that the region A is given by the inequalities

$$a \leqq x \leqq b \qquad \text{and} \qquad g_1(x) \leqq y \leqq g_2(x)$$

in the same manner as we studied before in Chapter VII, §2.

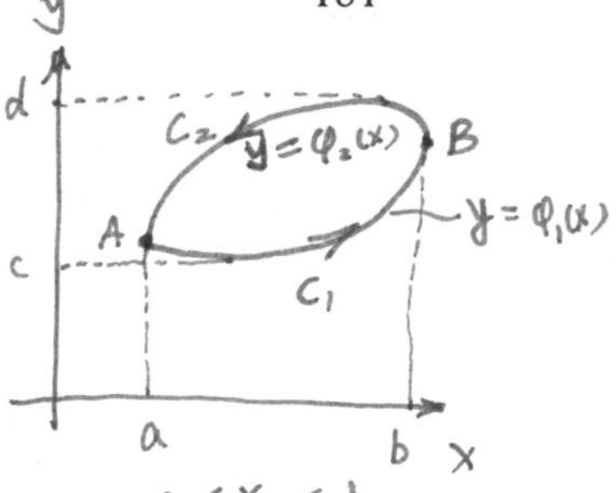

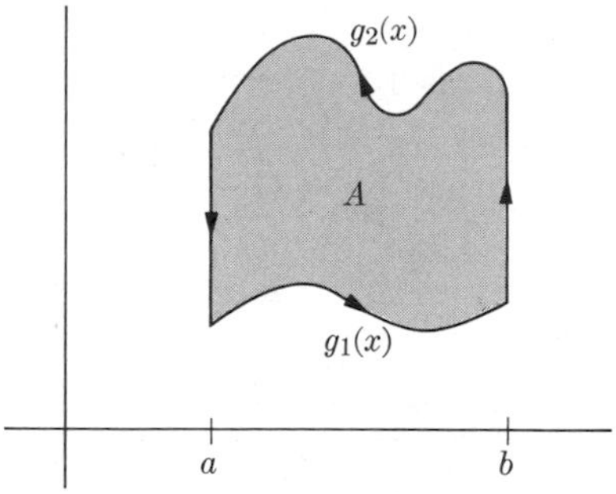

Figure 3

The boundary of A then consists of four pieces, the two vertical segments, and the pieces parametrized by

$$C_1(t) = (t, g_1(t)), \qquad a \leqq t \leqq b,$$
$$C_2(t) = (t, g_2(t)), \qquad a \leqq t \leqq b.$$

Then we can prove one-half of Green's theorem, namely

$$\int_C p\, dx = \iint_A -\frac{\partial p}{\partial y}\, dy\, dx.$$

Proof We have

$$\iint_A \frac{\partial p}{\partial y}\, dy\, dx = \int_a^b \int_{g_1(x)}^{g_2(x)} D_2 p(x, y)\, dy\, dx$$

$$= \int_a^b \left(p(x, y)\big|_{g_1(x)}^{g_2(x)} \right)\, dx$$

$$= \int_a^b \left[p(x, g_2(x)) - p(x, g_1(x)) \right]\, dx$$

$$= \int_{C_2} p\, dx - \int_{C_1} p\, dx.$$

However, the boundary of A, oriented counterclockwise, consists of four pieces,

$$C_1, C_2^-, C_3, C_4,$$

where C_2^- is the opposite curve to C_2, and C_3, C_4 are the vertical segments.

Handwritten annotations (right margin and bottom):

$a \leqslant x \leqslant b$

$\varphi_1(x) \leqslant \varphi_2(x)$

应用二重积分的公式我们得到

$$\iint \frac{\partial P}{\partial y}\, dx\, dy$$

$$= \int_a^b dx \int_{\varphi_1(x)}^{\varphi_2(x)} \frac{\partial P}{\partial y}\, dy$$

$$= \int_a^b \left[P(x, \varphi_2(x)) - P(x, \varphi_1(x)) \right]\, dx$$

再运用曲线积分的性质及计算公式

$$\oint_c p\, dx = \int_{C_1} p\, dx + \oint_{C_2} p\, dx$$

$$= \int_a^b P(x, \varphi_1(x))\, dx +$$

$$\left[-\int_b^a P(x, \varphi_2(x))\, dx \right]$$

$$= -\int_b^a P(x, \varphi_2(x))\, dx$$

$$- P(x, \varphi_1(x))\, dx$$

$$\boxed{\int p\, dx + Q\, dy = \iint_A \left(\frac{\partial Q}{\partial x} - \frac{\partial P}{\partial y} \right) dy\, dx} \qquad \iint \frac{\partial P}{\partial y}\, dx\, dy = -\int_c p\, dx$$

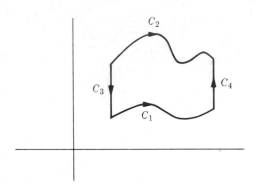

Figure 4

The integrals over the vertical segments are equal to 0. This is easily seen as follows. Consider the right vertical segment parametrized by

$$C_4(t) = (b, t), \quad \text{with} \quad g_1(b) \leqq t \leqq g_2(b).$$

Then $x = b$ (constant!) on this vertical segment, so $dx = 0$. Hence

$$\int_{C_4} p \, dx = 0,$$

thus showing that the interval over this vertical segment is 0. A similar argument applies to the integral over the other vertical segment, and this concludes the proof of Green's theorem in the present case.

Case 2. Suppose that the region is given by similar inequalities as in Case 1, but with respect to the y-axis. In other words, the region A is defined by inequalities

$$c \leqq y \leqq d \quad \text{and} \quad g_1(y) \leqq x \leqq g_2(y).$$

Then we prove the other half of Green's theorem, namely

$$\iint_A \frac{\partial q}{\partial x} \, dy \, dx = \int_C q \, dy.$$

Proof We take the integral with respect to x first:

$$\iint_A \frac{\partial q}{\partial x} \, dx \, dy = \int_c^d \left[\int_{g_1(y)}^{g_2(y)} D_1 q(x, y) \, dx \right] dy$$

$$= \int_c^d \left[q(g_2(y), y) - q(g_1(y), y) \right] dy.$$

In this case, the integral of $q \, dy$ over the horizontal segments is equal to 0 because y is constant on the horizontal segments, and so $dy = 0$. This proves Green's theorem in this second case.

Verify
Green's
Ther

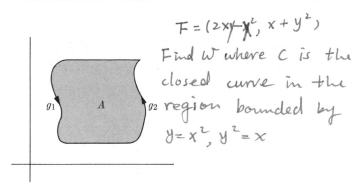

$F = (2xy - x^2, x + y^2)$
Find W where C is the
closed curve in the
region bounded by
$y = x^2, y^2 = x$

Figure 5

In particular, if a region is of a type satisfying both the preceding conditions, then the full theorem follows. Examples of such regions are rectangles and
triangles and interiors of circles:

Evaluate

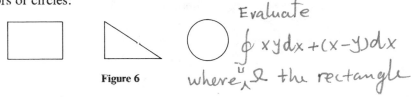

$\oint xy\,dx + (x - y)\,dx$
where $\mathscr{R}$ the rectangle

Figure 6

Other regions of this same type can also be drawn as follows.

$\{(x, y) : 0 \leq x \leq 1$
$1 \leq y \leq 3\}$

Figure 7

We have therefore proved Green's theorem in these cases.
We shall omit the proof of Green's theorem in complete generality.

Example 1. Find the integral of the vector field

$$F(x, y) = (y + 3x, 2y - x)$$

counterclockwise around the ellipse $4x^2 + y^2 = 4$.
   Let $p(x, y) = y + 3x$ and $q(x, y) = 2y - x$. Then

$$\partial q/\partial x = -1 \qquad \text{and} \qquad \partial p/\partial y = 1.$$

By Green's theorem, we get

$$\int_C p\,dx + q\,dy = \iint_A (-2)\,dy\,dx = -2\,\text{Area}(A),$$

where Area(A) is the area of the ellipse, which is known to be 2π ($= \pi ab$ when
the ellipse is in the form $x^2/a^2 + y^2/b^2 = 1$). Hence

$$\int_{\text{ellipse}} F = -4\pi.$$

Example 2. Let $F(x, y) = (3xy, x^2)$. Find the integral of F around the rectangle as shown on the figure, counterclockwise.

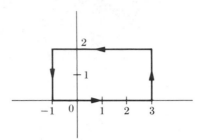

Figure 8

By Green's theorem, the desired integral is

$$\int_C p\ dx + q\ dy = \iint_R \left(\frac{\partial q}{\partial x} - \frac{\partial p}{\partial y} \right) dy\ dx$$

$$= \int_{-1}^{3} \int_0^2 (2x - 3x)\ dy\ dx$$

$$= \int_{-1}^{3} (-x)2\ dx$$

$$= -x^2 \Big|_{-1}^{3} = -(9 - 1) = -8.$$

It is clear that we could compute the integral of F over the boundary of the rectangle easier by using Green's theorem than by parametrizing all four sides and then adding the four integrals over these four sides.

We shall not prove Green's theorem other than in these special cases. In any case, the version stated above is insufficient to cover all applications, and we shall state a somewhat more general version which does suffice.

Suppose we have a region A whose boundary consists of a finite number of curves, which meet only in their end points. Let C_1 be one of these curves, so that A lies either to the right or to the left of C_1, as shown on the figure (Fig. 9).

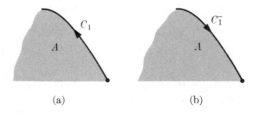

(a)   (b)

Figure 9

We have drawn the curve C_1 and its reverse curve C_1^-. In Fig. 9(a) the region A lies to the *left* of C_1. If we reverse the orientation of C_1 to obtain C_1^-, then A lies to the *right* of C_1^-.

Green's Theorem, General Version. *Let A be a region in the plane whose boundary consists of a finite number of curves. Assume that each curve of the boundary is oriented so that A lies to the left of the curve. Let p, q be functions on A. Let*

$$C = \{ C_1, \ldots, C_m \}$$

be the curves forming the boundary of A. Then ➚ Rotation .

$$\int_C p \, dx + q \, dy = \iint_A \left(\frac{\partial q}{\partial x} - \frac{\partial p}{\partial y} \right) dy \, dx.$$

Remark 1. In the first version of Green's theorem, making the assumption that the closed path forming the boundary is parametrized counterclockwise amounts to the assumption made in the general version that the region lies to the *left* of the curves forming its boundary. Some sort of assumption on the orientation of these curves must be made for the formula of the theorem to come out correctly.

Remark 2. We do not assume that the curves $C_1, \ldots, C_m$ are necessarily connected, i.e. form a path in the sense we used that word previously. In applications, these curves may be disconnected, as in the following example.

Example 3. Let A be the region between two concentric circles C_1, C_2 as shown, both with counterclockwise orientation (Fig. 10).

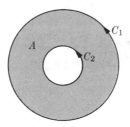

Figure 10

The boundary of the region A consists of the two circles C_1, C_2, which have both been shown with counterclockwise orientation. Then A lies to the *left* of C_1 but to the *right* of C_2. Therefore, if we wish to apply our version of Green's theorem, we must use

$$C = \{ C_1, C_2^- \}$$

as the curves describing the boundary, where C_2^- is the circle with *clockwise orientation*. Then A lies to the *left* of C_2^-. Hence Green's formula gives

$$\int_{C_1} p\,dx + q\,dy + \int_{C_2^-} p\,dx + q\,dy = \iint_A \left(\frac{\partial q}{\partial x} - \frac{\partial p}{\partial y} \right) dy\,dx.$$

Since

$$\int_{C_2^-} p\,dx + q\,dy = -\int_{C_2} p\,dx + q\,dy$$

we may also rewrite this formula in the form

$$\int_{C_1} - \int_{C_2} p\,dx + q\,dy = \iint_A \left(\frac{\partial q}{\partial x} - \frac{\partial p}{\partial y} \right) dy\,dx.$$

An important special case arises when $F = (p, q)$ is a vector field on A satisfying the additional assumption that

$$\frac{\partial p}{\partial y} = \frac{\partial q}{\partial x}, \qquad \text{that is,} \quad D_2 p = D_1 q.$$

Then the right-hand side in the above relation is equal to 0, and consequently we see that the integral of F over C_1 is equal to the integral of F over C_2, in other words

$$\int_{C_1} p\,dx + q\,dy = \int_{C_2} p\,dx + q\,dy.$$

Of course, if F is the gradient of a function, then both these integrals are 0. However, we saw previously that there exist vector fields satisfying the condition $\partial p/\partial y = \partial q/\partial x$, but not having potential functions, e.g.

$$F(x, y) = \left(\frac{-y}{x^2 + y^2}, \frac{x}{x^2 + y^2} \right).$$

Example 4. Let $F(x, y)$ be the above vector field. We wish to find the integral of F over the path γ shown on Fig. 11.

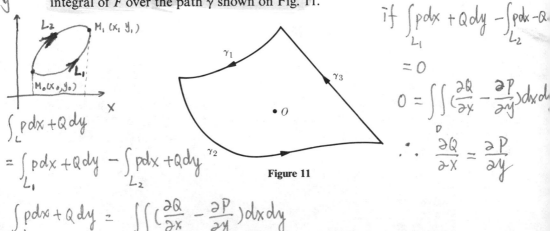

Figure 11

This path γ consists of the three curves γ_1, γ_2, γ_3. It is a mess. But we can use Green's theorem to simplify our problem. We draw a small circle C_1 around the origin O, oriented counterclockwise. We let A be the region between the circle and the path.

$$\int_C p\,dx + Q\,dy = \iint \left(\frac{\partial Q}{\partial x} - \frac{\partial P}{\partial y}\right) dx\,dy \quad \cdots \textcircled{2}$$

Green's 的公式把平面上的曲
线积分和二重积分联系起来

3. 用以求平面上的区域 D
而且还可以通过线积分来计
算。事实上，如果令 (2) 式中的
$p = -y$　$Q = x$ 就得到

$$A = \iint_D dx\,dy = \frac{1}{2}\oint_C x\,dy - y\,dx$$

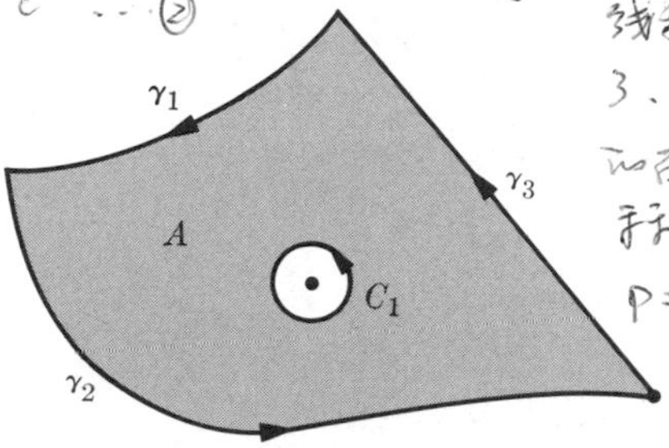

Figure 12

Then the boundary of A consists of the curves $\{\gamma_1, \gamma_2, \gamma_3, C_1^-\}$. Note that if we use C_1^- instead of C_1 then the region A lies to the left of each one of the curves. Therefore by Green's theorem, we get

$$\int_\gamma F + \int_{C_1^-} F = \iint_A \left(\frac{\partial q}{\partial x} - \frac{\partial p}{\partial y}\right) dy\,dx = 0$$

because our vector field satisfies the property $D_2 p = D_1 q$. Hence

$$\int_\gamma F = \int_{C_1} F.$$

It is now easy to find the integral of F over C_1, and was done in Chapter VI, where you found 2π. This is the answer.

EXERCISES

1. Use Green's theorem to find the integral $\int_C y^2\,dx + x\,dy$ when C is the following curve (taken counterclockwise).
   (a) The square with vertices $(0, 0)$, $(2, 0)$, $(2, 2)$, $(0, 2)$.
   (b) The square with vertices $(\pm 1, \pm 1)$.
   (c) The circle of radius 2 centered at the origin.
   (d) The circle of radius 1 centered at the origin.
   (e) The square with vertices $(\pm 2, 0)$, $(0, \pm 2)$.
   (f) The ellipse $x^2/a^2 + y^2/b^2 = 1$.

2. (a) Use Green's theorem to find the integral

$$\int_C y^2\,dx - x\,dy$$

   counterclockwise around the triangle whose vertices are at $(0, 0)$, $(0, 1)$, $(1, 0)$.

(b) Let C be the closed curve consisting of the graphs of

$$y = \sin x \quad \text{and} \quad y = 2 \sin x \quad \text{for} \quad 0 \leq x \leq \pi,$$

and oriented counterclockwise. Find

$$\int_C (1 + y^2)\, dx + y\, dy$$

both directly, and by using Green's theorem.

3. Find the integral

$$\int_C y\, dx + x^2\, dy$$

over the paths shown in Fig. 13.

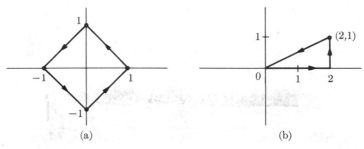

(a)   (b)

Figure 13

4. Let A be a region, which is the interior of a closed curve C oriented counterclockwise. Show that the area of A is given by

(a) $\text{Area}(A) = \dfrac{1}{2} \int_C -y\, dx + x\, dy$

(b) $\text{Area}(A) = \int_C x\, dy$.

5. Assume that the function f satisfies Laplace's equation,

$$\frac{\partial^2 f}{\partial x^2} + \frac{\partial^2 f}{\partial y^2} = 0,$$

on a region A which is the interior of a curve C, oriented counterclockwise. Show that

$$\int_C \frac{\partial f}{\partial y}\, dx - \frac{\partial f}{\partial x}\, dy = 0.$$

6. Find the integral

$$\int_{C_1} \frac{-y}{x^2 + y^2}\, dx + \frac{x}{x^2 + y^2}\, dy$$

when C_1 is each one of the following two paths.

(a) Let C_1 be the closed path consisting of the vertical segment on the line $x = 2$, and the piece of the parabola

$$y^2 = 2(x + 2)$$

lying to the left of this segment, as shown on Fig. 14(a). We assume that C_1 is oriented counterclockwise.

(b) Let C_1 be the square oriented counterclockwise as in Fig. 14(b).

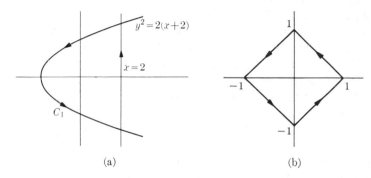

(a)   (b)

Figure 14

7. Find the integral of the vector field

$$F(x, y) = \left(\frac{-y + x}{x^2 + y^2}, \ \frac{x + y}{x^2 + y^2} \right)$$

over the same paths C_1 as in Exercise 6, in both cases (a) and (b).

§2. THE DIVERGENCE AND THE ROTATION OF A VECTOR FIELD

We shall investigate two quantities associated with a vector field

$$F = (p, q).$$

The divergence of F,

$$\operatorname{div} F = D_1 p + D_2 q,$$

which in (x, y)-coordinates also reads

$$(\operatorname{div} F)(x, y) = \frac{\partial p}{\partial x} + \frac{\partial q}{\partial y}.$$

The rotation of F,

$$\operatorname{rot} F = D_1 q - D_2 p,$$

curl F

which in (x, y)-coordinates also reads

$$(rot\ F)(x, y) = \frac{\partial q}{\partial x} - \frac{\partial p}{\partial y}.$$

We note that the rotation of F is exactly the expression which comes under the double integral sign in Green's theorem. So far these quantities have been defined purely algebraically, but in this section we shall derive physical interpretations for them by applying Green's theorem.

Since we have already formulated Green's theorem in the previous section, we begin with the discussion of the rotation. We shall see that the name is deserved, because it measures how much the vector field rotates. If we think in terms of a flow of fluid under the influence of the force field, we can interpret this rotation in terms of how much the fluid rotates.

Let us repeat Green's theorem. We have

$$\iint_A rot\ F\ dy\ dx = \int_a^b F(C(t)) \cdot C'(t)\ dt$$

if A is a region inside a curve C, oriented counterclockwise.

The norm of the velocity vector is the speed, i.e.

$$\|C'(t)\| = \frac{ds}{dt},$$

where $s = s(t)$ is the distance traveled. Let $\mathbf{u}$ be a unit vector in the tangential direction of the curve. We may write

$$C'(t) = \mathbf{u}(t)\frac{ds}{dt}.$$

It is then useful to rewrite the expression on the right in Green's theorem in terms of the unit vector, so that Green's theorem then reads

$$\boxed{\iint_A rot\ F\ dy\ dx = \int_C F \cdot \mathbf{u}\ ds.}$$

We shall apply the formula to a special case to derive the following result.

Corollary of Green's Theorem. *Let D_r be the disc of radius r centered at a point P. Let C_r be the circle of radius r which forms the boundary of D_r, oriented counterclockwise. Let F be a vector field on the closed disc, and let*

$$A(r) = \pi r^2$$

be the area of the disc. Let $\mathbf{u}$ be the unit tangent vector to the circle. Then

$$(rot\ F)(P) = \lim_{r \to 0} \frac{1}{A(r)} \int_{C_r} F \cdot \mathbf{u}\ ds.$$

Proof For an arbitrary point $X = (x, y)$ in the disc, let us write

$$\text{rot } F(X) = \text{rot } F(P) + h(X),$$

where

$$\lim_{X \to P} h(X) = 0.$$

By Green's theorem, we get

$$\frac{1}{A(r)} \int_{C_r} F \cdot u \, ds = \frac{1}{A(r)} \iint_{D_r} \text{rot } F \, dy \, dx$$

$$= \frac{1}{A(r)} \iint_{D_r} (\text{rot } F)(P) \, dy \, dx + \frac{1}{A(r)} \iint_{D_r} h(x, y) \, dy \, dx.$$

Observe that $(\text{rot } F)(P)$ is constant, and can therefore be taken out of the first integral. Since

$$\iint_{D_r} dy \, dx = \text{area of disc of radius } r,$$

we find that the first term on the right of (*) is equal to

$$\frac{1}{A(r)} (\text{rot } F)(P) \iint_{D_r} dy \, dx = (\text{rot } F)(P).$$

Thus to prove the corollary, we need only show that the second term approaches 0 as r approaches 0. This is done as follows. The function $h(x, y)$ approaches 0, and the integral on the right can be estimated as follows.

$$\left| \frac{1}{A(r)} \iint_{D_r} h(x, y) \, dy \, dx \right| \leq \max |h(x, y)| \frac{1}{A(r)} \iint_{D_r} dy \, dx$$

$$\leq \max |h(x, y)|,$$

where the maximum of $|h(x, y)|$ is taken over all points of the disc D_r. This maximum approaches 0 as r approaches 0 by assumption, and the corollary is proved.

This leads to the desired physical interpretation. The dot product

$$F \cdot u$$

is the component of F in the tangential direction of the circle, as shown on Fig. 15.

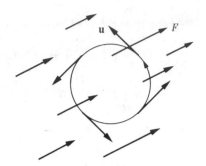

Figure 15

The integral

$$\int_{C_r} F \cdot \mathbf{u} \, ds$$

can be interpreted as the rotation of F around this circle. Dividing by the area of the disc, we obtain this rotation per unit area. Thus we get the interpretation for the rotation of F:

> *The rotation* (rot F)(P) *is the rate at which F rotates per unit area per unit time at P.*

We shall now turn to a similar discussion of the divergence of F. We need first to make some remarks on normal vectors. Let

$$C(t) = (x(t), y(t)), \qquad a \leqq t \leqq b$$

be a curve. Define the **right normal vector** at t to be the vector

$$N(t) = \left(\frac{dy}{dt}, \; -\frac{dx}{dt} \right).$$

It is easily verified that $N(t)$ is a vector perpendicular to the curve (Exercise 1). The picture looks as drawn on Fig. 16.

$$\frac{\partial f}{\partial x} - \frac{\partial f}{\partial y}, \qquad N(t) \qquad (-dy\,i, \; dx\,j) \; \longrightarrow \; \frac{d\varphi}{\partial x} + \frac{\partial f}{\partial y}$$

$C(t)$ $C(t)$

T
$c'(t)$

$dx\,i + dy\,j$ **Figure 16**

(dx, dy)

The word "right" is inserted in the definition above because $N(t)$ points to the right of the curve.

Example. Consider a circle

$$C(\theta) = (\cos \theta, \sin \theta).$$

Then

$$N(\theta) = (\cos \theta, \sin \theta).$$

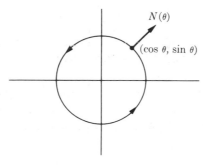

Figure 17

We see that $N(\theta)$ points in the same direction as the position vector $C(\theta)$, and thus points to the right of the circle.

Previously we integrated a vector field F along the curve by forming the dot product with the velocity (tangent) vector,

$$F(C(t)) \cdot C'(t),$$

giving the tangential component of the force, in the direction of the curve. Now we shall form the dot product

$$F(C(t)) \cdot N(t)$$

with the right normal vector giving the component in the perpendicular direction. If we want to abbreviate this by eliminating the reference to the variable t, we simply write

$$F \cdot N.$$

We then have:

Divergence Theorem. *Let A be a region which is the interior of a simple closed curve oriented counterclockwise. Let F be a vector field on A. Then*

$$\iint_A (\operatorname{div} F) \, dy \, dx = \int_a^b F \cdot N \, dt.$$

Proof Exercise 2.

The integral on the right-hand side is of course supposed to read in full

$$\int_a^b F(C(t)) \cdot N(t) \, dt.$$

Since $C'(t) = \left(\dfrac{dx}{dt}, \dfrac{dy}{dt} \right)$, it is immediate that

$$\|N(t)\| = \|C'(t)\| = v(t).$$

In other words, $N(t)$ and the velocity vector $C'(t)$ have the same norm, namely the speed of the curve. Since the distance traveled is given by the integral

$$s(t) = \int v(t) \, dt, \qquad \text{so} \quad \frac{ds}{dt} = v(t),$$

it is customary to rewrite the integral in the divergence theorem in terms of the variable s. Let

$$\mathbf{n}(t) = \frac{N(t)}{\|N(t)\|}$$

be the unit vector in the direction of the normal $N(t)$. Then

$$N(t) = \|N(t)\| \, \mathbf{n}(t) = \frac{ds}{dt} \mathbf{n}(t),$$

and the formula in the divergence theorem may be rewritten:

$$\boxed{\iint_A (\operatorname{div} F) \, dy \, dx = \int_C F \cdot \mathbf{n} \, ds.}$$

Of course, the right-hand side means

$$\int_a^b F(C(t)) \cdot \mathbf{n}(t) \frac{ds}{dt} \, dt.$$

The divergence theorem has an interesting corollary, which will allow us to give a physical interpretation for the divergence of a vector field.

Corollary of the divergence theorem. *Let D_r be the disc of radius r centered at a point P in the plane. Let C_r be the circle of radius r which forms the boundary of D_r, oriented counterclockwise. Let F be a vector field on the closed disc, and let*

$$A(r) = \pi r^2$$

be the area of the disc. Let $\mathbf{n}$ denote the unit right normal vector on the circle.

Then

$$(\text{div } F)(P) = \lim_{r \to 0} \frac{1}{A(r)} \int_{C_r} F \cdot \mathbf{n} \; ds.$$

Proof Let $g = \text{div } F = D_1 p + D_2 q$. Our vector fields are assumed continuous, so g is continuous, and we can write

$$g(X) = g(P) + h(X)$$

where

$$\lim_{X \to P} h(X) = 0.$$

By the divergence theorem, we get

$$\frac{1}{A(r)} \int_{C_r} F \cdot \mathbf{n} \; ds = \frac{1}{A(r)} \iint_{D_r} \text{div } F \; dy \; dx$$

(*) $$= \frac{1}{A(r)} \iint_{D_r} g(P) \; dy \; dx + \frac{1}{A(r)} \iint_{D_r} h(x, y) \; dy \; dx.$$

Observe that $g(P)$ is constant, and can therefore be taken out of the first integral. Since

$$\iint_{D_r} dy \; dx = \text{area of disc of radius } r,$$

we find that the first term on the right of (*) is equal to

$$\frac{1}{A(r)} g(P) \iint_{D_r} dy \; dx = g(P) = (\text{div } F)(P).$$

Thus to prove the corollary, we need only show that the second term approaches 0 as r approaches 0. This is done exactly in the same way that we handled the similar situation previously for the rotation, and concludes the proof.

We now give the physical interpretation for the divergence quite analogously to that of the rotation.

The dot product

$$F \cdot \mathbf{n}$$

is the component of F along the right normal vector, pointing outward. The integral

$$\int_{C_r} F \cdot \mathbf{n} \; ds$$

can be interpreted as the flow going outside the circle per unit time, in the direction of the unit outward normal vector. Dividing by the area of the disc, we obtain the mass per unit area flowing out of the disc. Thus we get the **interpretation for the divergence**:

> *The divergence of F at P is the rate of outward flow per unit area per unit time at P.*

An analogous theorem will be proved in Chapter X, §5 for the divergence in 3-space, and also in Chapter X, §6 for a similar interpretation of the curl. The patterns of proofs will also be quite analogous.

EXERCISES

1. Verify that $N(t)$ is perpendicular to the curve, i.e. perpendicular to $C'(t)$.

2. Prove the divergence theorem, by applying Green's theorem to the vector field $G = (-q, p)$.

3. Let $F(x, y) = (y, -x)$. Let C be the circle of radius 1 oriented counterclockwise. Show that

$$\int_C F \cdot \mathbf{n} \, ds = 0.$$

4. Let A be a region which is the interior of a closed curve C oriented counterclockwise. Let f, g be two functions on A. Define

$$\Delta f = \frac{\partial^2 f}{\partial x^2} + \frac{\partial^2 f}{\partial y^2}.$$

Let $\mathbf{n}$ be the unit right normal vector along the curve. Define

$$D_{\mathbf{n}} f = (\text{grad } f) \cdot \mathbf{n},$$

so that for any value of the parameter t, we have

$$(D_{\mathbf{n}} f)(t) = \text{grad } f(C(t)) \cdot \mathbf{n}(t).$$

This is called the **right normal derivative of f along the curve**. It is the directional derivative of F in the direction of $\mathbf{n}$.

Prove **Green's formulas:**

(a) $\iint_A [(\text{grad } f) \cdot (\text{grad } g) + g \, \Delta f] \, dx \, dy = \int_C g D_{\mathbf{n}} f \, ds$

(b) $\iint_A (g \, \Delta f - f \, \Delta g) \, dx \, dy = \int_C (g D_{\mathbf{n}} f - f D_{\mathbf{n}} g) \, ds$

[*Hint*: Apply the divergence theorem to the vector fields f grad g and g grad f. For instance, let

$$F = g \text{ grad } f = (gD_1 f, gD_2 f) = \left(g\frac{\partial f}{\partial x}, g\frac{\partial f}{\partial y} \right).$$

In computing the divergence of F, use the rule for the derivative of a product.]

5. Prove the following theorem.

Let f be a harmonic function on the disc of radius 1, *that is, assume that f is differentiable as needed, and satisfies Laplace's equation*

$$\frac{\partial^2 f}{\partial x^2} + \frac{\partial^2 f}{\partial y^2} = 0.$$

Then

$$f(0, 0) = \frac{1}{2\pi} \int_0^{2\pi} f(r \cos \theta, r \sin \theta) \, d\theta.$$

[*Hint*: For $0 < r < 1$, let

$$\varphi(r) = \frac{1}{2\pi} \int_0^{2\pi} f(r \cos \theta, r \sin \theta) \, d\theta.$$

Take the derivative $\varphi'(r)$ by differentiating under the integral sign, with respect to r. Using the divergence theorem, you will find

$$\varphi'(r) = \frac{1}{2\pi r} \iint_{D_r} \text{div grad } f(x, y) dy \, dx$$

$$= \frac{1}{2\pi r} \iint_{D_r} \left(\frac{\partial^2 f}{\partial x^2} + \frac{\partial^2 f}{\partial y^2} \right) dy \, dx$$

$$= 0.$$

Hence φ is constant. Then substitute $r = 0$ in the definition of φ to get what you want.]

The theorem in this exercise is sometimes called the mean value theorem for harmonic functions. It says that the value of the function at $(0, 0)$ is obtained by averaging the function over a circle (of any radius) centered at 0.

Evaluate $\iiint\limits_{S} 2x^3 y^2 z\, dx\, dy\, dz$ when s is region

$\{(x,y,z): 0 \le x \le 1, \ x^2 \le y \le x, \ x-y \le z \le x+y\}$

$$\int_0^1 \int_{x^2}^{x} \int_{x-y}^{x+y} 2x^3 y^2 z\, dz\, dy\, dx = \frac{4}{117}$$

Volume $x + \frac{y}{2} + \frac{z}{4} = 1$

coordinate planes

$$\int_0^1 \int_0^{2(1-x)} \int_0^{4-4x-2y} dz\, dy\, dx$$

$\left(\frac{4}{3}\right)$

(020)

(1p,0)

$$\int_0^2 \int_0^{4-2y} \int_0^{1-\frac{y}{2}-\frac{z}{4}} dx\, dz\, dy$$

① Find Volume of the 3-dim region bounded by $x+y+z = a$

$(a>0), \ x=0, \ y=0, \ z=0$

②. Find the volume $(\iiint\limits_{S} dv)$ when $S = \{(x,y,z): a \le x \le 1$

$x^2 \le y \le x, \ x-y \le z \le x+y\}$

Find the mass of a solid bounded by the cylinder $r \sin\theta$, the planes $z=0$, $\theta=0$ $\theta = \frac{\pi}{3}$ and the cone $z=r$ if the density is given by $\rho(r,\theta,z) = 4r$

Find the mass of a solid bounded by the paraboloid $z = x^2+y^2$ and the plane $z=4$ if the density at any point is proportional to the distance from the point to the z-axis

$$\int_{\theta_1}^{\theta_2} \int_{\phi_1}^{\phi_2} \int_{\rho_1}^{\rho_2} \rho^2 \sin\phi\, d\rho\, d\phi\, d\theta$$

Calculate the volume enclosed by the sphere $x^2+y^2+z^2 = a^2$

$$\int_{-a}^{a} \int_{-\sqrt{a^2-x^2}}^{\sqrt{a^2-x^2}} \int_{-\sqrt{a^2-x^2-y^2}}^{\sqrt{a^2-x^2-y^2}} dz\, dx\, dy$$

$$V = \int_0^{2\pi} \int_0^\pi \int_0^a \rho^2 \sin\phi \, d\rho \, d\phi \, d\theta$$

$$\int_0^{2\pi} \int_0^\pi \frac{a^3}{3} \sin\phi \, d\phi \, d\theta = \frac{a^3}{3} \int_0^{2\pi} -\cos\phi \Big|_0^\pi \, d\theta = \frac{2a^3}{3} \int_0^{2\pi} d\theta$$

$$= \frac{4\pi a^3}{3}$$

CHAPTER IX

Triple Integrals

Rectangular   Cylindrical (r, θ, z) Sherical

$\iiint dv$

$dv = dx\,dy\,dz$

(x, y, z)

$\iiint r\,dr\,d\theta\,dz$

In this chapter we carry out the analogue in 3-dimensional space of the integration theory developed in Chapter VII for 2-dimensional space.

§1. TRIPLE INTEGRALS

The entire discussion concerning 2-dimensional integrals generalizes to higher dimensions. We discuss briefly the 3-dimensional case.

A 3-dimensional rectangular box (rectangular parallelepiped) can be written as a product of three intervals:

$$R = [a_1, b_1] \times [a_2, b_2] \times [a_3, b_3].$$

It looks like this (Fig. 1).

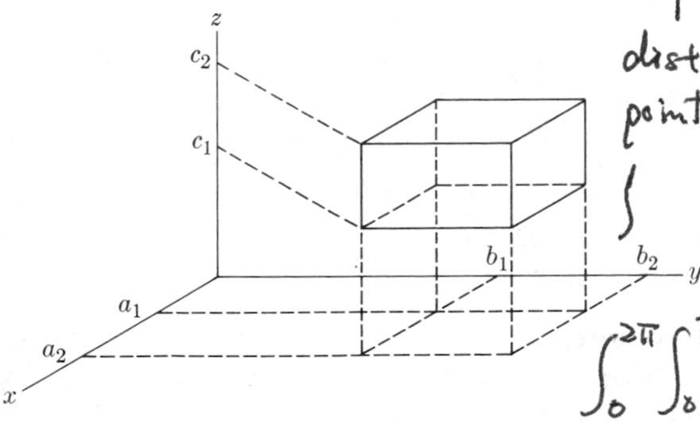

Find the mass of the sphere if the density is proportional to the distance from the point to the origin.

$$\int \int \int \alpha \sqrt{x^2 + y^2 + z^2} \, dV$$

Figure 1

$$\int_0^{2\pi} \int_0^\pi \int_0^a \alpha \rho (\rho^2 \sin\phi) d\rho \, d\phi \, d\theta$$

$$= \pi a^4 \alpha$$

A partition P of R is then determined by partitions P_1, P_2, P_3 of the three intervals respectively, and partitions R into 3-dimensional subrectangles, which we denote again by S.

If f is a bounded function on R, we may then form upper and lower sums. Indeed, we define the volume of the rectangle R above to be the 3-dimensional volume

$$\text{Vol}(R) = (b_3 - a_3)(b_2 - a_2)(b_1 - a_1)$$

and similarly for the subrectangles of the partition. Then we have

$$L(P, f) = \sum_S (\text{glb}_S f)\text{Vol}(S),$$

$$U(P, f) = \sum_S (\text{lub}_S f)\text{Vol}(S).$$

As before, every lower sum is less than or equal to every upper sum. A function f is called **integrable** if there exists a unique number which is $\geq$ every lower sum and $\leq$ every upper sum. If that is the case, this number is called the **integral of** f, and is denoted by

$$\iiint_R f = \iiint_R f(x, y, z)\, dz\, dy\, dx.$$

If $f \geq 0$, then we interpret this integral as the 4-dimensional volume of the 4-dimensional region lying in 4-space, bounded from below by R, and from above by the graph of f. Of course, we cannot draw this figure because it is in 4-space, but the terminology goes right over.

The basic theorems of Chapter VII are still valid here. We repeat them.

If f, g are integrable, then so is $f + g$ and kf for any constant k, and we have:

$$\iiint_R (f + g) = \iiint_R f + \iiint_R g, \qquad \iiint_R kf = k \iiint_R f.$$

In two variables, we stated that a function is integrable if it is bounded and continuous except at a finite number of smooth curves. We have also an analogue for this, except that instead of curves, we have to allow for surfaces.

Let R be a 3-dimensional rectangular box, and let f be a function defined on R, bounded and continuous except possibly at the points lying on a finite number of smooth surfaces. Then f is integrable on R.

Again we can integrate over a more general region than a rectangle, provided such a region A has a boundary which is contained in a finite number of smooth surfaces. If A denotes a 3-dimensional region and f is a function on A, we define

$$f(X) = 0 \quad \text{if} \quad X \quad \text{is not a point of} \quad A.$$

We always assume our regions are bounded, so we can find a suitably large rectangular box R which contains A. We define the integral of f over A to be the integral of the function over R, i.e. we define

$$\iiint_A f = \iiint_R f$$

or also in terms of the variables

$$\iiint_A f(x, y, z)\, dz\, dy\, dx = \iiint_R f(x, y, z)\, dz\, dy\, dx.$$

Since $f(x, y, z) = 0$ if (x, y, z) is not a point of A, the integral on the right represents the desired notion.

If we view A as a solid piece of material, and f is interpreted as a density distribution over A, then the integral of f over A may be interpreted as the **mass** of A.

To compute multiple integrals in the 3-dimensional case, we have the same situation as in the 2-dimensional case.

The theorem concerning the relation with repeated integrals holds, so that if R is the rectangular box given by

$$R = [a_1, b_1] \times [a_2, b_2] \times [a_3, b_3],$$

then

$$\iiint_R f = \int_{a_1}^{b_1} \left[\int_{a_2}^{b_2} \left(\int_{a_3}^{b_3} f(x, y, z)\, dz \right) dy \right] dx.$$

Of course, the repeated integral can be evaluated in any order.

Example 1. Find the integral of the function $f(x, y, z) = \sin x$ over the rectangular box

$$0 \le x \le \pi, \quad 2 \le y \le 3, \quad \text{and} \quad -1 \le z \le 1.$$

The integral is equal to

$$\int_0^\pi \int_2^3 \int_{-1}^1 \sin x\, dz\, dy\, dx.$$

If we first integrate with respect to z, we get

$$\int_{-1}^1 dz = z \Big|_{-1}^1 = 2.$$

Next with respect to y, we get

$$\int_2^3 dy = y \Big|_2^3 = 1.$$

(handwritten, right margin)

① Find the volume of the region inside both the sphere $x^2 + y^2 + z^2 = 4$ and the cylinder $(x-1)^2 + (y)^2 = 1$

② Find the volume of the solid inside the cone $z^2 = x^2 + y^2$

(handwritten, lower left)

$$\frac{16}{3}\left[\pi - \frac{4}{3}\right]$$

$$\frac{16\sqrt{2}}{3}\pi$$

We are then reduced to the integral

$$\int_0^\pi 2 \sin x \, dx = -2 \cos x \Big|_0^\pi = -2(\cos \pi - \cos 0) = 4.$$

We also have the integral over regions determined by inequalities.

Rectangular coordinates. *Let a, b be numbers, $a \leqq b$. Let g_1, g_2 be two smooth functions defined on the interval $[a, b]$ such that*

$$g_1(x) \leqq g_2(x),$$

and let $h_1(x, y) \leqq h_2(x, y)$ be two smooth functions defined on the region consisting of all points (x, y) such that

$$a \leqq x \leqq b \quad and \quad g_1(x) \leqq y \leqq g_2(x).$$

Let A be the set of points (x, y, z) such that

$$a \leqq x \leqq b, \qquad g_1(x) \leqq y \leqq g_2(x),$$

and

$$h_1(x, y) \leqq z \leqq h_2(x, y).$$

Let f be continuous on A. Then

$$\iiint_A f = \int_a^b \left[\int_{g_1(x)}^{g_2(x)} \left(\int_{h_1(x, y)}^{h_2(x, y)} f(x, y, z) dz \right) dy \right] dx$$

For simplicity, the integral on the right will also be written without the brackets.

Example 2. Consider the tetrahedron T spanned by 0 and the three unit vectors (Fig. 2).

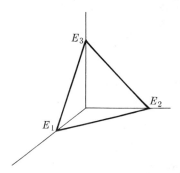

Figure 2

This tetrahedron is described by the inequalities:

$$0 \leqq x \leqq 1, \qquad 0 \leqq y \leqq 1 - x, \qquad 0 \leqq z \leqq 1 - x - y.$$

Hence if f is a function on the tetrahedron, its integral over T is given by

$$\iiint_T f = \int_0^1 \int_0^{1-x} \int_0^{1-x-y} f(x, y, z) \, dz \, dy \, dx.$$

For the constant function 1, the integral gives you the volume of the tetrahedron, and you should have no difficulty in evaluating it, finding the value

$$\mathrm{Vol}(T) = \iiint_T 1 \, dz \, dy \, dx = \frac{1}{6}.$$

EXERCISES

1. Find the volume of the region spanned by the following inequalities:

$$0 \leq x \leq 1, \quad 0 \leq y \leq \sqrt{1 - x^2}, \quad 0 \leq z \leq \sqrt{1 - x^2 - y^2}.$$

2. Find the integral

$$\int_0^\pi \int_0^{\sin \theta} \int_0^{\rho \cos \theta} \rho^2 \, dz \, d\rho \, d\theta.$$

3. Find the integral of the following functions over the indicated region, in 3-space.
   (a) $f(x, y, z) = x^2$ over the tetrahedron bounded by the plane

$$12x + 20y + 15z = 60,$$

   and the coordinate planes.
   (b) $f(x, y, z) = y$ over the tetrahedron as in (a).

4. Let A be the region in R^3 bounded by the planes

$$y = 1, \quad y = -x, \quad x = 0, \quad z = 0, \quad \text{and} \quad z = -x.$$

Find

$$\iiint_A e^{x+y+z} dx \, dy \, dz.$$

§2. CYLINDRICAL AND SPHERICAL COORDINATES

Cylindrical Coordinates

Analogously to the polar coordinates in the plane, we consider cylindrical coordinates in 3-space, given by

$$x = r \cos \theta$$
$$y = r \sin \theta$$
$$z = z.$$

We shall abbreviate the association

$$(r, \theta, z) \longmapsto (x, y, z)$$

by the symbols

$$(x, y, z) = G(r, \theta, z) = (r \cos \theta, r \sin \theta, z).$$

We also call G a **mapping**, or **transformation**, from the (r, θ, z)-space to the (x, y, z)-space. The numbers (r, θ, z) are called the **cylindrical coordinates** of the point (x, y, z), and are represented on the following figure (Fig. 3).

Find the centroid of ice-cream cone shaped region below to sphere $x^2 + y^2 + z^2 = z$ and above the $z^2 = x^2 + y^2$

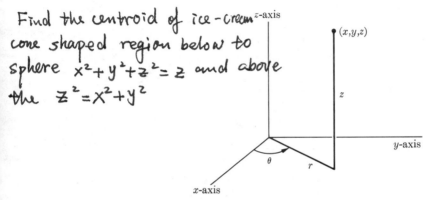

Figure 3

The cylindrical coordinates of a region are usually taken with values of (r, θ, z) such that

$$0 \leq r$$
$$0 \leq \theta \leq 2\pi$$
$$z \quad \text{is arbitrary.}$$

Consider the **elementary cylindrical region** shown on Fig. 4(b).

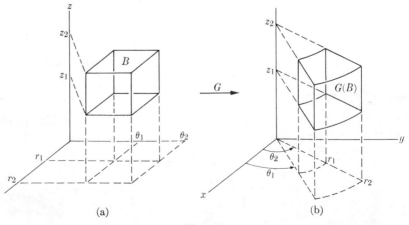

(a)   (b)

Figure 4

It is the transform of the rectangular box B in Fig. 4(a). It is the set of all points whose cylindrical coordinates satisfy the inequalities

$$0 \leq \theta_1 \leq \theta \leq \theta_2 \leq 2\pi,$$
$$0 \leq r_1 \leq r \leq r_2,$$
$$z_1 \leq z \leq z_2.$$

The volume of the elementary cylindrical region $G(B)$ is equal to the area of the base times the height. The height is $(z_2 - z_1)$. The area of the base is the area of a piece of a sector, which we already found when dealing with polar coordinates in the plane. Consequently, the volume of $G(B)$ is given by the formula:

$$(z_2 - z_1)\left(\frac{r_2^2 - r_1^2}{2} \right)(\theta_2 - \theta_1).$$

This expression can be rewritten in the form

$$\bar{r}(z_2 - z_1)(r_2 - r_1)(\theta_2 - \theta_1),$$

where

$$\bar{r} = \frac{r_2 + r_1}{2}.$$

Forming upper and lower sums with respect to partitions of the r-axis, θ-axis, and z-axis, we are then led to the formula analogous to the formula for integration with respect to polar coordinates, as follows.

Suppose A is some region in the (x, y, z)-space, and let A^* be the region of the (r, θ, z)-space corresponding to A under the cylindrical coordinates. Then

$$\iiint_A f(x, y, z) \, dz \, dy \, dx = \iiint_{A^*} f(r \cos \theta, r \sin \theta, z) r \, dz \, dr \, d\theta.$$

Indeed, the same kind of argument applies as with polar coordinates.

In practice, the region A^* is described by the same type of inequalities as with polar coordinates, and we state the relevant theorem as follows:

Let A be the region in the (x, y, z)-space consisting of points whose cylindrical coordinates (r, θ, z) satisfy inequalities

$$a \leq \theta \leq b \qquad (and \quad b \leq a + 2\pi),$$
$$0 \leq g_1(\theta) \leq r \leq g_2(\theta),$$
$$h_1(\theta, r) \leq z \leq h_2(\theta, r),$$

where g_1, g_2, h_1, h_2 are smooth functions. Let A^* be the corresponding region of points (z, r, θ) satisfying these inequalities in the (r, θ, z)-space. Let

$$f^*(\theta, r, z) = f(r \cos \theta, r \sin \theta, z).$$

Then

$$\iiint_A f = \int_a^b \int_{g_1(\theta)}^{g_2(\theta)} \int_{h_1(\theta,\, r)}^{h_2(\theta,\, r)} f^*(\theta, r, z) r \, dz \, dr \, d\theta.$$

The function which we denote by f^* may be viewed as the function f in terms of the cylindrical coordinates.

Example 1. Find the mass of a solid bounded by the polar coordinates $-\pi/3 \leqq \theta \leqq \pi/3$ and $r = \cos \theta$ and by $z = 0$, $z = r$, if the density is given by the function

$$f^*(r, \theta, z) = 3r.$$

The mass is given by the integral

$$\int_{-\pi/3}^{\pi/3} \int_0^{\cos \theta} \int_0^r 3r \cdot r \, dz \, dr \, d\theta.$$

Integrating the inner integral with respect to z yields $3r^2 r = 3r^3$. Integrating with respect to r between 0 and $\cos \theta$ yields

$$\left. \frac{3r^4}{4} \right|_0^{\cos \theta} = \frac{3 \cos^4 \theta}{4}.$$

Finally we integrate with respect to θ, using elementary techniques of integration: $\cos^2 \theta = (1 + \cos 2\theta)/2$ so that

$$\cos^4 \theta = \frac{1}{4}(1 + 2 \cos 2\theta + \cos^2 2\theta)$$

$$= \frac{1}{4}\left(1 + 2 \cos 2\theta + \frac{1 + \cos 4\theta}{2}\right).$$

We can now integrate this between the given limits, and we find

$$\frac{3}{4} \int_{-\pi/3}^{\pi/3} \cos^4 \theta \, d\theta = \frac{3}{16}\left(\frac{\pi}{3} - \sqrt{3} + \frac{\pi}{6} + \frac{\sqrt{3}}{8}\right).$$

Note. In the above example, the function is already given in terms of (r, θ, z). It corresponds to the function $f(x, y, z) = 3\sqrt{x^2 + y^2}$. Indeed, taking $f(r \cos \theta, r \sin \theta, z)$ yields $3r$.

Example 2. Let us find the volume of the region inside the cylinder $r = 4 \cos \theta$, bounded above by the sphere $r^2 + z^2 = 16$, and below by the plane $z = 0$. In the (x, y)-plane, the equation $r = 4 \cos \theta$ is that of a circle, with $-\pi/2 \leqq \theta \leqq \pi/2$. The region is then defined by means of the other two inequalities

$$0 \leqq z \leqq \sqrt{16 - r^2} \qquad \text{and} \qquad 0 \leqq r \leqq 4 \cos \theta.$$

Therefore the desired volume V is the integral

$$V = \int_{-\pi/2}^{\pi/2} \int_0^{4 \cos \theta} \int_0^{\sqrt{16 - r^2}} r \, dz \, dr \, d\theta$$

$$= \int_{-\pi/2}^{\pi/2} \int_0^{4 \cos \theta} r\sqrt{16 - r^2} \, dr \, d\theta$$

$$= -\frac{64}{3} \int_{-\pi/2}^{\pi/2} |\sin^3 \theta| - 1 \, d\theta = \frac{64\pi}{3} - \frac{64.4}{9}.$$

Spherical coordinates

We consider the region in coordinates (ρ, θ, φ) described by

$$0 \leqq \rho, \qquad 0 \leqq \varphi \leqq \pi, \qquad 0 \leqq \theta \leqq 2\pi.$$

These coordinates can be used to describe a point in 3-space as shown on the following picture.

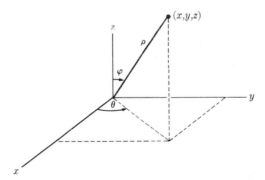

Figure 5

In fact, we let

$$\rho = \sqrt{x^2 + y^2 + z^2} .$$

We denote this by ρ to distinguish it from the polar coordinate r in the (x, y)-plane. Then ρ is the distance of (x, y, z) from the origin. Furthermore, θ is the same angle as with polar coordinates. We have a new coordinate φ which denotes the angle with respect to the z-axis.

We see that

$$x^2 + y^2 = \rho^2 - z^2 = \rho^2 \sin^2 \varphi$$

so that the polar r is given by

$$r = \sqrt{x^2 + y^2} = \rho \sin \varphi.$$

In taking the square root, we do not need to use the absolute value $|\sin \varphi|$ because we take $0 \leq \varphi \leq \pi$ so that $\sin \varphi \geq 0$ for our values of φ.

From the formulas $x = r \cos \theta$ and $y = r \sin \theta$, we then obtain the relationship between (x, y, z) and (ρ, θ, φ), namely:

$$
\begin{aligned}
x &= \rho \sin \varphi \cos \theta, \\
y &= \rho \sin \varphi \sin \theta, \\
z &= \rho \cos \varphi.
\end{aligned}
$$

We can also say that we have a transformation $G: \mathbf{R}^3 \rightarrow \mathbf{R}^3$ given by

$$G(\rho, \theta, \varphi) = (\rho \sin \varphi \cos \theta, \rho \sin \varphi \sin \theta, \cos \varphi).$$

Let R be the 3-dimensional box in the (ρ, θ, φ)-space described by the inequalities:

$$
\begin{aligned}
\theta_1 &\leq \theta \leq \theta_2, \quad (\theta_2 \leq \theta_1 + 2\pi), \\
0 &\leq \rho_1 \leq \rho \leq \rho_2, \\
0 &\leq \varphi_1 \leq \varphi \leq \varphi_2 \leq \pi.
\end{aligned}
$$

The image of R under the transformation G is then an elementary spherical region $G(R)$ as shown in Fig. 6.

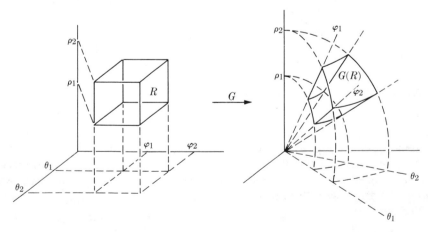

Figure 6

The volume of the elementary spherical region $G(R)$ *just described is equal to*

$$\left(\frac{\rho_2^3}{3} - \frac{\rho_1^3}{3}\right)(\cos\varphi_1 - \cos\varphi_2)(\theta_2 - \theta_1).$$

In order to see this, we shall find the volume of a slightly simpler region, namely that lying above a cone and inside a sphere as shown on the next figure (Fig. 7).

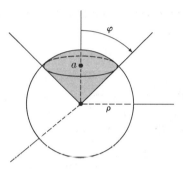

Figure 7

The radius of the sphere is ρ, and the angle of the cone is φ, as shown on the figure. We let a be the height at which the cone meets the sphere. The volume of this region consists of two pieces. The first is the volume of a cone of height a, and whose base is

$$b = \rho \sin\varphi.$$

Observe that $a = \rho \cos\varphi$. The volume of this cone is therefore equal to

$$\frac{\pi}{3}\rho^3 \sin^2\varphi \cos\varphi.$$

The other piece lies below the spherical dome, and can be obtained as a volume of revolution of the curve $x^2 + y^2 = \rho^2$, letting x range between a and ρ. You should know how to do this, and you will find the answer

$$\pi\left(\tfrac{2}{3}\rho^3 - \rho^3 \cos\varphi + \tfrac{1}{3}\rho^3 \cos^3\varphi\right).$$

Adding our two volumes together, and noting that

$$\cos^3\varphi = \cos^2\varphi \cos\varphi,$$

we find that the volume of the region lying above the cone and inside the sphere is equal to

$$\tfrac{2}{3}\pi\rho^3 - \tfrac{2}{3}\pi\rho^3 \cos\varphi.$$

The volume of this region lying between angles φ_1 and φ_2 is obtained by subtracting, and is equal to

$$\tfrac{2}{3}\pi\rho^3(\cos\varphi_1 - \cos\varphi_2).$$

Considering only the part lying between the spheres of radii ρ_1 and ρ_2, we obtain its volume again by subtraction, and get

$$\tfrac{2}{3}\pi(\rho_2^3 - \rho_1^3)(\cos\varphi_1 - \cos\varphi_2).$$

Finally, we have to take that part lying between angles θ_1 and θ_2, that is, take the fraction

$$\frac{\theta_2 - \theta_1}{2\pi}$$

of this last volume. In this way, we obtain precisely the desired volume of the elementary spherical region of Fig. 6.

Using the mean value theorem, we find that

$$\frac{\rho_2^3}{3} - \frac{\rho_1^3}{3} = \bar{\rho}^2(\rho_2 - \rho_1),$$

for some number $\bar{\rho}$ between ρ_1 and ρ_2. Again by the mean value theorem, we find that

$$\cos\varphi_1 - \cos\varphi_2 = \sin\bar{\varphi}(\varphi_2 - \varphi_1).$$

Hence:

> **The volume of the elementary spherical region $G(R)$ is equal to**
>
> $$\bar{\rho}^2\sin\bar{\varphi}(\rho_2 - \rho_1)(\varphi_2 - \varphi_1)(\theta_2 - \theta_1).$$

By forming Riemann sums we already had in polar coordinates, it is therefore reasonable that the triple integral of a function f over a region A in the (x, y, z)-space which corresponds to a region A^* in the (ρ, θ, φ)-space of spherical coordinates is given by the formula:

> $$\iiint_A f(x, y, z)\, dz\, dy\, dx = \iiint_{A^*} f^*(\rho, \theta, \varphi)\rho^2\sin\varphi\, d\rho\, d\varphi\, d\theta.$$

As usual, $f^*(\rho, \theta, \varphi)$ is the value of the function at the given point (x, y, z) in terms of the spherical coordinates of the point (ρ, θ, φ), namely,

$$f^*(\rho, \theta, \varphi) = f(\rho\sin\varphi\cos\theta,\ \rho\sin\varphi\sin\theta,\ \rho\cos\varphi).$$

We can also abbreviate this with the notation

$$f^*(\rho, \theta, \varphi) = f(G(\rho, \theta, \varphi)).$$

Symbolically, it is convenient to use a notation which does not contain variables when expressing an integral. Thus we sometimes write

$$\iiint_A f \, dV$$

where dV means, in the various coordinates:

$$dV = dz \, dy \, dx = z \, dz \, dr \, d\theta = \rho^2 \sin \varphi \, d\rho \, d\varphi \, d\theta.$$

Example 3. As a check, let us apply the general formula directly to see if it gives us the same answer for the volume of the elementary spherical region $G(R)$. We are supposed to evaluate the integral

$$\iiint_{G(R)} dz \, dy \, dx = \int_{\theta_1}^{\theta_2} \int_{\varphi_1}^{\varphi_2} \int_{\rho_1}^{\rho_2} \rho^2 \sin \varphi \, d\rho \, d\varphi \, d\theta.$$

In this case, the repeated 3-fold integral splits into separate integrals with respect to ρ, φ, θ independently. These integrals are of course very simple to evaluate. In this case, the limits of integration are constant. Integrating with respect to ρ yields the factor $\frac{1}{3}(\rho_2^3 - \rho_1^3)$. Integrating with respect to φ yields the factor $(\cos \varphi_1 - \cos \varphi_2)$. Integrating with respect to θ yields the factor $(\theta_2 - \theta_1)$. Thus the evaluation of the integral checks with the arguments given previously.

Example 4. Find the volume above the cone $z^2 = x^2 + y^2$ and inside the sphere $x^2 + y^2 + z^2 = z$ (Fig. 9.8).

Figure 8

Handwritten notes (left):

using the triple integrals find the volume

Inside $x^2 + y^2 = 9$ above $z = 0$ and below $x + z = 4$ (36π unit)

bounded by the coordinate planes and $6x + 4y + 3z = 12$ 4 unit

inside $x^2 + y^2 = 4x$ above $z = 0$ and below $x^2 + y^2 = 4z$ 6π unit

cut from the cone $\phi = \frac{1}{4}\pi$ by the sphere $\rho = 2a\cos\phi$ πa^3

Handwritten notes (right, Chinese):

母线平行于坐标轴的柱面方程与图形

定义：平行于定直线 L 的动直线 L，沿着定曲线 C（C 与 L 不在同一平面上）运动而形成的曲面叫柱面，动直线 L 叫母线，定曲线 C 叫准线

$x^2 + y^2 = R^2$ 是圆柱方程，

这里值得注意的是，在 OXY 平面直角坐标系中 $x^2 + y^2 = R^2$ 是表示一个圆；而在空间坐标系中 $x^2 + y^2 = R^2$ 不再表示一个圆，而是表示一个圆柱面，就象 $y = 3$ 在 OXY 平面直角坐标系中是表示一条直线，而在空间直角坐标系中就是表示一个平面一样。 $\frac{x^2}{a^2} + \frac{y^2}{b^2} = 1$ 椭圆柱面 $\frac{x^2}{a^2} - \frac{y^2}{b^2} = 1$ 双曲柱面，$x^2 - 2py = 0$ 抛物柱面

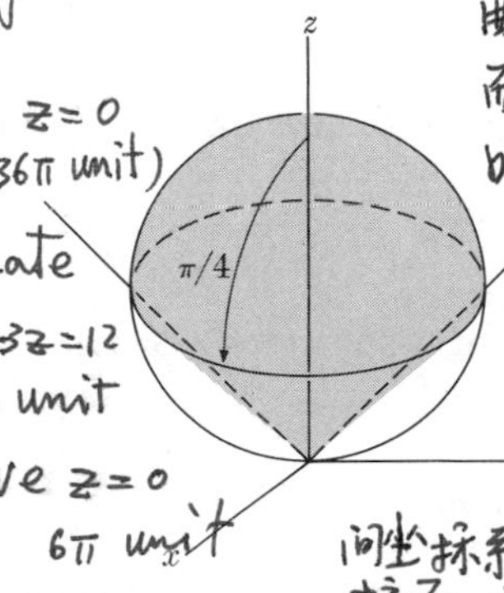

上面我们讨论了母线平行于 z 轴的柱面方程，这种方程的特征是不含变量 z，同理，如果方程不含变量 x 或者 y，那么这方程在空间就表示母线平行于 x 轴或 y 轴的柱面。

The equation for the sphere in spherical coordinates is obtained by the values

$$\rho^2 = x^2 + y^2 + z^2$$

and

$$z = \rho \cos \varphi.$$

Thus the sphere is given by the equation

$$\rho = \cos \varphi.$$

The cone is given by $\cos^2 \varphi = \sin^2 \varphi$, and since $0 \leq \varphi \leq \pi$ this is the same as $\varphi = \pi/4$. Thus the region of the integration is the set of points whose spherical coordinates satisfy the inequalities

$$0 \leq \theta \leq 2\pi, \qquad 0 \leq \varphi \leq \pi/4, \qquad 0 \leq \rho \leq \cos \varphi.$$

Hence our volume is equal to the integral

$$\iiint_A 1 = \int_0^{2\pi} \int_0^{\pi/4} \int_0^{\cos \varphi} \rho^2 \sin \varphi \, d\rho \, d\varphi \, d\theta.$$

The inside integral with respect to ρ is equal to

$$(\sin \varphi) \frac{\rho^3}{3} \Big|_0^{\cos \varphi} = \frac{1}{3} \cos^3 \varphi \sin \varphi.$$

This is now easily integrated with respect to φ, using

$$u = \cos \varphi, \quad du = -\sin \varphi \, d\varphi,$$

and yields

$$\int_0^{\pi/4} \frac{1}{3} \cos^3 \varphi \sin \varphi \, d\varphi = \frac{1}{3} \frac{-\cos^4 \varphi}{4} \Big|_0^{\pi/4} = \frac{1}{12} \left(-\frac{1}{4} + 1 \right) = \frac{3}{48}.$$

Finally, we integrate with respect to θ, and the final answer is therefore equal to

$$\tfrac{3}{48} \cdot 2\pi = \tfrac{1}{8}\pi.$$

Example 5. Find the mass of a solid body S determined by the inequalities of spherical coordinates:

$$0 \leq \theta \leq \frac{\pi}{2}, \qquad \frac{\pi}{4} \leq \varphi \leq \arctan 2, \qquad 0 \leq \rho \leq \sqrt{6},$$

if the density, given as a function of the spherical coordinates (ρ, θ, φ), is equal to $1/\rho$.

To find the mass, we have to integrate the given function over the region. The integral is given by

$$\int_0^{\pi/2} \int_{\pi/4}^{\arctan 2} \int_0^{\sqrt{6}} \frac{1}{\rho} \rho^2 \sin \varphi \, d\rho \, d\varphi \, d\theta.$$

Performing the repeated integral, we obtain

$$\frac{3\pi}{2} \left(\frac{1}{\sqrt{2}} - \frac{1}{\sqrt{5}} \right).$$

We note that in the present example, the limits of integration are constants, and hence the repeated integral is equal to a product of the integrals

$$\int_0^{\pi/2} d\theta \cdot \int_{\pi/4}^{\arctan 2} \sin \varphi \, d\varphi \cdot \int_0^{\sqrt{6}} \rho \, d\rho.$$

Each integration can be performed separately. Of course, this does not hold when the limits of integration are non-constant functions.

As before, we have a similar integral when the boundaries of integration are not constant. We state the result:

Let A be a region in the (x, y, z)-space which consists of all points whose spherical coordinates (ρ, θ, φ) satisfy the inequalities:

$$a \leqq \theta \leqq b,$$
$$g_1(\theta) \leqq \varphi \leqq g_2(\theta),$$
$$h_1(\theta, \varphi) \leqq \rho \leqq h_2(\theta, \varphi)$$

where:

a, b are numbers such that $0 \leqq b - a \leqq 2\pi$;

$g_1(\theta)$, $g_2(\theta)$ are smooth functions of θ, defined on the interval $a \leqq \theta \leqq b$ such that

$$0 \leqq g_1(\theta) \leqq g_2(\theta) \leqq \pi;$$

h_1, h_2 are functions of two variables, defined and smooth on the region consisting of all points (θ, φ) such that

$$a \leqq \theta \leqq b,$$
$$g_1(\theta) \leqq \varphi \leqq g_2(\theta)$$

and such that $0 \leqq h_1(\theta, \varphi) \leqq h_2(\theta, \varphi)$ for all (θ, φ) in this region.

Let f be a continuous function on A, and let

$$f^*(\rho, \theta, \varphi) = f(G(r, \theta, \varphi))$$

be the corresponding function of (ρ, θ, φ). Then

$$\iiint_A f = \int_a^b \int_{g_1(\theta)}^{g_2(\theta)} \int_{h_1(\theta,\,\varphi)}^{h_2(\theta,\,\varphi)} f^*(\theta, \varphi, \rho)\rho^2 \sin \varphi \, d\rho \, d\varphi \, d\theta.$$

EXERCISES

1. Find the volume inside the sphere

$$x^2 + y^2 + z^2 = a^2,$$

by using spherical coordinates.

2. Find the volume inside the cone

$$\sqrt{x^2 + y^2} \leq z \leq 1$$

by using spherical coordinates.

3. (a) Find the mass of a spherical ball of radius $a > 0$ if the density at any point is equal to a constant k times the distance of that point to the center.
    (b) Find the integral of the function

$$f(x, y, z) = \frac{1}{\sqrt{x^2 + y^2 + z^2}}$$

over the spherical shell of inside radius a and outside radius 1. Assume $0 < a < 1$. What is the limit of this integral as $a \to 0$?

4. Find the mass of a spherical shell of inside radius a and outside radius b if the density at any point is inversely proportional to the distance from the center.

5. Find the integral of the function

$$f(x, y, z) = x^2$$

over that portion of the cylinder

$$x^2 + y^2 = a^2$$

lying between the planes

$$z = 0 \quad \text{and} \quad z = b > 0.$$

6. Find the mass of a sphere of radius a if the density at any point is proportional to the distance from a fixed plane passing through a diameter.

7. Find the volume of the region bounded by the cylinder $y = \cos x$, and the planes

$$z = y, \quad x = 0, \quad x = \pi/2, \quad \text{and} \quad z = 0.$$

② 锥面的标准方程与图形：定义通过定点O, 沿动直线 L, 沿着定曲线 C (C5 Q, 不在同一平面内) 运动而形成的曲面叫做锥面, 定点O, 叫做顶点, 动直线 L 叫做母线 定曲线 C 叫做准线

⑧ ... $\dfrac{x^2}{a^2} + \dfrac{y^2}{b^2} = \dfrac{z^2}{c^2}$ 叫做椭圆锥面

如果 $a=b$, 就叫圆锥面, 它的准线 是圆. 上式可以化为

$$AX^2 + BY^2 + CZ^2 = 0 \quad \cdots ⑨$$

如果其中三个系数有两个同号, 一个异号 时, 那么方程⑨就表示顶点在坐标 标原点的锥面, 这种锥面叫做二 次锥面, 如果A, B, C 同号方程⑨只 表示一个点 $(0,0,0)$

8. Find the volume of the region bounded above by the sphere

$$x^2 + y^2 + z^2 = 1$$

and below by the surface

$$z = x^2 + y^2.$$

9. Find the volume of that portion of the sphere $x^2 + y^2 + z^2 = a^2$, which is inside the cylinder $r = a \sin \theta$, using cylindrical coordinates.

10. Find the volume above the cone $z^2 = x^2 + y^2$ and inside the sphere $\rho = 2a \cos \varphi$ (spherical coordinates). [Draw a picture. What is the center of the sphere? What is the equation of the cone in spherical coordinates?]

11. Find the volumes of the following regions, in 3-space.
    (a) Bounded above by the plane $z = 1$, and below by the top half of $z^2 = x^2 + y^2$.
    (b) Bounded above and below by $z^2 = x^2 + y^2$, and on the sides by

$$x^2 + y^2 + z^2 = 1.$$

    (c) Bounded above by $z = x^2 + y^2$, below by $z = 0$, and on the sides by

$$x^2 + y^2 = 1.$$

    (d) Bounded above by $z = x$, and below by $z = x^2 + y^2$.

12. Find the integral of the function $f(x, y, z) = 7yz$ over the region on the positive side of the (x, z)-plane, bounded by the planes $y = 0$, $z = 0$, and $z = a$ (for some positive number a), and the cylinder $x^2 + y^2 = b^2$ ($b > 0$).

13. Find the volume of the region bounded by the cylinder $r^2 = 16$, by the plane $z = 0$, and below the plane $y = 2z$.

14. Let n be a positive integer, and let $f(x, y, z) = 1/\rho^n$, where

$$\rho = \sqrt{x^2 + y^2 + z^2}\,.$$

    (a) Find the integral of the function

$$f(x, y, z) = 1/\rho^n$$

    over the region contained between two spheres of radii a and b respectively, with $0 < a < b$.
    (b) For which values of n does this integral approach a limit as $a \to 0$? Compare with the similar result which you may have worked out in Chapter VIII for a function of two variables.

§3. CENTER OF MASS

Double and triple integrals have an application to finding the center of mass of a body in the plane or in 3-space. Let A be such a body, say in the plane, and let f be its density function, giving the density at every point. Let m be the total mass. Let $(\bar{x}, \bar{y})$ be the coordinates of the center of mass. Then they are given by the integrals:

$$\bar{x} = \frac{1}{m} \iint_A xf(x, y) \, dy \, dx = \frac{\displaystyle\iint_A xf(x, y) \, dy \, dx}{\displaystyle\iint_A f(x, y) \, dy \, dx}.$$

$$\bar{y} = \frac{1}{m} \iint_A yf(x, y) \, dy \, dx = \frac{\displaystyle\iint_A yf(x, y) \, dy \, dx}{\displaystyle\iint_A f(x, y) \, dy \, dx}.$$

In 3-space, we would of course use the triple integral of $xf(x, y, z)$ and $yf(x, y, z)$ over the body. For instance, the third coordinate of the center of mass of a body of total mass m in 3-space is given by

$$\bar{z} = \frac{1}{m} \iiint_A zf(x, y, z) \, dx \, dy \, dz.$$

Example 1. Let us find the center of mass of the part of the first quadrant lying in the disc of radius 1, as shown on Fig. 9. We assume in this case that the density is uniform, say equal to 1.

The total mass m is equal to $\pi/4$, and

$$\bar{x} = \frac{1}{m} \iint_A x \, dy \, dx.$$

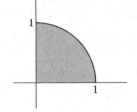

Figure 9

The integral is best evaluated by changing variables, i.e. using polar coordinates. The first quadrant consists of the points whose polar coordinates satisfy the inequalities

$$0 \leqq \theta \leqq \pi/2 \qquad \text{and} \qquad 0 \leqq r \leqq 1.$$

Thus we find:

$$\iint_A x \, dy \, dx = \int_0^{\pi/2} \int_0^1 r \cos \theta r \, dr \, d\theta = \tfrac{1}{3}.$$

Hence

$$\bar{x} = \frac{4}{3\pi}.$$

Similarly, or by symmetry, we have $\bar{y} = \dfrac{4}{3\pi}$ also.

Example 2. Let us find the z-coordinate of the center of mass of the part of the unit ball consisting of all points (x, y, z) whose coordinates are ≥ 0. If A denotes this part of the ball, then we have

$$\bar{z} = \frac{1}{m} \iiint_A z \, dx \, dy \, dz.$$

The region A consists of those points whose spherical coordinates satisfy the inequalities

$$0 \leq \theta \leq \pi/2 \qquad 0 \leq \varphi \leq \pi/2, \qquad 0 \leq \rho \leq 1.$$

By using spherical coordinates, the integral is equal to

$$\int_0^{\pi/2} \int_0^{\pi/2} \int_0^1 \rho \cos \varphi \rho^2 \sin \varphi \, d\rho \, d\varphi \, d\theta.$$

Again we easily find the value $\pi/16$. We also know that the mass of the total ball is $\frac{4}{3}\pi$. Hence the mass of our part of the ball is $\dfrac{1}{8} \cdot \dfrac{4\pi}{3} = \dfrac{\pi}{6}$, so that

$$\bar{z} = \frac{\pi}{16} \cdot \frac{6}{\pi} = \frac{3}{8}.$$

EXERCISES

In each of the following cases, find the center of mass of the given body, assuming that the density is equal to 1.

1. The triangle whose vertices are $(0, 0)$, $(3, 0)$, and $(0, 5)$.

2. The region enclosed by the parabola $y = 6x - x^2$ and the line $y = x$.

3. The upper half of the region enclosed by the ellipse as shown on Fig. 10.

① Find the centroid of the plane bounded
by the parabola $y = 6x - x^2$ and the line

$y = x$ ($m_x = \frac{625}{6}$ $m_y = \frac{625}{12}$ $A = \frac{125}{6}$)

$$\frac{x^2}{a^2} + \frac{y^2}{b^2} = 1.$$

$\bar{x} = \frac{5}{2}$ $\bar{y} = 5$

② Find the centroid of the plane area
bounded by the parabolas   **Figure 10**
$y = 2x - x^2$ and $y = 3x^2 - 6x$

$A = \frac{16}{3}$ $m_y = \frac{16}{5}$ $m_x = -\frac{64}{15}$ $\bar{x} = 1$ $\bar{y} = -\frac{4}{5}$

4. The region enclosed by the parabolas $y = 2x - x^2$ and $y = 3x^2 - 6x$.

5. The region enclosed by one arch of the curve $y = \sin x$.

6. The region bounded by the curves $y = \sin x$ and $y = \cos x$, for $0 \leq x \leq \pi/4$.

7. The region bounded by $y = \log x$ and $y = 0$, $1 \leq x \leq a$.

8. The inside of a cone of height h and base radius a, as shown on Fig. 11.

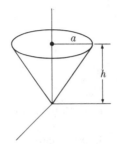

Figure 11

9. Find (a) mass and (b) the center of mass of a plate bounded by the upper half of the curve $r = 2(1 + \cos \theta)$ (in polar coordinates) if the density is proportional to the distance from the origin. The plate is drawn on Fig. 12.

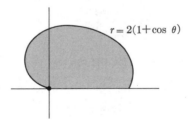

$r = 2(1 + \cos \theta)$

Figure 12

10. Find (a) the mass and (b) the center of mass of a right circular cylinder of radius a and height h if the density is proportional to the distance from the base.

11. (a) Find the mass of a circular plate of radius a whose density is proportional to the distance from the center.
    (b) Find the center of mass of this plate.
    (c) Find the center of mass of one quadrant of this plate.

12. Find the mass of a circular cylinder of radius a and height h if the density is proportional to the square of the distance from the axis.

13. Find the center of mass of a cone of height h and radius of the base equal to a, if the density is proportional to the distance from the base.

CHAPTER X

Surface Integrals

We assume that you are acquainted with the cross product of Chapter I, §7. Read that section if you have not already done so.

§1. PARAMETRIZATION, TANGENT PLANE, AND NORMAL VECTOR

Let us first recall that a curve can be described by an equation, like

$$x^2 + y^2 = 1,$$

or it can be given parametrically, as when we set

$$x = \cos \theta,$$
$$y = \sin \theta,$$

with $0 \leq \theta \leq 2\pi$. A similar situation will occur for surfaces, and we consider first the parametric representation.

Let R be a region in the plane, whose variables are denoted by (t, u). Let us associate to each pair (t, u) of R a point $X(t, u)$ in 3-space which can be written in terms of its coordinate functions

$$X(t, u) = (x_1(t, u), x_2(t, u), x_3(t, u)),$$

where x_1, x_2, x_3 are functions from R into the real numbers. We say that such an association is a **mapping** from R into $\mathbf{R}^3$, or also a **parametrization.** This is a higher dimensional analogue of parametrizing curves in space. A curve $C(t)$ depends only on one variable t. Here, the parametrization $X(t, u)$ depends on the two variables (t, u).

219

If each coordinate function is differentiable, and if its partial derivatives are continuous, we may view X as parametrizing a surface in $\mathbf{R}^3$, as shown on Fig. 1. *We shall always assume that our parametrizations satisfy all needed assumptions of differentiability and continuity, without usually repeating such assumptions.*

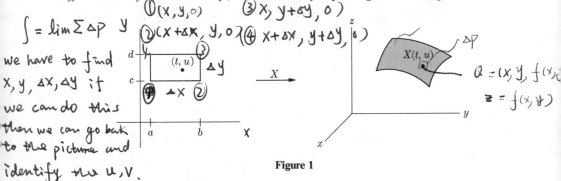

[handwritten annotations:]
$\int = \lim \Sigma \Delta p$
we have to find
$x, y, \Delta x, \Delta y$ if
we can do this
then we can go back
to the picture and
identify the u, V.

$\textcircled{1}(x, y, 0)$ $\textcircled{3}x, y+\Delta y, 0)$
$\textcircled{2}(x+\Delta x, y, 0)$ $\textcircled{4}x+\Delta x, y+\Delta y, 0)$
$X(t, u)$
$Q = (x, y, f(x, y)$
$z = f(x, y)$

Figure 1

If x, y, z are the three coordinates of $\mathbf{R}^3$, then we also write the parametrization of our surface in the form

$$x = f_1(u, v) \quad \text{or} \quad x(u, v),$$
$$y = f_2(u, v) \quad \text{or} \quad y(u, v),$$
$$z = f_3(u, v) \quad \text{or} \quad z(u, v).$$

Example 1. We parametrize the **sphere of radius** ρ by means of spherical coordinates, as studied in Chapter IX, namely

$$x = \rho \sin \varphi \cos \theta,$$
$$y = \rho \sin \varphi \sin \theta,$$
$$z = \rho \cos \varphi.$$

The region R in $\mathbf{R}^2$ is the rectangle described by the inequalities

$$0 \leqq \varphi \leqq \pi$$

and

$$0 \leqq \theta < 2\pi.$$

Our mapping "wraps" this rectangle around the sphere. If we evaluate

$$x^2 + y^2 + z^2,$$

and use relations like $\sin^2 \theta + \cos^2 \theta = 1$, we get the value ρ^2. This kind of technique shows us how to get back the equation in rectangular coordinates from the parametrization.

Example 2. A **torus** (i.e. a doughnut-shaped surface) can be parametrically by the functions:

$$x = (a + b \cos \varphi) \cos \theta,$$
$$y = (a + b \cos \varphi) \sin \theta,$$
$$z = b \sin \varphi.$$

The torus is centered at the origin, and $a > 0$ is the distance from the origin to the center of a cross section, as shown on Fig. 2. The variables φ, θ satisfy inequalities

$$0 \leqq \varphi < 2\pi$$

and

$$0 \leqq \theta < 2\pi.$$

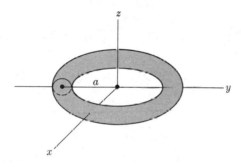

Figure 2

The number $b > 0$ is the radius of a cross section. The angle φ determines the rotation of a point in this cross section, as shown in Fig. 3.

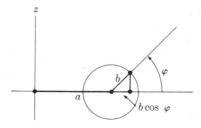

Figure 3

It is clear from this picture that the elevation z of a point is given by $b \sin \varphi$. If we project the point on the (x, y)-plane, then the distance of this projection from the origin is exactly

$$a + b \cos \varphi.$$

To get the x-coordinate of this projection, we have to multiply the projection with $\cos \theta$, and to get the y-coordinate of this projection, we have to multiply the projection with $\sin \theta$, as shown on Fig. 4.

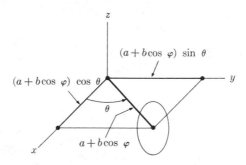

Figure 4

Let R be a region in $\mathbf{R}^2$, and let $X\,(t, u)$ be the parametrization of a surface. If

$$X(t, u) = (x_1(t, u), x_2(t, u), x_3(t, u))$$

is represented by coordinates, then for each value of u we may consider the curve

$$C_1(t) = X(t, u)$$

as a curve parametrized by the variable t, and for each value of t, we may also consider a second curve

$$C_2(u) = X(t, u)$$

as a curve parametrized by the variable u. These curves lie on the surface. We may then take the partial derivatives

$$A_1 = \frac{\partial X}{\partial t} \qquad \text{and} \qquad A_2 = \frac{\partial X}{\partial u}$$

giving the tangent vectors (velocity vectors) of each one of these curves. They may be viewed as tangent vectors to the surface, as shown on Fig. 5.

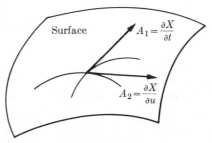

Figure 5

We shall say that (t, u) is a **regular** point if the two vectors A_1, A_2 span a plane in $\mathbf{R}^3$. The translation of this plane to the point $X(t, u)$ is called the **tangent plane** of the surface at the given point. This is illustrated on Fig. 6. It is the plane passing through the point $X(t, u)$, parallel to the vectors $A_1 = \partial X / \partial t$ and $A_2 = \partial X / \partial u$.

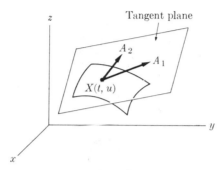

Figure 6

We now assume that you have read the section on the cross product in Chapter I. Then you realize that if A, B are non-zero vectors in $\mathbf{R}^3$, and are not parallel, their cross product

$$A \times B = (a_2 b_3 - a_3 b_2, a_3 b_1 - a_1 b_3, a_1 b_2 - a_2 b_1)$$

is perpendicular to both of them, as illustrated on Fig. 7.

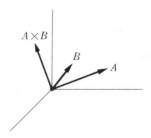

Figure 7

If we want a vector of norm 1 perpendicular to both A and B, all we have to do is divide $A \times B$ by its norm.

In the case of a parametrized surface, we can do this with the two vectors A_1 and A_2 as above. Of course, $B \times A = -A \times B$ is also perpendicular to both A and B, but has opposite direction. We use the notation

$$\boxed{N = \frac{\partial X}{\partial t} \times \frac{\partial X}{\partial u}}$$

whenever the surface is given parametrically by $X(t, u)$. Then

$$N = N(t, u)$$

is a vector perpendicular to the surface, as shown on Fig. 8.

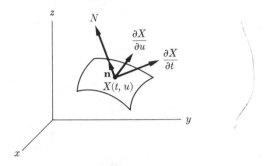

Figure 8

If we have chosen the orientation, i.e. the order of t, u, such that N points outwards from the surface, and if we denote by **n** the **outward unit normal vector** to the surface, then we have

$$\mathbf{n} = \frac{N}{\|N\|} = \frac{\dfrac{\partial X}{\partial t} \times \dfrac{\partial X}{\partial u}}{\left\| \dfrac{\partial X}{\partial t} \times \dfrac{\partial X}{\partial u} \right\|}.$$

Example 3. We compute the above quantities in the case of the parametrization of the sphere given above. We get very easily

$$\frac{\partial X}{\partial \varphi} = (\rho \cos \varphi \cos \theta, \rho \cos \varphi \sin \theta, -\rho \sin \varphi)$$

and

$$\frac{\partial X}{\partial \theta} = (-\rho \sin \varphi \sin \theta, \rho \sin \varphi \cos \theta, 0).$$

Hence

$$N(\varphi, \theta) = \frac{\partial X}{\partial \varphi} \times \frac{\partial X}{\partial \theta} = (\rho^2 \sin^2 \varphi \cos \theta, \rho^2 \sin^2 \varphi \sin \theta, \rho^2 \sin \varphi \cos \varphi)$$

$$= \rho \sin \varphi X(\varphi, \theta).(\rho \sin \varphi \cos \theta, \rho \sin \varphi \sin \theta, \rho \cos \varphi)$$

Since $\sin \varphi$ and ρ are ≥ 0, we see that N has the same direction as the position vector $X(\varphi, \theta)$, and therefore points outward. Taking the square root of the sum of the squares of the coordinates, we find

$$\left\| \frac{\partial X}{\partial \varphi} \times \frac{\partial X}{\partial \theta} \right\| = \rho^2 |\sin \varphi| = \rho^2 \sin \varphi.$$

Hence **for the sphere,**

$$\mathbf{n} = \frac{1}{\rho} X(\varphi, \theta).$$

EXERCISES

1. Compute the coordinates of the vectors $\partial X / \partial \theta$ and $\partial X / \partial \varphi$, when X is the mapping parametrizing the torus as in Example 2. Compute the norms of these vectors.

In each one of the following exercises, where you are given a parametrization $X(t, u)$, compute the tangent vectors $\dfrac{\partial X}{\partial t}, \dfrac{\partial X}{\partial u}$, their cross product, and the norm of this cross product. In each case, get an equation in cartesian coordinates for the surface parametrized by X. Draw the picture of the surface.

2. **The cone.** Let α be a fixed number, $0 < \alpha < \pi/2$. Let

$$X(\theta, z) = (z \sin \alpha \cos \theta, z \sin \alpha \sin \theta, z \cos \alpha),$$

$0 \leq \theta < 2\pi$ and $0 \leq z \leq h \sec \alpha$. Describe how you get a cone of height h.

3. **Paraboloid.** Let $X(t, \theta) = (at \cos \theta, at \sin \theta, t^2)$, with

$$0 \leq \theta < 2\pi \qquad \text{and} \qquad 0 \leq t \leq h.$$

4. **Ellipsoid.** Let $a, b, c > 0$. Let $0 \leq \varphi \leq \pi, 0 \leq \theta < 2\pi$, and

$$X(\varphi, \theta) = (a \sin \varphi \cos \theta, b \sin \varphi \sin \theta, c \cos \varphi).$$

5. **Cylinder.** Let $a > 0$. Let

$$X(\theta, z) = (a \cos \theta, a \sin \theta, z),$$

with $0 \leq \theta < 2\pi$, and $h_1 \leq z \leq h_2$.

6. **Surface of revolution** (around the z-axis). Let f be a function of one variable r, defined for $r_1 \leq r \leq r_2$. Let $0 \leq \theta < 2\pi$, and let

$$X(r, \theta) = (r \cos \theta, r \sin \theta, f(r)).$$

§2. SURFACE AREA

Let A, B be a non-zero vectors in $\mathbf{R}^3$, and assume that they are not parallel. Then they span a parallelogram, as shown on Fig. 9, and this parallelogram is contained in a plane.

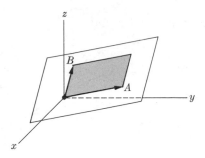

Figure 9

If θ is the angle between A and B, then the area of this parallelogram is precisely equal to

$$\|A\| \, \|B\| \, |\sin \theta|,$$

as one sees at once from Fig. 10, and as we already mentioned in Chapter I.

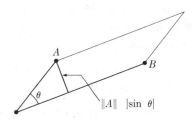

Figure 10

We observe that $\|A\| \, \|B\| \, |\sin \theta|$ is precisely the norm of $A \times B$. Thus in 3-space, we may say that the area of the parallelogram spanned by A and B is equal to

$$\|A \times B\|.$$

We apply this to a surface, parametrized by $X(t, u)$ as before. Then the two tangent vectors

$$A = \frac{\partial X}{\partial t} \quad \text{and} \quad B = \frac{\partial X}{\partial u}$$

span a parallelogram. By the preceding remark, the area of this parallelogram is equal to

$$\left\| \frac{\partial X}{\partial t} \times \frac{\partial X}{\partial u} \right\|.$$

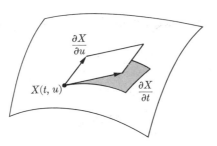

Figure 11

We don't want the parametrization $X(t, u)$ to be degenerate. Hence we have to make some assumption that it represents the surface in a non-degenerate way. To phrase this assumption we need a new word. We say that the parametrization X is **injective** if it satisfies the condition:

If $(t_1, u_1) \neq (t_2, u_2)$, then $X(t_1, u_1) \neq X(t_2, u_2)$.

In other words, two different values of the parameters correspond to two different points on the surface.

Assume that X is defined on a region R, and that the parametrization $X(t, u)$ is injective, except possibly for a finite number of smooth curves in R. Also assume that the coordinate functions of $X(t, u)$ are continuously differentiable, and that all points of R are regular, except for a finite number of smooth curves. It is then reasonable to define the **area of the parametrized surface** to be the integral

$$\text{Area} = \iint_S d\sigma = \iint_R \left\| \frac{\partial X}{\partial t} \times \frac{\partial X}{\partial u} \right\| dt\, du.$$

We write symbolically

$$d\sigma = \left\| \frac{\partial X}{\partial t} \times \frac{\partial X}{\partial u} \right\| dt\, du.$$

Example 1. Let us compute the area of a sphere, whose parametrization was given in §1. We had already computed that

$$\left\| \frac{\partial X}{\partial \varphi} \times \frac{\partial X}{\partial \theta} \right\| = \rho^2 \sin \varphi.$$

Consequently for our **parametrization of the sphere**, we can write

$$\boxed{d\sigma = \rho^2 \sin \varphi \, d\varphi \, d\theta.}$$

Hence the area of the sphere is equal to

$$\int_0^{2\pi} \int_0^{\pi} \rho^2 \sin \varphi \, d\varphi \, d\theta.$$

Since ρ^2 is constant, we take it out of the integral. It is a trivial matter to carry out the integration, and we find that the desired area is equal to $4\pi\rho^2$.

   Graph of a function. Sometimes a surface is given by the graph of a function

$$z = f(x, y),$$

defined over some region R of the (x, y)-plane. In this case, we use $t = x$ and $u = y$ as the parameters, so that

$$X(x, y) = (x, y, f(x, y)).$$

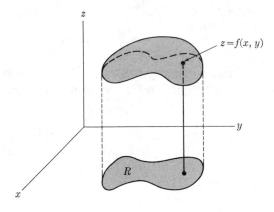

Figure 12

Thus the case when a surface is so defined is a special case of the general parametrization. In this special case, we find

$$\frac{\partial X}{\partial x} = \left(1, 0, \frac{\partial f}{\partial x}\right) \quad \text{and} \quad \frac{\partial X}{\partial y} = \left(0, 1, \frac{\partial f}{\partial y}\right).$$

Consequently

$$\frac{\partial X}{\partial x} \times \frac{\partial X}{\partial y} = \left(-\frac{\partial f}{\partial x}, \, -\frac{\partial f}{\partial y}, \, 1\right),$$

and

$$\left\| \frac{\partial X}{\partial x} \times \frac{\partial X}{\partial y} \right\| = \sqrt{1 + \left(\frac{\partial f}{\partial x} \right)^2 + \left(\frac{\partial f}{\partial y} \right)^2}.$$

The **area of the surface** is given by the integral

$$\iint_R \sqrt{1 + \left(\frac{\partial f}{\partial x} \right)^2 + \left(\frac{\partial f}{\partial y} \right)^2} \; dx \, dy.$$

Symbolically we may write in this case

$$d\sigma = \sqrt{1 + \left(\frac{\partial f}{\partial x} \right)^2 + \left(\frac{\partial f}{\partial y} \right)^2} \; dx \, dy.$$

Example 2. Find the area of the paraboloid

$$z = x^2 + y^2, \quad \text{with} \quad 0 \leq z \leq 2.$$

The surface looks as on the figure (Fig. 13).

Figure 13

Handwritten annotations:

Find the area of the portion of the cone $x^2 + y^2 = 3z^2$ and inside the cylinder $x^2 + y^2 = 4x$

Evaluate $\iint_S u(x, y, z) \, ds$ where S is the surface of the paraboloid $z = 2 - (x^2 + y^2)$ above the xy plane. let $u(x, y, z) = 1$

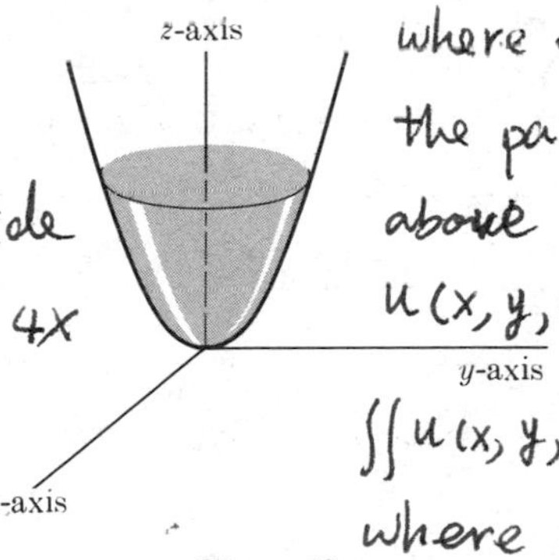

$\iint u(x, y, z) \sqrt{\left(\frac{\partial f}{\partial x}\right)^2 + \left(\frac{\partial f}{\partial y}\right)^2 + 1} \, dx \, dy$ where R is the projection of S on the xy plane give by $x^2 + y^2 = 2$

Here $f(x, y) = x^2 + y^2$, and the region R in the (x, y)-plane is the disc of radius $\sqrt{2}$. Hence

$$\text{Area of paraboloid} = \iint_R \sqrt{1 + (2x)^2 + (2y)^2} \; dx \, dy$$

$$= \iint_R \sqrt{1 + 4x^2 + 4y^2} \; dx \, dy.$$

Changing to polar coordinates, this yields

$$\int_0^{2\pi} \int_0^{\sqrt{2}} \sqrt{1 + 4r^2} \; r \; dr \; d\theta,$$

which you should know how to integrate by substitution. Let

$$u = 1 + 4r^2 \quad \text{and} \quad du = 8r \; dr.$$

The answer comes out $13\pi/3$.

Example 3. It may also happen that a surface is defined by an equation

$$g(x, y, z) = 0,$$

and that over a certain region R of the (x, y)-plane, we can then solve for z by a function

$$z = f(x, y),$$

satisfying this equation, that is

$$g(x, y, f(x, y)) = 0.$$

Taking the partials with respect to x and y, we find the relations:

$$\frac{\partial f}{\partial x} = -\frac{\partial g/\partial x}{\partial g/\partial z} \quad \text{and} \quad \frac{\partial f}{\partial y} = -\frac{\partial g/\partial y}{\partial g/\partial z}.$$

We can now use the formula for the area obtained in the preceding example, and thus obtain a formula for the area just in terms of the given g, namely:

$$\iint_R \frac{\sqrt{(\partial g/\partial x)^2 + (\partial g/\partial y)^2 + (\partial g/\partial z)^2}}{|\partial g/\partial z|} \; dx \; dy.$$

Example 4. Take the special case of this formula arising from the equation of a sphere

$$x^2 + y^2 + z^2 - a^2 = 0,$$

where $a > 0$ is the radius. Then $g(x, y, z)$ is the expression on the left, and the partials are trivially computed:

$$\frac{\partial g}{\partial x} = 2x, \qquad \frac{\partial g}{\partial y} = 2y, \qquad \frac{\partial g}{\partial z} = 2z.$$

We can solve for z explicitly in terms of x, y by letting

$$z = \sqrt{a^2 - x^2 - y^2} = f(x, y),$$

$$(u \times v) \cdot k$$
$$= \Delta x \Delta y \left(-f_x(x,y) i - f_y(x,y) + k \right) \cdot k$$
$$= \Delta A$$

where (x, y) ranges over the points in the disc of radius a in the plane. The surface is then the upper hemisphere.

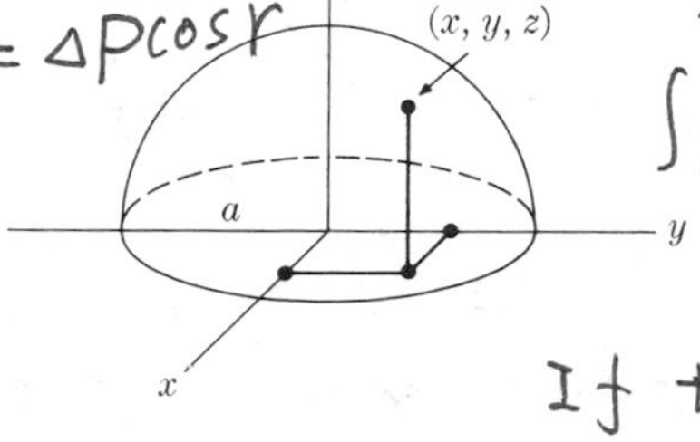

$$(u \times v) \cdot k = |u \times v||k| \cos r$$
$$= \Delta P \cos r$$

(x, y, z)

a

Figure 14

$\Delta A = \Delta P \cos r$

$\text{Surface} = \iint \frac{1}{\cos r} dA$

$= \iint \sec r \, dA$

If the equation of s is

$F(x, y, z) = 0$ N at (x, y, z)

$= \nabla F = \dfrac{\partial f}{\partial x} i + \dfrac{\partial f}{\partial y} j + \dfrac{\partial f}{\partial z} k$

Then

$$\text{Area of hemisphere} = \iint_R \frac{\sqrt{4x^2 + 4y^2 + 4z^2}}{|2z|}\, dx\, dy = \iint_R \frac{a}{z}\, dx\, dy,$$

$\nabla F \cdot k = |\nabla F||k| \cos r$

using the fact that $x^2 + y^2 + z^2 = a^2$. The region R is the disc of radius a in the (x, y)-plane. Using polar coordinates, we know how to evaluate this last integral. We get

$$\text{Area of hemisphere} = a \int_0^{2\pi} \int_0^a \frac{1}{\sqrt{a^2 - r^2}}\, r\, dr\, d\theta.$$

$F_z = \sqrt{(f_x)^2 + (f_y)^2 + (f_z)^2}$

Integrating 1 with respect to θ between 0 and 2π yields 2π. The integral with respect to r is reducible to the form

$\cdot \cos r$

$$\int \frac{1}{\sqrt{u}}\, du,$$

$\dfrac{F_z}{\sqrt{(f_x)^2 + (f_y)^2 + (f_z)^2}} = \cos r$

and is therefore easily found. Thus, finally we obtain the value

$$2\pi a^2$$

for the area of the hemisphere. Naturally, this jibes with the answer found from the parametrization by means of spherical coordinates.

Remark. Just as in the case of curves, it can be shown that the area of a surface is independent of the parametrization selected. This amounts to a change of variables in a 2-dimensional integral, but we shall omit the proof.

surface $= \displaystyle\iint \frac{\sqrt{(f_x)^2 + (f_y)^2 + (f_z)^2}}{F_z}\, dx\, dy$

EXERCISES

Compute the following areas.

1. (a) A cone as shown on the following figure.

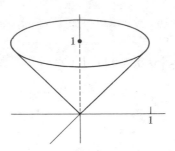

Figure 15

(b) The cone of height h obtained by rotating the line $z = 3x$ around the z-axis.

2. The surface $z = x^2 + y^2$ lying above the disc of radius 1 in the (x, y)-plane.

3. The surface $2z = 4 - x^2 - y^2$ over the disc of radius $\sqrt{2}$ in the (x, y)-plane.

4. $z = xy$ over the disc of radius 1.

5. The surface given parametrically by

$$X(t, \theta) = (t \cos \theta, t \sin \theta, \theta),$$

with $0 \leq t \leq 1$ and $0 \leq \theta \leq 2\pi$. [*Hint*: Use $t = \sinh u = (e^u - e^{-u})/2$.]

6. The surface given parametrically by

$$X(t, u) = (t + u, t - u, t),$$

with $0 \leq t \leq 1$ and $0 \leq u \leq 2\pi$.

7. The part of the sphere $x^2 + y^2 + z^2 = 1$ between the planes $z = 1/\sqrt{2}$ and $z = -1/\sqrt{2}$.

8. The part of the sphere $x^2 + y^2 + z^2 = 1$ inside the upper part of the cone $x^2 + y^2 = z^2$.

9. The torus, using the parametrization in §1, assuming that the cross section has radius 1.

§3. SURFACE INTEGRALS

Integral of a function over a surface.

Let R be a region in the plane, and let $X(t, u)$ be the parametrization of a surface by a smooth mapping X. Let S be the image of X, i.e. the surface, and let ψ be a function on S. Then when ψ is sufficiently smooth, we define the **integral of ψ over S** by the formula

$$\iint_S \psi \, d\sigma = \iint_R \psi(X(t, u)) \left\| \frac{\partial X}{\partial t} \times \frac{\partial X}{\partial u} \right\| dt \, du.$$

When ψ is the constant 1, then our formula expresses simply the area of the parametrized surface.

Example 1. Let S be the surface defined by

$$z = x^2 + y,$$

with x, y satisfying the inequalities

$$0 \leq x \leq 1 \quad \text{and} \quad -1 \leq y \leq 1.$$

Find the integral

$$\iint_S x \, d\sigma.$$

The surface is here given as the graph of a function, so we use the formula for $d\sigma$ given in the preceding section. We let R be the region of points (x, y) satisfying the above inequalities. Then:

$$\iint_S x \, d\sigma = \iint_R x\sqrt{1 + (2x)^2 + 1^2} \, dx\,dy$$

$$= \int_{-1}^{1} \int_0^1 x\sqrt{2 + 4x^2} \, dx\,dy.$$

volume

area

The inner integral

$$\int_0^1 x\sqrt{2 + 4x^2} \, dx$$

work

surface

flux

can be evaluated by substitution, putting

$$u = 2 + 4x^2 \quad \text{and} \quad du = 8x \, dx.$$

Thus

$$\int_0^1 x\sqrt{2 + 4x^2} \, dx = \frac{1}{8} \int_0^1 \sqrt{2 + 4x^2} \, (8x) \, dx = \frac{1}{12}(6^{3/2} - 2^{3/2}).$$

Hence finally

$$\iint_S x \, d\sigma = \int_{-1}^{1} \frac{1}{12}(6^{3/2} - 2^{3/2}) \, dy = \frac{1}{6}(6^{3/2} - 2^{3/2}).$$

Heat flux. Suppose that the function ψ is interpreted as a temperature. Then the integral

$$\iint_S \psi \, d\sigma$$

is called the **heat flux** across the surface.

Density and mass. Suppose that ψ is the function representing a **positive density** of the surface. Then the integral above is interpreted as the **mass** m of the surface, corresponding to this density.

Let ψ be a density as above, and m the mass. The integrals

$$\bar{x} = \frac{1}{m} \iint_S x\psi(x, y, z) \, d\sigma,$$

$$\bar{y} = \frac{1}{m} \iint_S y\psi(x, y, z) \, d\sigma,$$

$$\bar{z} = \frac{1}{m} \iint_S z\psi(x, y, z) \, d\sigma$$

give the coordinates $(\bar{x}, \bar{y}, \bar{z})$ of the center of mass of the surface.

Example 2. Let us find the center of mass of a hemisphere of radius a, having constant density c. We use the spherical coordinate parametrization of §1. The hemisphere is the one lying above the (x, y)-plane as in Fig. 16.

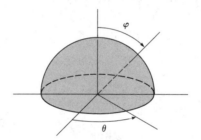

Figure 16

By symmetry, it is easy to see that $\bar{x} = \bar{y} = 0$. We have $z = a \cos \varphi$. The third coordinate $\bar{z}$ is given by the integral

$$\bar{z} = \frac{c}{m} \iint_S z \, d\sigma = \frac{c}{m} \int_0^{2\pi} \int_0^{\pi/2} a \cos \varphi \cdot a^2 \sin \varphi \, d\varphi \, d\theta,$$

which is easily evaluated to be

$$\bar{z} = ca^3\pi / m.$$

The total mass is equal to the density times the area, since the density is constant, and we know that the area of the hemisphere is $2\pi a^2$. Hence we find

$$\bar{z} = a/2.$$

Integral of a vector field over a surface. Let $X(t, u)$ parametrize a surface, and suppose that the image of X, that is the surface, is contained in some open set U in $\mathbf{R}^3$. Let F be a vector field on U, so to each point X of U, F associates a vector $F(X)$ in $\mathbf{R}^3$. **We assume that F is as smooth as needed.** We define the **integral of the vector field along the surface** in a manner similar to the integral a vector field along a curve in the lower dimensional case. Namely, <u>let $\mathbf{n}$ be the outward normal unit vector to the surface</u>, it being assumed that we have agreed on an orientation of the surface which determines its outside and inside. Then

$$F \cdot \mathbf{n}$$

is the projection of the vector field along the normal to the surface, and we **define the above integral** by the formula

$$\iint_S F \cdot \mathbf{n} \, d\sigma = \iint_R F \cdot \mathbf{n} \left\| \frac{\partial X}{\partial t} \times \frac{\partial X}{\partial u} \right\| dt \, du.$$

By definition, we have

$$\mathbf{n} \left\| \frac{\partial X}{\partial t} \times \frac{\partial X}{\partial u} \right\| = \frac{\partial X}{\partial t} \times \frac{\partial X}{\partial u}.$$

Hence our integral for F over the surface can be rewritten

$$\iint_S F \cdot \mathbf{n} \, d\sigma = \iint_R F(X(t, u)) \cdot \left(\frac{\partial X}{\partial t} \times \frac{\partial X}{\partial u} \right) dt \, du.$$

Example 3. Consider a fluid flow, subject to a force field G, so that we may interpret G as a vector field. Let ψ be the function representing the density of the fluid, so that $\psi(x, y, z)$ is the density at a given point (x, y, z), and is a number. We call

$$F(x, y, z) = \psi(x, y, z) G(x, y, z)$$

the **force field** of the flow, and visualize it as in Fig. 17.

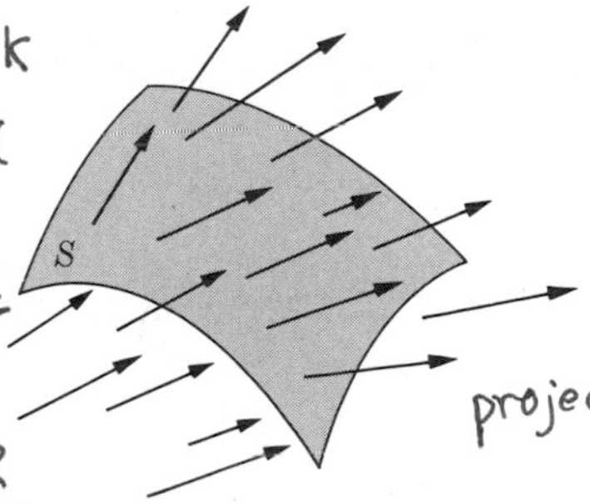

Figure 17

let $\vec{F} = P\vec{i} + Q\vec{j} + R\vec{k}$

$\vec{F} \cdot n = (P\vec{i} + Q\vec{j} + R\vec{k})(\dfrac{\vec{i}(f_x) + \vec{j}(f_y) - 1}{\sqrt{f_{(x)}^2 + (f_y)^2 + 1}})$

$\vec{F} \cdot n = \dfrac{Pf_x + Qf_y - R}{\sec \gamma}$

$\iint \sec \gamma \, dx \, dy = \iint \dfrac{Pf_{(x)} + Qf_y - R}{\sec \gamma} \sec \gamma \, dx \, dy$

$= \iint Pf_{(x)} + Qf_y - R \, dx \, dy$.

(normal unit)

$n \ (\vec{N})$ V A

B

$c_B = \dfrac{A \cdot B}{B \cdot B}$

projection $n = \dfrac{V \cdot n}{h \cdot n} \cdot n$

$= \dfrac{V \cdot n}{|n|^2} \cdot n$

$= \dfrac{V \cdot n}{|n|} \cdot \dfrac{n}{|n|}$

$= \vec{F} \cdot \vec{n}$

The amount of fluid passing through the surface per unit time is then called the **flux**, and is given by the integral of the force field over the surface, namely

$$\text{Flux} = \iint_S F \cdot \mathbf{n} \, d\sigma,$$

where F is the force field.

It is not true that all surfaces can be oriented so that we can define an outside and an inside. The well-known Moebius strip gives an example when this cannot be done. In all the applications that we deal with, however, it is geometrically clear what is meant by the inside and outside. It is fairly difficult to give a definition in general, and so we don't go into this.

Observe that when we give a parametrization $X(t, u)$, we could interchange the role of t, u as the first and second variable, respectively. Thus, for instance, if

$$X(t, u) = (t, u, t^2 + u^2),$$

we could let

$$Y(u, t) = (t, u, t^2 + u^2).$$

Then

$$\frac{\partial Y}{\partial u} \times \frac{\partial Y}{\partial t} = - \frac{\partial X}{\partial t} \times \frac{\partial X}{\partial u}.$$

Interchanging the variables amounts to changing the orientation. The two normal vectors corresponding to these two parametrizations have opposite direction. In finding the integral of a vector field with respect to a given parametrization, one must therefore agree on what is the "inside" and what is the "outside" of the surface, and check that the normal vector obtained from the cross product of the two partial derivatives points to the outside.

Example 4. Compute the integral of the vector field

$$F(x, y) = (x, y)$$

over the sphere $x^2 + y^2 = a^2$ $(a > 0)$. We use the parametrization of §1. Then

$$N(\varphi, \theta) = \frac{\partial X}{\partial \varphi} \times \frac{\partial X}{\partial \theta} = a \sin \varphi \, X(\varphi, \theta).$$

Thus $N(\varphi, \theta)$ is a positive multiple of the position vector X, because

$$0 \leq \varphi \leq \pi,$$

and hence points outward. So we get

$$F(X(\varphi, \theta)) \cdot N(\varphi, \theta) = (a \sin \varphi)\left[(a \sin \varphi \cos \theta)^2 + (a \sin \varphi \sin \theta)^2\right],$$

and

$$\iint_R F \cdot N \, d\varphi \, d\theta = a^3 \int_0^{2\pi} \int_0^{\pi} \sin^3 \varphi \, d\varphi \, d\theta = \frac{8\pi a^3}{3}.$$

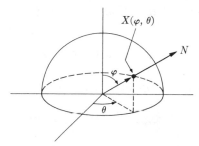

Figure 18

Example 5. Let S be the paraboloid defined by the equation

$$z = x^2 + y^2.$$

We can use x, y as parameters, and represent S parametrically by

$$X(x, y) = (x, y, x^2 + y^2).$$

Then

$$N(x, y) = (1, 0, 2x) \times (0, 1, 2y)$$
$$= (-2x, -2y, 1).$$

Thus with the parametrization as given, we see from Fig. 19 that N points inside the paraboloid.

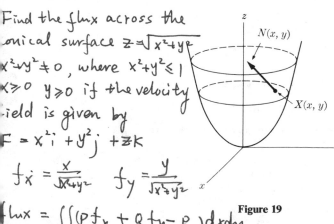

Find the flux across the conical surface $z = \sqrt{x^2 + y^2}$ $x^2 + y^2 \neq 0$, where $x^2 + y^2 \leq 1$ $x \geq 0$ $y \geq 0$ if the velocity field is given by

$$F = x^2 i + y^2 j + z k$$

$$f_x = \frac{x}{\sqrt{x^2 + y^2}} \quad f_y = \frac{y}{\sqrt{x^2 + y^2}}$$

$$\text{flux} = \iint (P f_x + Q f_y - R) \, dx \, dy$$

$$= \iint \frac{x^3}{\sqrt{x^2 + y^2}} + \frac{y^3}{\sqrt{x^2 + y^2}} - \sqrt{x^2 + y^2} \, dx \, dy$$

$$\int_0^{\frac{\pi}{2}} \int_0^{1} \left(\frac{r^3 \cos^3 \theta}{r} + \frac{r^3 \sin^3 \theta}{r} - r \right) r \, dr \, d\theta$$

Figure 19

For instance, when x, y are positive, say equal to 1, then

$$N(1, 1) = (-2, -2, 1),$$

which points inward. Consequently, if we want the integral of a vector field F with respect to the outward orientation, then we have to take minus the integral, namely

$$-\iint F \cdot N \, dx \, dy.$$

Example 6. Compute the integral of the vector field

$$F(x, y, z) = (y, -x, z^2)$$

over the paraboloid

$$z = x^2 + y^2 \quad \text{with} \quad 0 \leq z \leq 1.$$

We have

$$F(X(x, y)) \cdot N(x, y) = -2xy + 2xy + z^2 = z^2$$
$$= (x^2 + y^2)^2.$$

Hence

$$\iint_S F \cdot \mathbf{n} \, d\sigma = -\iint_R (x^2 + y^2)^2 \, dx \, dy,$$

where R is the unit disc in the (x, y)-plane. Changing to polar coordinates, it is easy to evaluate this integral,

$$\iint_S F \cdot \mathbf{n} \, d\sigma = -\int_0^{2\pi} \int_0^1 r^4 r \, dr \, d\theta = -\pi/3.$$

Note that in the present case, we have

$$\mathbf{n} = -\frac{N}{\|N\|}.$$

EXERCISES

Integrate the following function over the indicated surface.

1. (a) The function $x^2 + y^2$ over the same upper hemisphere of radius a as in Example 2 of this section.
   (b) The function $(x^2 + y^2)z$ over this same hemisphere.
   (c) The function $(x^2 + y^2)z^2$ over this same hemisphere.
   (d) The function $z(x^2 + y^2)^2$ over this same hemisphere.

2. The function z^2 over the unit sphere

$$x^2 + y^2 + z^2 = 1.$$

3. The function z over the upper hemisphere of radius a.

4. The function z over the surface

$$z = x^2 + y^2 \qquad \text{with} \qquad x^2 + y^2 \leq 1.$$

5. The function z over the surface

$$z = 1 - x^2 - y^2, \qquad z \geq 0.$$

   (Use polar coordinates and sketch the surface.)

6. The function x over the cone $x^2 + y^2 = z^2, 0 \leq z \leq a$.

7. The function x over the part of the sphere $x^2 + y^2 + z^2 = a^2$ contained inside the cone of Exercise 6.

8. The function x^2 over the cylinder defined by $x^2 + y^2 = a^2$, and $0 \leq z \leq 1$, excluding its top and bottom.

9. The same function x^2 over the top and bottom of the cylinder.

10. **Theorem of Pappus.** Let C be the parametrization of a smooth curve in the plane, defined as an interval $[a, b]$, say

$$C(t) = (f(t), z(t)).$$

   We view $C(t)$ as lying in the (x, z)-plane, as shown on Fig. 20. We assume that $f(t) \geq 0$. Let $\bar{x}$ be the x-coordinate of the center of mass of this curve in the (x, z)-plane. Prove that the area of the surface of revolution of this curve is equal to

$$2\pi\bar{x}L,$$

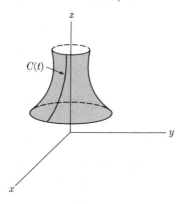

Figure 20

where L is the length of the curve. [*Hint*: Parametrize the surface of revolution by the mapping

$$X(t, \theta) = (f(t) \cos \theta, f(t) \sin \theta, z(t)).]$$

What is θ in Fig. 20? Recall that $\bar{x}$ is given by

$$\bar{x} = \frac{1}{L} \int_a^b f(t) \|C'(t)\| \, dt.$$

How does this apply to get the area of torus in a simple way?

11. Let S be the sphere of radius a and centered at O. Let P be a fixed point, either inside or outside the sphere, but not on S. Let

$$r(X) = \|X - P\|.$$

Show that

$$\iint_S \frac{1}{r} \, d\sigma = \begin{cases} 4\pi a & \text{if } P \text{ is inside the sphere} \\ \dfrac{4\pi a^2}{\|P\|} & \text{if } P \text{ is outside the sphere.} \end{cases}$$

[*Hint*: You may assume that the point P is on the z-axis. This will simplify the direct computation.]

Find the integrals of the following vector fields over the given surfaces.

12. $F(x, y, z) = \dfrac{1}{\sqrt{x^2 + y^2}} (y, -y, 1)$ over the paraboloid

$$z = 1 - x^2 - y^2, \qquad 0 \le z \le 1.$$

(Draw the picture.)

13. The same vector field as in Exercise 12, over the lower hemisphere of a sphere centered at the origin, of radius 1. Note: $\pi/2 \le \varphi \le \pi$.

14. The vector field $F(x, y, z) = (y, -x, 1)$ over the surface

$$X(t, \theta) = (t \cos \theta, t \sin \theta, \theta),$$

$0 \le t \le 1$ and $0 \le \theta \le 2\pi$.

15. The vector field $F(x, y, z) = (x^2, y^2, z^2)$ over the surface

$$X(t, u) = (t + u, t - u, t),$$

$0 \le t \le 2$ and $1 \le u \le 3$.

16. The vector field $F(X) = X$, over the part of the sphere $x^2 + y^2 + z^2 = 1$ between the planes $z = 1/\sqrt{2}$ and $z = -1/\sqrt{2}$.

17. The vector field $F(x, y, z) = (x, 0, 0)$ over the part of the unit sphere inside the upper part of the cone $x^2 + y^2 = z^2$.

18. The vector field $F(x, y, z) = (x, y^2, z)$ over the triangle determined by the plane $x + y + z = 1$, and the coordinate planes.

19. The vector field $F(x, y, z) = (x, y, z^2)$ over the cylinder defined by $x^2 + y^2 = a^2$, $0 \leq z \leq 1$,
    (a) excluding the top and bottom,
    (b) including the top and bottom.

20. The vector field $F(x, y, z) = (xy, y^2, y^3)$ over the boundary of the unit cube

$$0 \leq x \leq 1, \qquad 0 \leq y \leq 1, \qquad 0 \leq z \leq 1.$$

21. The vector field $F(x, y, z) = (xz, 0, 1)$ over the upper hemisphere of radius 1.

22. Let an electric field be given by

$$F(x, y, z) = (2x, 2y, 2z).$$

Find the electric flux across the *closed* surface consisting of the hemisphere

$$x^2 + y^2 + z^2 = 1, \qquad z \geq 0$$

together with the base

$$x^2 + y^2 \leq 1 \qquad \text{and} \qquad z = 0.$$

[Compute the desired integral over the two surfaces separately.]

23. The force field of a fluid is given by

$$F(x, y, z) = (1, x, z),$$

measured in meters/second. Find how may cubic meters of fluid per second cross the upper hemisphere

$$x^2 + y^2 + z^2 = 1, \qquad z \geq 0.$$

§4. CURL AND DIVERGENCE OF A VECTOR FIELD

Let U be an open set in $\mathbf{R}^3$, and let F be a vector field on U. Thus F associates a vector to each point of U, and F is given by three coordinate functions,

$$F(x, y, z) = (f_1(X), f_2(X), f_3(X)).$$

We assume that F is as differentiable as needed.
    We define the **divergence** of F to be the *function*

$$\text{div } F = \frac{\partial f_1}{\partial x} + \frac{\partial f_2}{\partial y} + \frac{\partial f_3}{\partial z}.$$

$$\nabla \cdot F = \frac{\partial P}{\partial x} + \frac{\partial Q}{\partial y}$$

$$\text{divergence}.$$

$$\nabla \times F = \frac{\partial Q}{\partial x} - \frac{\partial P}{\partial y}$$

$$\text{curl}.$$

$$\nabla F = \frac{\partial F}{\partial x} i + \frac{\partial F}{\partial y} j + \frac{\partial F}{\partial z} k$$

$$F(x, y, z) = P(x, y, z) i + Q(x, y, z) j + R(x, y, z) k$$

$$\text{div } F = \frac{\partial P}{\partial x} + \frac{\partial Q}{\partial y} + \frac{\partial R}{\partial z}$$

$$\text{curl: } \nabla \times F = \begin{vmatrix} \bar{i} & \bar{j} & \bar{k} \\ \frac{\partial}{\partial x} & \frac{\partial}{\partial y} & \frac{\partial}{\partial z} \\ P & Q & R \end{vmatrix} = \left(\frac{\partial R}{\partial y} - \frac{\partial Q}{\partial z}\right)i + \left(\frac{\partial P}{\partial z} - \frac{\partial R}{\partial x}\right)j + \left(\frac{\partial Q}{\partial x}\right)$$

Thus the divergence is the sum of the partial derivatives of the coordinate functions, taken with respect to the corresponding variables. It is scalar valued.

div F is scalar

curl F is a vector.

Example 1. Let $F(x, y, z) = (\sin xy, e^{xz}, 2x + yz^4)$. Then

$$(\text{div } F)(x, y, z) = y \cos xy + 0 + 4yz^3$$
$$= y \cos xy + 4yz^3.$$

As a matter of notation, one sometimes writes symbolically

$$\nabla = \left(\frac{\partial}{\partial x}, \frac{\partial}{\partial y}, \frac{\partial}{\partial z}\right) = (D_1, D_2, D_3),$$

where D_1, D_2, D_3 are the partial derivative operators with respect to the corresponding variables. Then one also writes

$$\text{div } F = \nabla \cdot F = D_1 f_1 + D_2 f_2 + D_3 f_3.$$

We shall interpret the divergence geometrically later. Similarly, we now define the **curl** of F, and interpret it geometrically later. We define

$$\text{curl } F = \left(\frac{\partial f_3}{\partial y} - \frac{\partial f_2}{\partial z}, \frac{\partial f_1}{\partial z} - \frac{\partial f_3}{\partial x}, \frac{\partial f_2}{\partial x} - \frac{\partial f_1}{\partial y}\right)$$

$$= (D_2 f_3 - D_3 f_2, D_3 f_1 - D_1 f_3, D_1 f_2 - D_2 f_1).$$

The curl of F is therefore also a *vector* field.

Again, we use the symbolic notation

$$\text{curl } F = \nabla \times F.$$

Example 2. Let F be the same vector field as in the preceding example. Then

$$\text{curl } F = (z^4 - xe^{xz}, 0 - 2, ze^{xy} - x \cos xy)$$
$$= (z^4 - xe^{xz}, -2, ze^{xy} - x \cos xy).$$

Remark on notation. If you look at Chapter XV, §2, giving the expansion of a 3×3 determinant according to the first row, then you see that we may write the curl symbolically as a "determinant"

$$\text{curl } F = \begin{vmatrix} E_1 & E_2 & E_3 \\ D_1 & D_2 & D_3 \\ f_1 & f_2 & f_3 \end{vmatrix}.$$

As in R^2, the curl of a vector field F represents the circulation per unit area at the point (x, y, z) if curl $F = 0$, for every (x, y, z) in some region w in R^3 then the fluid flow is called irrotational. The divergence of F at a point (x, y, z) in w represents the net rate flow away from (x, y, z)

If div $F = 0$, for every (x, y, z) in W, then the flow is called *incompressable*

① when the [line] intergral is dependent of the path taken, we say that F is exact.

Indeed, expanding symbolically this determinant, we find

$$E_1(D_2 f_3 - D_3 f_2) + E_2(D_1 f_3 - D_3 f_1) + E_3(D_1 f_2 - D_2 f_1)$$

which yields exactly the expression of the definition of curl F. Writing the curl in this fashion makes it easier to remember in which order the indices occur in the components.

② when the differentiable vector filed has a curl $F = 0$, we say that F is exact.

Compute the divergence and curl of $F(x, y, z)$
$= (x y)i + (z^2 - 2y)j + (\cos yz)k$

EXERCISES

Compute the divergence and the curl of the following vector fields.

1. $F(x, y, z) = (x^2, xyz, yz^2)$

$div = y - z - y \sin yz$

2. $F(x, y, z) = (y \log x, x \log y, xy \log z)$

3. $F(x, y, z) = (x^2, \sin xy, e^x yz)$

4. $F(x, y, z) = (e^{xy} \sin z, e^{xz} \sin y, e^{yz} \cos x)$

5. Let φ be a smooth function. Prove that curl grad $\varphi = O$.

6. Prove that div curl $F = 0$.

7. Let $\nabla^2 = \nabla \cdot \nabla = D_1^2 + D_2^2 + D_3^2 = \left(\dfrac{\partial}{\partial x}\right)^2 + \left(\dfrac{\partial}{\partial y}\right)^2 + \left(\dfrac{\partial}{\partial z}\right)^2$. A function f is said to be **harmonic** if $\nabla^2 f = 0$. Prove that the following functions are harmonic.

(a) $\dfrac{1}{\sqrt{x^2 + y^2 + z^2}}$ (b) $x^2 - y^2 + 2z$

(c) If f is harmonic, prove that div grad $f = 0$.

8. Let $F(X) = c \dfrac{X}{\|X\|^3}$, where c is constant. Prove that div $F = 0$ and that curl $F = O$.

9. Prove that div$(F \times G) = G \cdot$ curl $F - F \cdot$ curl G, if F, G are vector fields.

10. Prove that div (grad $f \times$ grad $g) = 0$, if f, g are functions.

§5. DIVERGENCE THEOREM IN 3-SPACE

In this section, we let U be a 3-dimensional region in $\mathbf{R}^3$, whose boundary is a closed surface which is smooth, except for a finite number of smooth curves. For instance, a 3-dimensional rectangular box is such a region. The inside of a sphere, or of an ellipsoid is such a region. The region bounded by the plane $z = 2$, and inside the paraboloid $z = x^2 + y^2$ is such a region, illustrated in Fig. 21.

Let W be a solid in $\mathbf{R}^3$ totally enclosed by a smooth surface. Let F be a smooth vector field on W and let $\vec{n}$ denote the outward unit nomal to S Then

$$\int\int_{} F \cdot n \, d\sigma = \int\int\int_W \text{div} F \, dx \, dy \, dz$$

Compute $\iint_S F \cdot n \, d\sigma$ where $F(x, y, z) = x^2 i + 2y j + 4 z^2 k$ and s is the surface of the cylinder $x^2 + y^2 \leq 4$ $0 \leq z \leq 2$

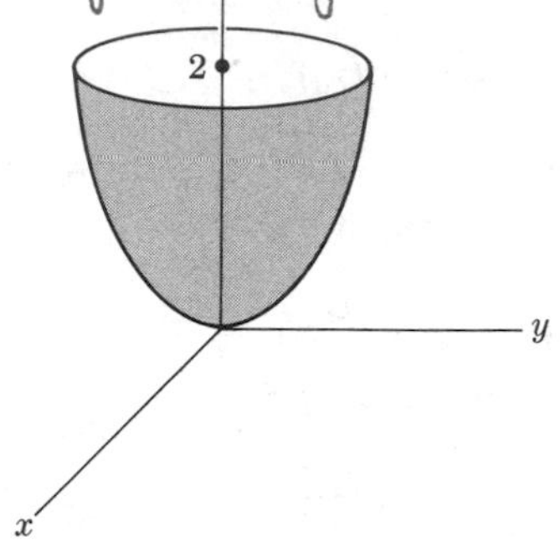

Figure 21

Note that the boundary consists of two pieces, the surface of the paraboloid and the disc on top, each of which can be easily parametrized.

Divergence Theorem. *Let U be a region in 3-space, forming the inside of a surface S which is smooth, except for a finite number of smooth curves. Let F be a vector field on an open set containing U and S. Let* **n** *be the unit outward normal vector to S. Then*

$$\iint_S F \cdot \mathbf{n} \, d\sigma = \iiint_U \text{div } F \, dV,$$

where the expression on the right is simply the triple integral of the function div F over the region U.

It is not easy to give a proof of the divergence theorem in general, but we shall give it in a special case of a rectangular box. This makes the general case very plausible, because we could reduce the general case to the special case by the following steps:

(i) Analyze how surface integrals change (or rather do not change) when we change the variables.

(ii) Reduce the theorem to a "local one" where the region admits one parametrization from a rectangular box. This can be done by various chopping-up processes, some of which are messy, some of which are neat, but all of which take up a fair amount of space to establish fully.

(iii) Combine the first and second steps, reducing the local theorem concerning the region to the theorem concerning a box, by means of the change of variables formula.

We now prove the theorem for a box, expressed as a product of intervals:

$$[a_1, b_1] \times [a_2, b_2] \times [a_3, b_3],$$

and illustrated in Fig. 22.

Verify the divergence theorem for $F = (2x - z)i + (x^4 y)j$
taken over the region bounded $x = 0$ $- (xz^2)k$
$x = 1, y = 0, y = 1, z = c_2, z = $

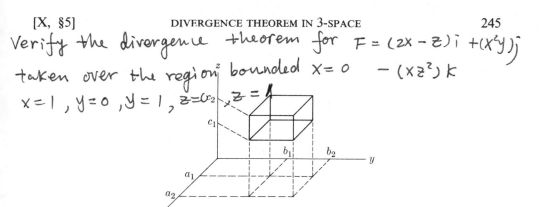

Figure 22

The surface surrounding the box consists of six sides, so that the integral over S will be a sum of six integrals, each one taken over one of the sides.

Let S_1 be the front face. We can parametrize S_1 by

$$X(y, z) = (a_2, y, z),$$

with y, z satisfying the inequalities

$$b_1 \leqq y \leqq b_2 \quad \text{and} \quad c_1 \leqq z \leqq c_2.$$

Let $\mathbf{n}_1$ be the unit outward normal vector on S_1. Then

$$\mathbf{n}_1 = (1, 0, 0).$$

If $F = (f_1, f_2, f_3)$, then $F \cdot \mathbf{n}_1 = f_1$, and hence

$$\iint_{S_1} F \cdot \mathbf{n} \, d\sigma = \int_{c_1}^{c_2} \int_{b_1}^{b_2} f_1(a_2, y, z) \, dy \, dz.$$

Similarly, let S_2 be the back face, parametrized by

$$X(y, z) = (a_1, y, z),$$

with y, z satisfying the same inequalities as above. Then

$$\mathbf{n}_2 = -(1, 0, 0),$$

the geometric interpretation being that the outward unit normal vector points to the back of the box drawn on Fig. 22. Hence

$$\iint_{S_2} F \cdot \mathbf{n} \, d\sigma = \int_{c_1}^{c_2} \int_{b_1}^{b_2} - f_1(a_1, y, z) \, dy \, dz.$$

Adding the integrals over S_1 and S_2 yields

$$\iint_{S_1} + \iint_{S_2} F \cdot \mathbf{n} \, d\sigma = \int_{c_1}^{c_2} \int_{b_1}^{b_2} \left[f_1(a_2, y, z) - f_1(a_1, y, z) \right] dy \, dz$$

$$= \int_{c_1}^{c_2} \int_{b_1}^{b_2} \int_{a_1}^{a_2} D_1 f_1(x, y, z) \, dx \, dy \, dz$$

$$= \iiint_U D_1 f_1 \, dV.$$

We now carry out a similar argument for the right side and the left side, as well as the top side and the bottom side. We find that the sums of the surface integral taken over these pairs of sides are equal to

$$\iiint_U D_2 f_2 \, dV$$

and

$$\iiint_U D_3 f_3 \, dV,$$

respectively. Adding all three volume integrals yields

$$\iint_S F \cdot \mathbf{n} \, d\sigma = \iiint_U (D_1 f_1 + D_2 f_2 + D_3 f_3) \, dV,$$

which is precisely the integral of the divergence, thus proving what we wanted.

Example 1. Let us compute the integral of the vector field

$$F(x, y, z) = (x^2, y^2, z^2)$$

over the unit cube by using the divergence theorem. The divergence of F is equal to $2x + 2y + 2z$, and hence the integral is equal to

$$\int_0^1 \int_0^1 \int_0^1 (2x + 2y + 2z) \, dx \, dy \, dz,$$

which is easily evaluated to give the value 3.

Example 2. Let us compute the integral of the vector field

$$F(x, y, z) = (x, y, z),$$

that is $F(X) = X$ over the sphere of radius a. The divergence of F is equal to

$$\frac{\partial x}{\partial x} + \frac{\partial y}{\partial y} + \frac{\partial z}{\partial z} = 3.$$

The ball B is the inside of the sphere. By the divergence theorem, we get

$$\iint_S F \cdot \mathbf{n} \, d\sigma = \iiint_B 3 \, dV = 3 \cdot \frac{4}{3} \pi a^3 = 4\pi a^3.$$

Note that the volume integral over the ball B of radius a is the integral of the constant 3, and hence is equal to 3 times the volume of the ball.

The divergence theorem has an interesting application, which can be used to interpret the divergence geometrically. It is the 3-dimensional analogue of the interpretation given in Chapter VIII, §2 for the 2-dimensional case, and the proof will be entirely similar.

Corollary. *Let $B(t)$ be the solid ball of radius $t > 0$, centered at a point P in $\mathbf{R}^3$. Let $S(t)$ denote the boundary of the ball, i.e. the sphere of radius t, centered at P. Let F be a C^1 vector field, and let $V(t)$ denote the volume of $B(t)$. Let $\mathbf{n}$ denote the unit normal vector pointing out from the spheres. Then*

$$(\operatorname{div} F)(P) = \lim_{t \to 0} \frac{1}{V(t)} \iint_{S(t)} F \cdot \mathbf{n} \, d\sigma.$$

Proof. Let $g = \operatorname{div} F$. Since g is continuous by assumption, we can write

$$g(X) = g(P) + h(X),$$

where

$$\lim_{X \to P} h(X) = 0.$$

Using the divergence theorem, we get

$$\frac{1}{V(t)} \iint_{S(t)} F \cdot \mathbf{n} \, d\sigma = \frac{1}{V(t)} \iiint_{B(t)} \operatorname{div} F \, dV$$

$$= \frac{1}{V(t)} \iiint_{B(t)} g(P) \, dV + \frac{1}{V(t)} \cdot \iiint_{B(t)} h \, dV.$$

Observe that $g(P) = (\operatorname{div} F)(P)$ is constant, and hence can be taken out of the first integral. The simple integral of dV over $B(t)$ yields the volume $V(t)$, which cancels, so that the first term is equal to $(\operatorname{div} F)(P)$, which is the desired answer.

There remains to show that the second term approaches 0 as t approaches 0. But this is clear: The function h approaches 0, and the integral on the right can be estimated as follows:

$$\left| \frac{1}{V(t)} \iiint_{B(t)} h \, dV \right| \leq \operatorname*{Max}_{\|X - P\| \leq t} |h(X)| \frac{1}{V(t)} \iiint_{B(t)} dV$$

$$\leq \operatorname*{Max}_{\|X - P\| \leq t} |h(X)|.$$

As $t \to 0$, the maximum of $h(X)$ for $\|X - P\| \leq t$ approaches 0, thus proving what we wanted.

The integral expression under the limit sign in the corollary can be interpreted as the flow going outside the sphere per unit time, in the direction of the unit outward normal vector. Dividing by the volume of the ball $B(t)$, we obtain the mass per unit volume flowing out of the sphere. Thus we get an interpretation:

> *The divergence of F at P is the rate of change of mass per unit volume per unit time at P.*

As in the case of Green's theorem, whose general form was stated for regions which are more general than interiors of closed curves, we have an analogue in the higher dimensional case for the divergence theorem.

Divergence Theorem, General Case. *Let U be an open set whose boundary consists of a finite number of surfaces,*

$$S = \{S_1, \ldots, S_m\}$$

oriented so that U lies to the left of each surface S_i. Let F be a vector field on an open set containing U and S. Let $\mathbf{n}$ be the unit outward normal vector to S. Then

$$\iint_S F \cdot \mathbf{n} \, d\sigma = \iiint_U \operatorname{div} F \, dV.$$

In the formula the integral over S is of course the sum of the integrals over the pieces S_i for $i = 1, \ldots, m$.

Example 3. Suppose that U is the region between two concentric spheres, S_1 and S_2^-, and that div $F = 0$. Then the integral on the right-hand side is 0. Hence

$$\iint_{S_1} F \cdot \mathbf{n} \, d\sigma + \iint_{S_2^-} F \cdot \mathbf{n} \, d\sigma = 0.$$

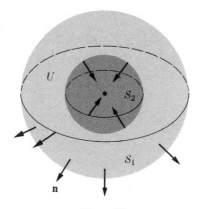

Figure 23

The outer sphere S_1 is oriented so that the unit outward normal vector points outward. The inner sphere has to be oriented so that unit normal vector points toward the common center, in order for the region between the spheres to lie to the left of the inner sphere. Thus, if S_2 denotes the inner sphere with its standard orientation, we have to take S_2^- with opposite orientation to apply the divergence theorem. Consequently, we find that

$$\iint_{S_1} F \cdot \mathbf{n} \, d\sigma = \iint_{S_2} F \cdot \mathbf{n} \, d\sigma.$$

Of course, we did not need to assume S_1 to be a sphere. The same argument proves the following corollary.

Corollary. *Let S_1, S_2 be closed surfaces such that S_1 is contained in the interior of S_2, and let U be the region between them. Let F be a vector field such that* div $F = 0$ *on a region containing U and its boundary. Then the integral of F over S_1 is equal to the integral of F over S_2.*

Example 4. **(Gauss' Law).** In 3-space, let q be a constant, and let

$$f(x, y, z) = \frac{q}{4\pi r} \qquad \text{where} \quad r = \|X\| = \sqrt{x^2 + y^2 + z^2} \, .$$

Let $E = -$ grad f. We interpret f as the potential function associated with a point charge of electricity q at the origin, and we interpret E as the corresponding electric field. Verify that (Exercise 16)

$$\text{div } E = 0.$$

Let S_1 be any closed surface whose interior contains the origin. The integral

$$\iint_{S_1} E \cdot \mathbf{n} \, d\sigma$$

is interpreted as the total electric flux over the surface, due to that point charge. Whereas it is probably difficult to evaluate the integral over S_1 directly, we can use the corollary which tells us that the flux can be computed as the integral

$$\iint_{S_2} E \cdot \mathbf{n} \, d\sigma = q$$

where S_2 is a small sphere centered at the origin. It is then easy to find the value q on the right-hand side (Exercise 16). Thus the flux is equal to the point charge of electricity. This is known as **Gauss' Law.**

EXERCISES

1. Compute explicitly the integrals over the top, bottom, right, and left sides of the box to check in detail the remaining steps of the proof of the divergence theorem, left to the reader in the text, as "similar arguments".

2. Let S be the boundary of the unit cube,

$$0 \leq x \leq 1, \qquad 0 \leq y \leq 1, \qquad 0 \leq z \leq 1.$$

Compute the integral of the vector field $F(x, y, z) = (xy, y^2, y^2)$ over the surface of this cube.

3. Calculate the integral

$$\iint_S (\mathrm{curl}\ F) \cdot \mathbf{n}\ d\sigma$$

where F is the vector field

$$F(x, y, z) = (-y, x^2, z^3),$$

and S is the surface

$$x^2 + y^2 + z^2 = 1, \qquad -\tfrac{1}{2} \leq z \leq 1.$$

Don't make things more complicated than they need be.

4. Find the integral of the vector field

$$F(X) = \frac{X}{\|X\|}$$

over the sphere of radius 4.

Find the integral of the following vector fields over the indicated surface.

5. (a) $F(x, y, z) = (yz, xz, xy)$ over the cube centered at the origin and with sides of length 2.
   (b) $F(x, y, z) = (x^2, y^2, z^2)$ over the same cube.
   (c) $F(x, y, z) = (x - y, y - z, x - y)$ over the same cube.
   (d) $F(X) = X$ over the same cube.

6. Let $F(x, y, z) = (2x, y^2, z^2)$. Compute the integral of F over the unit sphere.

7. Let $F(x, y, z) = (x^3, y^3, z^3)$. Compute the integral of F over the unit sphere.

8. Let $F(x, y, z) = (x, y, -z)$. Compute the integral of F over the unit cube, consisting of all points (x, y, z) with

$$0 \leq x \leq 1, \qquad 0 \leq y \leq 1, \qquad \text{and} \qquad 0 \leq z \leq 1.$$

9. $F(x, y, z) = (x + y, y + z, x + z)$ over the surface bounded by the paraboloid

$$z = 4 - x^2 - y^2,$$

and the disc of radius 2 centered at the origin in the (x, y)-plane.

10. $F(x, y, z) = (2x, 3y, z)$ over the surface bounding the region enclosed by the cylinder

$$x^2 + y^2 = 4$$

and the planes $z = 1$ and $z = 3$.

11. $F(x, y, z) = (x, y, z)$, over the surface bounding the region enclosed by the paraboloid $z = x^2 + y^2$, the cylinder $x^2 + y^2 = 9$, and the plane $z = 0$.

12. $F(x, y, z) = (x + y, y + z, x + z)$ over the surface bounding the region defined by the inequalities

$$0 \leq x^2 + y^2 \leq 9 \quad \text{and} \quad 0 \leq z \leq 5.$$

13. $F(x, y, z) = (3x^2, xy, z)$ over the tetrahedron bounded by the coordinate planes and the plane $x + y + z = 1$.

14. Let f be a harmonic function, that is a function satisfying

$$\frac{\partial^2 f}{\partial x^2} + \frac{\partial^2 f}{\partial y^2} + \frac{\partial^2 f}{\partial z^2} = 0.$$

Let S be a closed smooth surface bounding a region U in 3-space. Let f be a harmonic function on an open set containing the region and its boundary. If $\mathbf{n}$ is the unit normal vector to the surface pointing outward, let $D_{\mathbf{n}} f$ be the directional derivative of f in the direction of $\mathbf{n}$.
(a) Prove that

$$\iint_S D_{\mathbf{n}} f \, d\sigma = 0.$$

[*Hint:* Let $F = \text{grad } f$.]
(b) Prove that

$$\iint_S f D_{\mathbf{n}} f \, d\sigma = \iiint_U \|\text{grad } f\|^2 \, dV.$$

[*Hint:* Let $F = f \text{ grad } f$.]

15. (a) Let U be the interior of a closed surface S. Show that

$$\iint_S X \cdot \mathbf{n} \, d\sigma = 3 \text{ Vol } (U).$$

(b) Show that

$$\iint_S \frac{X \cdot \mathbf{n}}{r^2} \, d\sigma = \iiint_U \frac{1}{r^2} \, dV.$$

As usual in this exercise, $X = (x, y, z)$ and $r = \|X\| = \sqrt{x^2 + y^2 + z^2}$.

16. Let q be a constant, and let

$$f(X) = f(x, y, z) = \frac{q}{4\pi r} \quad \text{where} \quad r = \|X\|.$$

(a) Verify that div grad $f = 0$. [Cf. Exercise 7a of §4.]
(b) Compute the integral of $E = -\text{grad } f$ over a sphere centered at the origin to find the value stated in the text in the last example, namely q.

17. Let U be the interior of a closed surface S.
(a) Assume that the origin O does not lie in U or its boundary. Show that

$$\iint_S \frac{X \cdot \mathbf{n}}{r^3} \, d\sigma = 0.$$

As usual, $X = (x, y, z)$. How is this exercise related to Exercise 14?

(b) If the origin O is contained in U, show that

$$\iint_S \frac{X \cdot \mathbf{n}}{r^3} \, d\sigma = 4\pi.$$

How is this related to Exercise 16?

18. Let $P_1, \ldots, P_m$ be fixed points in 3-space, and let $q_1, \ldots, q_m$ be numbers, which we call charges. Let

$$f(X) = \sum_{j=1}^m \frac{q_j}{4\pi \|X - P_j\|}.$$

(This is interpreted as the potential function associated with the finite number of charges at the given points.) Let S be a closed surface not containing any of the points P_j. Let q be the sum of the charges inside S. Let $E = -\operatorname{grad} f$. Show that

$$\iint_S E \cdot \mathbf{n} \, d\sigma = q.$$

19. Let U be the interior of a closed surface S. Let f, g be functions. Prove the formulas known as **Green's identities**:

(a) $\displaystyle \iint_S f(\operatorname{grad} g) \cdot \mathbf{n} \, d\sigma = \iiint_U [f \, \nabla^2 g + \nabla f \cdot \nabla g] \, dV$

(b) $\displaystyle \iint_S (f \, \nabla g - g \, \nabla f) \cdot \mathbf{n} \, d\sigma = \iiint_U (f \, \nabla^2 g - g \, \nabla^2 f) \, dV.$

[Note: ∇f means grad f, and $\nabla^2 f = \operatorname{div} \operatorname{grad} f$ by definition.]

§6. STOKES' THEOREM

We recall Green's theorem in the plane. It stated that if S is a plane region bounded by a closed path C, oriented counterclockwise, and F is a vector field on some open set containing the region, $F = (f_1, f_2)$, then

$$\iint_S (D_1 f_2 - D_2 f_1) \, d\sigma = \int_C F \cdot dC.$$

Of course in the plane with variables (x, y), $d\sigma = dx \, dy$.

We can now ask for a similar theorem in 3-space, when the surface lies in 3-space, and the surface is bounded by a curve in 3-space. The analogous statement is true, and is called Stokes' theorem:

Stokes' Theorem. *Let S be a smooth surface in $\mathbf{R}^3$, bounded by a closed curve C. Assume that the surface is orientable, and that the boundary curve is oriented so that the surface lies to the left of the curve. Let F be a vector field* *in an open set containing the surface S and its boundary. Then*

$$\nabla \times F = \begin{vmatrix} i & j & k \\ \dfrac{\partial}{\partial x} & \dfrac{\partial}{\partial y} & \dfrac{\partial}{\partial z} \\ P & Q & R \end{vmatrix}$$

$$\iint_S (\operatorname{curl} F) \cdot \mathbf{n} \, d\sigma = \int_C F \cdot dC.$$

$$= \left(\frac{\partial R}{\partial y} - \frac{\partial Q}{\partial z}\right) i + \left(\frac{\partial P}{\partial z} - \frac{\partial R}{\partial x}\right) j$$

$$+ \left(\frac{\partial Q}{\partial x} - \frac{\partial P}{\partial y}\right) k$$

[handwritten] $F = Pi + Qj + Rk$ and the line integral can be this simple closed curve $x = f(t), \ y = g(t), \ z = h(t)$

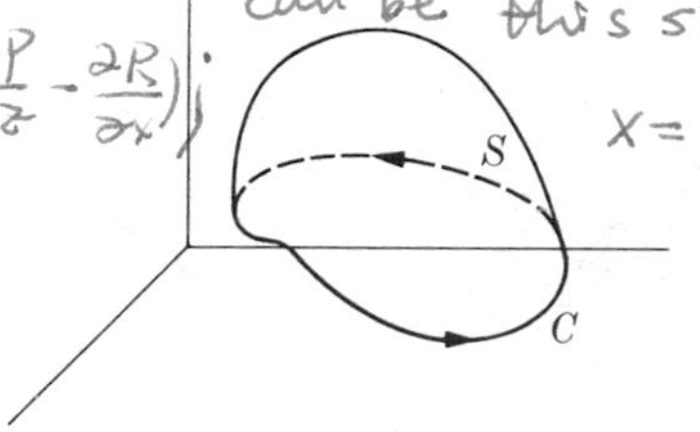

Figure 24

When the surface consists of a finite number of smooth pieces, and the boundary also consists of a finite number of smooth curves, then the analogous statement holds, by taking a sum over these pieces.

We shall not prove Stokes' theorem. The proof can be reduced to that of Green's theorem in the plane by making an analysis of the way both sides of the formula behave under changes of variables, i.e. changes of parametrization. Note that Green's theorem in the plane is a special case, because then the unit normal vector is simply (0, 0, 1), and the curl of F dotted with the unit normal vector is simply the third component of the curl, namely

$$D_1 f_2 - D_2 f_1.$$

Thus Green's theorem in the plane makes the 3-dimensional analogue quite plausible.

Example 1. Suppose that two surfaces S_1 and S_2 are bounded by a curve C, and lie on opposite sides of the curve, as on Fig. 25. Then

$$\iint_{S_1} (\operatorname{curl} F) \cdot \mathbf{n} \, d\sigma = -\iint_{S_2} (\operatorname{curl} F) \cdot \mathbf{n} \, d\sigma$$

because the integral over S_1 is equal to the integral of F over C, whereas the integral over S_2 is equal to the integral of F over C^-, which is the same as C but oriented in the opposite direction. We have also drawn separately the surfaces S_1 and S_2 having C as boundary. Observe that taken together, S_1 and S_2 bound the inside of a 3-dimensional region.

Let S be the hemisphere $z = \sqrt{4-x^2-y^2}$ $0 \leq x^2+y^2 \leq 4$ lying above the xy plane, with the center at the origin. The boundary of this circle· C : $z=0$, $x^2+y^2 = 4$, where $\vec{F} = y\hat{j} - x\hat{j}$ Find the line integral.

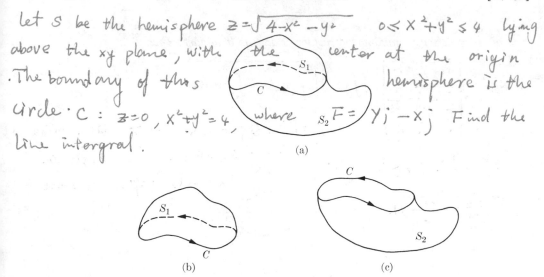

(a)

(b)   (c)

Figure 25

Example 2. Similarly, consider a ball, bounded by a sphere. The two hemispheres have a common boundary, namely the circle in the plane as on Fig. 26. Note that C is oriented so that S_1 lies to the left of C, but S_2 lies to the right of C.

Compute $\oint F\,dR$

$F(x, y, z) = (z-2y)\hat{i} + (3x-4y)\hat{j} + (z+3y)\hat{k}$ and C is the boundary of the triangle joining the points $(1,0,0)$, $(0,10)$ $(0,0,1)$.

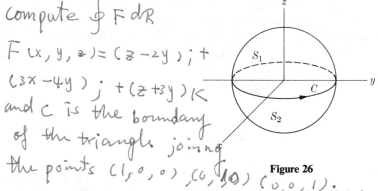

Figure 26

By the divergence theorem, we know that if S denotes the union of S_1 and S_2, then

$$\iint_S (\text{curl } F) \cdot \mathbf{n}\, d\sigma = \iiint_U \text{div curl } F\, dV.$$

However, div curl $F = 0$. Since

$$\iint_S = \iint_{S_1} + \iint_{S_2},$$

we obtain in another way that

$$\iint_{S_1} (\text{curl } F) \cdot \mathbf{n} \, d\sigma = -\iint_{S_2} (\text{curl } F) \cdot \mathbf{n} \, d\sigma.$$

Example 3. We shall verify Stokes' theorem for the vector field

$$F(x, y, z) = (z - y, x + z, -(x + y)),$$

and the surface bounded by the paraboloid

$$z = 4 - x^2 - y^2$$

and the plane $z = 0$, as on Fig. 27.

Stoke's thm (curl)

curl B = V

$\oint B \cdot T \, ds = \iint \text{curl } B \cdot n \, d\sigma$
$= \iint r \cdot n \, d\sigma$

The magnetic flux of an electric field through a surface S is equal to the circulation induced by V around the boundary S

Ampare's law

Figure 27

First we compute the integral over the boundary curve, which is just the circle

$$x^2 + y^2 = 4.$$

We parametrize the circle by $x = 2 \cos \theta$ and $y = 2 \sin \theta$ as usual. Then

$$\begin{aligned} F \cdot dC &= (z - y) \, dx + (x + z) \, dy - (x + y) \, dz \\ &= -2 \sin \theta(-2 \sin \theta \, d\theta) + 2 \cos \theta(2 \cos \theta) \, d\theta \\ &= 4 \, d\theta. \end{aligned}$$

Consequently,

$$\int_C F \cdot dC = \int_0^{2\pi} 4 \, d\theta = 8\pi.$$

Now we evaluate the surface integral. First we get the curl, namely

$$\text{curl } F = \begin{vmatrix} E_1 & E_2 & E_3 \\ D_1 & D_2 & D_3 \\ z - y & x + z & -x - y \end{vmatrix} = (-2, 2, 2).$$

use stoke's theorem to evaluate $\oint F \cdot dx$ where
$F(x, y, z) = (z - 2y)\,i + (3x - 4y)\,j + (z + 3y)\,k$
and C is the unit circle in the plane $z = 2$ (5π)

We can compute the normal vector as in §1, or by observing that the surface is defined by the equation

$$f(x, y, z) = z - 4 + x^2 + y^2 = 0,$$

and then finding

$$\text{grad } f(x, y, z) = (2x, 2y, 1),$$

so that

$$\mathbf{n} = \frac{1}{\sqrt{4x^2 + 4y^2 + 1}} (2x, 2y, 1).$$

Then

$$\iint_S \text{curl } F \cdot \mathbf{n} \, d\sigma = \iint_D (-4x + 4y + 2) \, dx \, dy,$$

where D is the disc defined by $x^2 + y^2 \leq 4$. This last integral is easily found to be equal to 8π, which is, of course, the same value as the integral of F over the curve in the first part of the example.

Remark. Green's and Stokes' theorems are special cases of higher dimensional theorems expressing a relation between an integral over a region in space, and another integral over the boundary of the region. To give a systematic treatment requires somewhat more elaborate foundations, and lies beyond the bounds of this course.

Stokes' theorem allows us to give an interpretation for the curl of a vector field similar to that given for the divergence.

Let P be a point on a surface S (smoothly parametrized), and for each small positive number r, let C_r be the closed curve consisting of the points on the surface at distance r from P. We assume without proof that this curve is smooth, and we take it with counterclockwise orientation, as shown on the figure (Fig. 28).

Verify stoke's theorem for the surface $z = x^2 + y^2$
$x^2 + y^2 \leq 1$ $F(x, y, z) = y^2 i + x j + z^2 k$

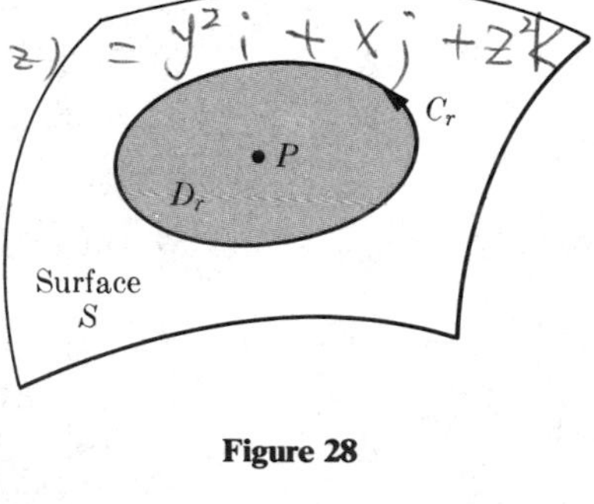

Figure 28

We let D_r be the portion of the surface in the interior of C_r. Then C_r and D_r constitute the analogue of a circle and a disc centered at P, but of course since the surface may bend in 3-space, C_r is not actually a circle, and D_r is not actually a disc. We let $A(r)$ be the surface area of D_r.

Corollary of Stokes' Theorem. *Let $\mathbf{n}_P$ be the unit normal vector to the surface at P. Then*

$$(\text{curl } F(P)) \cdot \mathbf{n}_P = \lim_{r \to 0} \frac{1}{A(r)} \int_{C_r} F \cdot dC.$$

Proof. Our vector fields are always assumed continuously differentiable, so we can write

$$(\text{curl } F)(X) = (\text{curl } F)(P) + H(X),$$

where

$$\lim_{X \to P} H(X) = 0.$$

Similarly, if $\mathbf{n}$ is the unit normal vector at some point in D_r, then

$$\mathbf{n} = \mathbf{n}_P + \mathbf{w}$$

where $\mathbf{n}_P$ is the unit normal at P, and $\mathbf{w}$ approaches 0 as r approaches 0. Then

$$(\text{curl } F) \cdot \mathbf{n} = ((\text{curl } F)(P) + H) \cdot (\mathbf{n}_P + \mathbf{w})$$
$$= (\text{curl } F(P)) \cdot \mathbf{n}_P + \text{three other terms,}$$

where the three other terms are obtained by distributivity, and each term contains some vector H or $\mathbf{w}$ which approaches 0 as r approaches 0. Let us abbreviate these three other terms by h. Substituting in the left-hand side of Stokes' theorem yields

$$\iint_{D_r} (\text{curl } F) \cdot \mathbf{n} \, d\sigma = \iint_{D_r} (\text{curl } F(P)) \cdot \mathbf{n}_P \, d\sigma + \iint_{D_r} h \, d\sigma$$

$$= (\text{curl } F(P)) \cdot \mathbf{n}_P \iint_{D_r} d\sigma + \iint_{D_r} h \, d\sigma$$

(because $(\text{curl } F(P)) \cdot \mathbf{n}_P$ is constant and can be taken out of the integral)

$$= A(r)(\text{curl } F(P)) \cdot \mathbf{n}_P + \iint_{D_r} h \, d\sigma$$

because the integral

$$\iint_{D_r} d\sigma = A(r)$$

is the area of the surface lying inside C_r.

Now apply Stokes' theorem, and divide by $A(r)$. We then find

$$(\text{curl } F(P)) \cdot \mathbf{n}_P + \frac{1}{A(r)} \iint_{D_r} h \, d\sigma = \frac{1}{A(r)} \int_{C_r} F \cdot dC.$$

Let r approach 0. The integral remaining on the left-hand side is bounded in absolute value by

$$\left| \iint_{D_r} h \, d\sigma \right| \le \left(\max_{D_r} |h| \right) \iint_{D_r} d\sigma = \left(\max_{D_r} |h| \right) A(r)$$

where $\max_{D_r} |h|$ is the maximum of the absolute value of h over the region D_r, and tends to 0 as r tends to 0. Hence

$$\lim_{r \to 0} \frac{1}{A(r)} \iint_{D_r} h \, d\sigma = 0.$$

This proves the corollary.

Physical Interpretation for the Curl. The curve integral

$$\int_{C_r} F \cdot dC$$

along the curve C_r represents the integral along C_r of the tangential component of F along the curve. This tangential component is interpreted as the amount by which F is rotating, or as we appropriately could say, curling around the point, rather than the normal component

$$F \cdot \mathbf{n},$$

which is the amount by which the vector field F points outward from the curve. Thus $F \cdot \mathbf{n}$ represents the flow outward from the curve, while $F \cdot dC$ represents the flow remaining inside the curve.

Dividing by $A(r)$ is a normalizing procedure, which determines the amount by which F is curling around the point per unit area. Hence the limit on the right-hand side, equal to the left-hand side, gives the following interpretation for the curl:

The curl $F(P)$ is the amount by which the vector field F (or the fluid flow determined by F) rotates (curls) around the point P.

This is illustrated on Fig. 29.

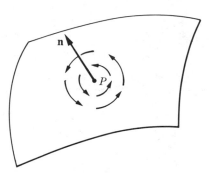

Figure 29

EXERCISES

Verify Stokes' theorem in each one of the following cases.

1. $F(x, y, z) = (z, x, y)$, S defined by $z = 4 - x^2 - y^2$, $z \geqq 0$.

2. $F(x, y, z) = (x^2 + y, yz, x - z^2)$ and S is the triangle defined by the plane

$$2x + y + 2z = 2$$

and $x, y, z \geqq 0$.

3. $F(x, y, z) = (x, z, -y)$ and the surface is the portion of the sphere of radius 2 centered at the origin, such that $y \geqq 0$.

4. $F(x, y, z) = (x, y, 0)$ and the surface is the part of the paraboloid $z = x^2 + y^2$ inside the cylinder $x^2 + y^2 = 4$.

5. $F(x, y, z) = (y + x, x + z, z^2)$, and the surface is that part of the cone $z^2 = x^2 + y^2$ between the planes $z = 0$ and $z = 1$.

Compute the integral $\iint_S \operatorname{curl} F \cdot \mathbf{n} \, d\sigma$ by means of Stokes' theorem.

6. $F(x, y, z) = (y, z, x)$ over the triangle with vertices at the unit points $(1, 0, 0)$, $(0, 1, 0)$, $(0, 0, 1)$.

7. $F(x, y, z) = (x + y, y - z, x + y + z)$ over the hemisphere

$$x^2 + y^2 + z^2 = a^2, \quad z \geqq 0.$$

8. (a) Let C be the curve given by

$$C(t) = (\cos t, \sin t, \sin t) \quad \text{with} \quad 0 \leqq t \leqq 2\pi.$$

Find

$$\int_C z \, dx + 2x \, dy + y^2 \, dz$$

directly from the definition of curve integrals.

(b) Find the integral of (a) by using Stokes' theorem.
[*Hint*: The curve C is the boundary of the graph of the function $f(x, y) = y$, defined on the disc of radius 1.]

9. Let $F(x, y, z) = (ye^z, xe^z, xye^z)$. Let C be a simple closed curve which is the boundary of a surface S. Show that the integral of F along C is equal to 0.

10. Let C be a closed curve which is the boundary of a surface S. Prove the following:

(a) $\int_C (f \operatorname{grad} g) \cdot dC = \int\int_S [(\operatorname{grad} f) \times (\operatorname{grad} g)] \cdot \mathbf{n} \, d\sigma$

(b) $\int_C (f \operatorname{grad} g + g \operatorname{grad} f) \cdot dC = 0.$

11. Let S be a surface bounded by a curve C. Let F be a vector field on an open set containing the surface and its boundary, and assume that F is perpendicular to the boundary (i.e. at every point of the boundary, the value of the vector field is perpendicular to the tangent line of the curve). Show that

$$\int\int_S (\operatorname{curl} F) \cdot \mathbf{n} \, d\sigma = 0.$$

求极值的步骤：设函数 $f(x,y)$ 有二阶连续偏微商，求 $f(x,y)$ 的极值的步骤如下

① 计算一阶偏微商，并令一阶偏微商等于 0 $\begin{cases} f'_x(x,y)=0 \\ f'_y(x,y)=0 \end{cases}$
解这个方程组，得诸组解 $(X_1,Y_1),(X_2,Y_2)$
···，每一组解叫做函数 $f(x,y)$ 的一个临界点

② 计算二阶偏微商 $f''_{xx}(x,y)$ $f''_{xy}(x,y)$, $f''_{yy}(x,y)$

③ 把①中求得的每一组解，例如 (X_1,Y_1) 代入三个二阶偏微商令
$A = f''_{xx}(X_1,Y_1)$, $B = f''_{xy}(X_1,Y_1)$, $C\ f''_{yy}(X_1,Y_1)$.

$D = B^2 - AC$

如果 $D<0$，那么函数在 (X_0,Y_0) 处有极值：当 $A<0$ 时，$f(X_0,Y_0)$ 是极大值，当 $A>0$ 时，$f(X_0,Y_0)$ 是极小值。如果 $D>0$，那么函数在 (X_0,Y_0) 处无极值。如果 $D=0$，那么还不能判断函数在 (X_0,Y_0) 处有无极值。

Ex: $f(x,y) = x^2 + 2xy + 2y^2 + 4x + 2y - 5$

解 $\begin{cases} f'_x(x,y) = 2x + 2y + 4 = 0 \\ f'_y(x,y) = 2x + 4y + 2 = 0 \end{cases}$ 解这个方程得 $X = -3$, $Y = 1$

$f''_{xx}(x,y) = 2$ $f''_{xy}(x,y) = 2$ $f''_{yy}(x,y) = 4$

它们已是常数，不必将 $X=-3$ $Y=1$ 代入于是有

$D = B^2 - AC = 2^2 - 2 \times 4 = -4 < 0$ $A > 0$ $A = 2 > 0$ 函数在 $(-3, 1)$

处有极小值. $f(-3, 1) = (-3)^2 + 2(-3)\cdot 1 + 2\cdot 1^2 + 4\cdot(-3) + 2\cdot 1 - 5 = -10$

$$f_{xx}(x_0, y_0) > 0 \text{ min}$$
$$f_{xx}(x_0, y_0) < 0 \text{ max}$$

CHAPTER XI

Maximum and Minimum

$$D = \begin{vmatrix} f_{xx} & f_{xy} \\ f_{yx} & f_{yy} \end{vmatrix}$$

If $D > 0$ $f_{xx}(x_0, y_0)$ min

$D \gtrless 0$ $f_{xx}(x_0, y_0)$ max

$D < 0 \longrightarrow$ saddle point

$D = 0$ you don't know

When we studied functions of one variable, we found maxima and minima by first finding critical points, i.e. points where the derivative is equal to 0, and then determining by inspection which of these are maxima or minima. We can carry out a similar investigation for functions of several variables. The condition that the derivative is equal to 0 must be replaced by the vanishing of all partial derivatives.

§1. CRITICAL POINTS

Let f be a differentiable function defined on an open set U. Let P be a point of U. If all partial derivatives of f are equal to 0 at P, then we say that P is a **critical point** of the function. In two variables, the point (x_0, y_0) is a critical point if and only if

$$D_1 f(x_0, y_0) = 0 \quad \text{and} \quad D_2 f(x_0, y_0) = 0.$$

In other words, the two partial derivatives

$$\frac{\partial f}{\partial x} \quad \text{and} \quad \frac{\partial f}{\partial y}$$

must be equal to 0 when evaluated at the point $P = (x_0, y_0)$.

In n variables, the condition reads

$$D_1 f(P) = 0, \ldots, D_n f(P) = 0.$$

or more concisely, grad $f(P) = O$.

Example 1. Find the critical points of the function $f(x, y) = e^{-(x^2+y^2)}$. Taking the partials, we see that

$$\frac{\partial f}{\partial x} = -2x e^{-(x^2+y^2)} \quad \text{and} \quad \frac{\partial f}{\partial y} = -2y e^{-(x^2+y^2)}.$$

The only value of (x, y) for which both these quantities are equal to 0 is $x = 0$ and $y = 0$. Hence the only critical point is $(0, 0)$.

A critical point of a function of one variable is a point where the derivative is equal to 0. We have seen examples where such a point need not be a local maximum or a local minimum, for instance as in the following picture (Fig. 1):

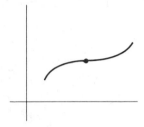

Figure 1

A fortiori, a similar thing may occur for functions of several variables. However, once we have found critical points, it is usually not too difficult to tell by inspection whether they are of this type or not.

Let f be any function (differentiable or not), defined on an open set U. We shall say that a point P of U is a **local maximum** for the function if there exists an open ball (of positive radius) B, centered at P, such that for all points X of B we have

$$f(X) \leq f(P).$$

As an exercise, define **local minimum** in an analogous manner.

In the case of functions of one variable, we took an open interval instead of an open ball around the point P. Thus our notion of local maximum in n-space is the natural generalization of the notion in 1-space.

Theorem 1. *Let f be a function which is defined and differentiable on an open set U. Let P be a local maximum for f in U. Then P is a critical point of f.*

Proof. The proof is exactly the same as for functions of one variable. In fact, we shall prove that the directional derivative of f at P in any direction is 0. Let H be a non-zero vector. For small values of t, $P + tH$ lies in the open set U, and $f(P + tH)$ is defined. Furthermore, for small values of t, tH is small, and hence $P + tH$ lies in our open ball such that

$$f(P + tH) \leq f(P).$$

Hence the function of one variable $g(t) = f(P + tH)$ has a local maximum at $t = 0$. Hence its derivative $g'(0)$ is equal to 0. By the chain rule, we obtain as usual:

$$\text{grad } f(P) \cdot H = 0.$$

This equation is true for every non-zero vector H, and hence

$$\text{grad } f(P) = O.$$

This proves what we wanted.

Just as in one-variable theory, a critical point may be a maximum, a minimum, or neither. Remember the possibilities for the graph of a function of one variable in these three cases, as shown on Fig. 2.

(a)   (b)   (c)

Figure 2

In several variables, we have exactly the same situations, and the three cases might look like this.

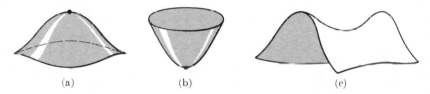

(a)   (b)   (c)

Figure 3

We shall study these possibilities more systematically in the next chapter. In the present chapter, we shall determine which possibilities occur by inspection.

EXERCISES

Find the critical points of the following functions.

1. $x^2 + 4xy - y^2 - 8x - 6y$ 2. $x + y \sin x$

3. $x^2 + y^2 + z^2$ 4. $(x + y)e^{-xy}$

5. $xy + xz$ 6. $\cos(x^2 + y^2 + z^2)$

7. $x^2 y^2$ 8. $x^4 + y^2$

9. $(x - y)^4$ 10. $x \sin y$

11. $x^2 + 2y^2 - x$ 12. $e^{-(x^2 + y^2 + z^2)}$

13. $e^{(x^2 + y^2 + z^2)}$

14. In each of the preceding exercises, find the minimum value of the given function, and give all points where the value of the function is equal to this minimum. [*Do this exercise after you have read §2.*]

§2. BOUNDARY POINTS

In considering intervals, we had to distinguish between closed and open intervals, We must make an analogous distinction when considering sets of points in space.

Let S be a set of points, in some n-space. Let P be a point of S. We shall say that P is an **interior point** of S if there exists an open ball B of positive radius, centered at P, and such that B is contained in S. The next picture illustrates an interior point (for the set consisting of the region enclosed by the curve).

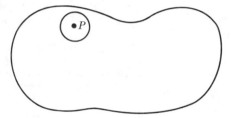

Figure 4

We have also drawn an open ball around P.

From the very definition, we conclude that the set consisting of all interior points of S is an open set.

A point P (not necessarily in S) is called a **boundary point** of S if every open ball B centered at P includes a point of S, and also a point which is not in S. We illustrate a boundary point in the following picture:

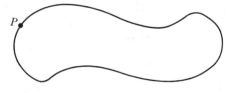

Figure 5

For example, the set of boundary points of the closed ball of radius $a > 0$ is the sphere of radius a. In 2-space, the plane, the region consisting of all points with $y > 0$ is open. Its boundary points are the points lying on the x-axis.

If a set contains all of its boundary points, then we shall say that the set is **closed**.

Finally, a set is said to be **bounded** if there exists a number $b > 0$ such that, for every point X of the set, we have

$$\|X\| \leqq b.$$

We are now in a position to state the existence of maxima and minima for continuous functions.

Theorem 2. *Let S be a closed and bounded set. Let f be a continuous function defined on S. Then f has a maximum and a minimum in S. In other words, there exists a point P in S such that*

$$f(P) \geqq f(X)$$

for all X in S, and there exists a point Q in S such that

$$f(Q) \leqq f(X)$$

for all X in S.

We shall not prove this theorem. It depends on an analysis which is beyond the level of this course.

When trying to find a maximum (say) for a function f, one should first determine the critical points of f in the interior of the region under consideration. If a maximum lies in the interior, it must be among these critical points.

Next, one should investigate the function on the boundary of the region. By parametrizing the boundary, one frequently reduces the problem of finding a maximum on the boundary to a lower-dimensional problem, to which the technique of critical points can also be applied.

Finally, one has to compare the possible maximum of f on the boundary and in the interior to determine which points are maximum points.

Example 1. In Example 1 in §1, we observe that the function

$$f(x, y) = e^{-(x^2 + y^2)}$$

becomes very small as x or y becomes large. Consider some big closed disc centered at the origin. We know by Theorem 2 that the function has a maximum in this disc. Since the value of the function is small on the boundary, it follows that this maximum must be an interior point, and hence that the maximum is a critical point. But we found in the Example in §1 that the only critical point is at the origin. Hence we conclude that the origin is *the* only maximum of the function $f(x, y)$. The value of f at the origin is $f(0, 0) = 1$. Furthermore, the function has no minimum, because $f(x, y)$ is always positive and approaches 0 as x and y become large.

Example 2. Find the maximum of the function

$$f(x, y) = x^2 e^{-x^4 - y^2}.$$

You should know from first year calculus that

$$\lim_{x \to \infty} x^2 e^{-x} = 0.$$

As x becomes large, x^4 is bigger than x^2, and so e^{-x^4} is smaller than e^{-x}. Consequently

$$x^2 e^{-x^4} \to 0 \quad \text{as } x \text{ becomes large.}$$

Since $y^2 \geq 0$, it follows that $e^{-y^2} \leq 1$. Hence as

$$r = \sqrt{x^2 + y^2}$$

becomes large, the function $f(x, y)$ approaches 0. Hence any maximum occurs in a bounded region of the plane.

To find it we find the critical points. We have:

$$\frac{\partial f}{\partial x} = e^{-y^2}\left[x^2(-4x^3)e^{-x^4} + 2xe^{-x^4}\right]$$

$$= e^{-x^4 - y^2}\left[-4x^5 + 2x\right].$$

$$\frac{\partial f}{\partial y} = x^2 e^{-x^4}(-2y)e^{-y^2} = -2x^2 ye^{-x^4 - y^2}.$$

The only points (x, y) where $\partial f/\partial y$ is 0 occur when $x = 0$ or $y = 0$. Among these, the only points (x, y) where $\partial f/\partial x$ is 0 are:

$$\text{when} \quad x = 0$$

or

$$\text{when} \quad -4x^4 + 2 = 0, \quad \text{that is} \quad x = (1/2)^{1/4}.$$

Hence the critical points are the points:

$$((1/2)^{1/4}, 0) \quad \text{and} \quad (0, y)$$

for an arbitrary value of y. But

$$f(0, y) = 0 \quad \text{and} \quad f((1/2)^{1/4}, 0) = \frac{1}{\sqrt{2}} e^{-1/2}.$$

Hence the maximum of the function is at $((1/2)^{1/4}, 0)$ and the maximum value is that given above.

Example 3. Find the maximum of the function

$$f(x, y) = x^2 y$$

on the square drawn in the figure (Fig. 6).

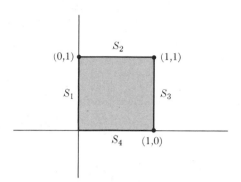

Figure 6

Let U be the interior of the square. We first find the critical points of f on U. We have:

$$\operatorname{grad} f(x, y) = (2xy, x^2).$$

Thus

$$\operatorname{grad} f(x, y) = (0, 0) \quad \text{if and only if} \quad (x, y) = (0, y)$$

with an arbitrary value of y. In particular, the x-coordinate of a critical point must be 0, and when that happens we have

$$f(0, y) = 0.$$

Hence the critical points do not occur in the interior of the square, only on a piece of boundary. Hence the maximum of the function must occur on the boundary.

This boundary consists of four segments, and we evaluate the function on these four segments to test where the maximum lies. The segments have been labeled S_1, through S_4.

The segment S_1 is the left vertical segment, with $x = 0$, and we have just seen that the value of f is 0 on this segment.

On the segment S_2, we have $y = 1$, and

$$f(x, 1) = x^2,$$

so the maximum occurs when $x = 1$, in which case $f(1, 1) = 1$.

On the segment S_3 we have $x = 1$, and

$$f(1, y) = y,$$

so the maximum occurs when $y = 1$, in which case $f(1, 1) = 1$ again.

On the segment S_4 we have $y = 0$, and

$$f(x, 0) = 0.$$

Putting it all together, we see that the maximum is at the point (1, 1), and the maximum value of f on the square is therefore

$$f(1, 1) = 1.$$

EXERCISES

Find the maximum and minimum points of the following functions in the indicated region

1. $x + y$ in the square with corners at $(\pm 1, \pm 1)$

2. (a) $x + y + z$ in the region $x^2 + y^2 + z^2 < 1$
   (b) $x + y$ in the region $x^2 + y^2 < 1$

3. $xy - (1 - x^2 - y^2)^{1/2}$ in the region $x^2 + y^2 \leq 1$

4. $x^3 y^2 (1 - x - y)$ in the region $x \geq 0$ and $y \geq 0$ (the first quadrant together with its boundary)

5. $(x^2 + 2y^2)e^{-(x^2 + y^2)}$ in the plane

6. (a) $(x^2 + y^2)^{-1}$ in the region $(x - 2)^2 + y^2 \leq 1$
   (b) $(x^2 + y^2)^{-1}$ in the region $x^2 + (y - 2)^2 \leq 1$

7. Which of the following functions have a maximum and which have a minimum in the whole plane?
   (a) $(x + 2y)e^{-x^2 - y^4}$ (b) e^{x-y}
   (c) $e^{x^2 - y^2}$ (d) $e^{x^2 + y^{10}}$
   (e) $(3x^2 + 2y^2)e^{-(4x^2 + y^2)}$ (f) $-x^2 e^{x^4 + y^{10}}$
   (g) $\begin{cases} \dfrac{x^2 + y^2}{|x| + |y|} & \text{if } (x, y) \neq (0, 0) \\ 0 & \text{if } (x, y) = (0, 0) \end{cases}$

8. Which is the point on the curve $(\cos t, \sin t, \sin(t/2))$ farthest from the origin?

In the following exercises, find the maximum of the function on the indicated square.

9. $f(x, y) = x^3 + xy$ on the square (Fig. 7):

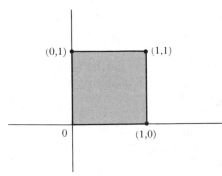

Figure 7

10. $f(x, y) = x^3 + xy$ on the square (Fig. 8):

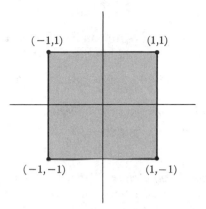

Figure 8

11. $f(x, y) = 3xy^3$ on the rectangle (Fig. 9):

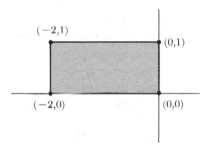

Figure 9

§3. LAGRANGE MULTIPLIERS

In this section, we shall investigate another method for finding the maximum or minimum of a function on some set of points. This method is particularly well adapted to the case when the set of points is described by means of an equation.

We shall work in 3-space. Let g be a differentiable function of three variables x, y, z. We consider the surface

$$g(X) = 0.$$

Let U be an open set containing this surface, and let f be a differentiable function defined for all points of U. We wish to find those points P on the surface $g(X) = 0$ such that $f(P)$ is a maximum or a minimum on the surface. In other words, we wish to find all points P such that $g(P) = 0$, and either

$$f(P) \geq f(X) \quad \text{for all } X \text{ such that } g(X) = 0,$$

or

$$f(P) \leqq f(X) \quad \text{for all } X \text{ such that} \quad g(X) = 0.$$

Any such point will be called an **extremum for f subject to the constraint g.**
In what follows, we consider only point P such that

$$g(P) = 0 \qquad \text{but grad } g(P) \neq O$$

Theorem 3. *Let g be a continuously differentiable function on an open set U. Let S be the set of points X in U such that $g(X) = 0$ but*

$$\text{grad } g(X) \neq O.$$

Let f be a continuously differentiable function on U and assume that P is a point of S such that P is an extremum for f on S. (In other words, P is an extremum for f, subject to the constraint g.) Then there exists a number λ such that

$$\text{grad } f(P) = \lambda \text{ grad } g(P).$$

Proof. Let $X(t)$ be a differentiable curve on the surface S passing through P, say $X(t_0) = P$. Then the function $f(X(t))$ has a maximum or a minimum at t_0. Its derivative

$$\frac{d}{dt} f(X(t))$$

is therefore equal to 0 at t_0. But this derivative is equal to

$$\text{grad } f(P) \cdot X'(t_0) = 0.$$

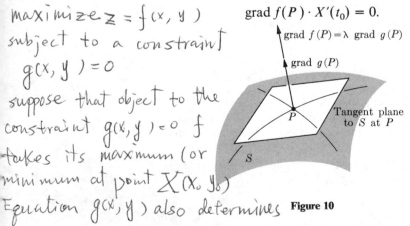

Figure 10

Hence grad $f(P)$ is perpendicular to every curve on the surface passing through P (Fig. 10). Under these circumstances, and the hypothesis that grad $g(P) \neq O$, there exists a number λ such that

$$\text{grad } f(P) = \lambda \text{ grad } g(P), \tag{1}$$

[Handwritten margin notes:]

maximize $z = f(x, y)$
subject to a constraint
$g(x, y) = 0$
suppose that object to the
constraint $g(x, y) = 0$ f
takes its maximum (or
minimum at point $X(x_0, y_0)$
Equation $g(x, y)$ also determines
a curve in the xy plane.
We know that $\nabla F(x_0, y_0)$ is orthogonal to c at (x_0, y_0), but
we also observe that $\nabla g(x_0, y_0)$ is also orthogonal to c
at (x_0, y_0) ∴ $\nabla g(x_0, y_0)$ and $\nabla f(x_0, y_0)$ are parallel

or in other words, grad $f(P)$ has the same, or opposite direction, as grad $g(P)$, provided it is not O. This is rather clear, since the direction of grad $g(P)$ is the direction perpendicular to the surface, and we have seen that grad $f(P)$ is also perpendicular to the surface.

Conversely, when we want to find an extremum point for f subject to the constraint g, we find all points P such that $g(P) = 0$, and such that relation (1) is satisfied. We can then find our extremum points among these by inspection.

(Note that this procedure is analogous to the procedure used to find maxima or minima for functions of one variable. We first determined all points at which the derivative is equal to 0, and then determined maxima or minima by inspection.)

Example 1. Find the maximum of the function $f(x, y) = x + y$ subject to the constraint $x^2 + y^2 = 1$.

Note: The constraint is the equation of a circle. Hence the problem can also be stated as: Find the maximum of the function $f(x, y) = x + y$ on the circle of radius 1.

We let $g(x, y) = x^2 + y^2 - 1$, so that S consists of all points (x, y) such that $g(x, y) = 0$. We have

$$\text{grad } f(x, y) = (1, 1),$$
$$\text{grad } g(x, y) = (2x, 2y).$$

Let (x_0, y_0) be a point for which there exists a number λ satisfying

$$\text{grad } f(x_0, y_0) = \lambda \text{ grad } g(x_0, y_0),$$

or in other words

$$1 = 2x_0\lambda \qquad \text{and} \qquad 1 = 2y_0\lambda.$$

Then $x_0 \neq 0$ and $y_0 \neq 0$. Hence $\lambda = 1/2x_0 = 1/2y_0$, and consequently $x_0 = y_0$. Since the point (x_0, y_0) must satisfy the equation $g(x_0, y_0) = 0$, we get the possibilities:

$$x_0 = \pm \frac{1}{\sqrt{2}} \qquad \text{and} \qquad y_0 = \pm \frac{1}{\sqrt{2}}.$$

It is then clear that $(1/\sqrt{2}, 1/\sqrt{2})$ is a maximum for f since the only other possibility $(-1/\sqrt{2}, -1/\sqrt{2})$ is a point at which f takes on a negative value, and $f(1/\sqrt{2}, 1/\sqrt{2}) = 2/\sqrt{2} > 0$.

Example 2. Find the extrema for the function $x^2 + y^2 + z^2$ subject to the constraint $x^2 + 2y^2 - z^2 - 1 = 0$. The function is the square of the distance from the origin, and the constraint defines a surface, so at a minimum for f, we are finding the point on the surface which is at minimum distance from the origin.

$$\therefore \quad \nabla f(x_0) = \lambda \nabla g(x_0)$$

Computing the partial derivatives of the functions f and g, we find that we must solve the system of equations

(a) $2x = \lambda \cdot 2x,$ (b) $2y = \lambda \cdot 4y,$
(c) $2z = \lambda \cdot (-2z),$ (d) $g(X) = x^2 + 2y^2 - z^2 - 1 = 0.$

Let (x_0, y_0, z_0) be a solution. If $z_0 \neq 0$, then from (c) we conclude that $\lambda = -1$. The only way to solve (a) and (b) with $\lambda = -1$ is that $x = y = 0$. In that case, from (d), we would get

$$z_0^2 = -1,$$

which is impossible. Hence any solution must have $z_0 = 0$.

If $x_0 \neq 0$, then from (a) we conclude that $\lambda = 1$. From (b) and (c) we then conclude that $y_0 = z_0 = 0$. From (d), we must have $x_0 = \pm 1$. In this manner, we have obtained two solutions satisfying our conditions, namely

$$(1, 0, 0) \quad \text{and} \quad (-1, 0, 0).$$

Similarly, if $y_0 \neq 0$, we find two more solutions, namely

$$\left(0, \sqrt{\tfrac{1}{2}}, 0\right) \quad \text{and} \quad \left(0, -\sqrt{\tfrac{1}{2}}, 0\right).$$

These four points are therefore the extrema of the function f subject to the constraint g.

If we ask for the minimum of f, then a direct computation shows that the last two points

$$\left(0, \pm\sqrt{\tfrac{1}{2}}, 0\right)$$

are the only possible solutions (because $1 > \tfrac{1}{2}$).

So far we have formulated the method of Lagrange multipliers in geometric terms, allowing us to find the extrema of a function on a surface. In some applications, e.g. economics, the problem is posed in different terms, as in the next example.

Example 3. Suppose a student has $90, with which he wants to buy lecture notes at $3 a piece, and also packs of beer costing $5 a pack. Suppose he buys x books and y packs of beer. To get maximum satisfaction out of his purchases, he wants the product xy to be maximum. How many of each should he buy?

The constraint imposed by the student's budget can be written down by the equation

(*) $3x + 5y = 90.$

So the problem is to maximize the function $f(x, y) = xy$ subject to the above constraint. For this we simply follow the previous pattern. Let

$$g(x, y) = 3x + 5y - 90.$$

Find the points on the sphere $x^2+y^2+z^2=1$ closed to and farthest from the point $(1,2,3)$

Then

$g(x,y,z)=x^2+y^2+z^2-1=0$

$f(x,y,z)=\sqrt{(x-1)^2+(y-2)^2+(z-3)^2}$

Find the maximum and minimum values of $f(x,y)=xy^2$ subject to the condition $x^2+y^2=1$

$$\operatorname{grad} g(x, y) = (3, 5)$$
$$\operatorname{grad} f(x, y) = (y, x).$$

$\left(-\frac{1}{\sqrt{3}},\sqrt{\frac{2}{3}}\right)$

The maximum occurs for values of λ such that

$$(y, x) = \lambda(3, 5) = (3\lambda, 5\lambda),$$

$\left(-\frac{1}{\sqrt{3}},-\sqrt{\frac{2}{3}}\right)$

so

$$y = 3\lambda \quad \text{and} \quad x = 5\lambda.$$

$\left(\frac{1}{\sqrt{3}},-\sqrt{\frac{2}{3}}\right)$

We substitute these values back in the constraint equation (*) to get

$$3 \cdot 5\lambda + 5 \cdot 3\lambda = 90.$$

$(1,0)$

Solving for λ yields $\lambda = 3$. Hence the extremum of f is at the point   *$(-1,0)$*

$$\lambda(5, 3) = 3(5, 3) = (15, 9).$$

The answer is that the student must buy 15 books and 9 packs of beer.

Note. The function $f(x, y) = xy$ which expresses the relation between how much satisfaction is derived from buying x units of one thing and y units of another is called the **utility function** by economists.

EXERCISES

Find the point on the plane $2x-3y+5z=19$ that is nearest the origin (using lagrange mult)

1. (a) Find the minimum of the function $x + y^2$ subject to the constraint

$$2x^2 + y^2 = 1.$$

   (b) Find its maximum.

2. Find the maximum value of $x^2 + xy + y^2 + yz + z^2$ on the sphere of radius 1.

3. Let $A = (1, 1, - 1)$, $B = (2, 1, 3)$, $C = (2, 0, - 1)$. Find the point at which the function

$$f(X) = (X - A)^2 + (X - B)^2 + (X - C)^2$$

reaches its minimum, and find the minimum value.

$f(x^2+y^2+z^2)$

$g=(x,y,z)$

$=2x-3y+5z+9$

4. Do Exercise 3 in general, for any three distinct vectors

$$A = (a_1, a_2, a_3), \quad B = (b_1, b_2, b_3), \quad C = (c_1, c_2, c_3).$$

5. Find the maximum of the function $3x^2 + 2\sqrt{2}\, xy + 4y^2$ on the circle of radius 3 in the plane.

6. Find the maximum of the functions xyz subject to the constraints $x \geq 0, y \geq 0,$ $z \geq 0$, and $xy + yz + xz = 2$.

来缚；掬束

7. Find the maximum and minimum distance from points on the curve

$$5x^2 + 6xy + 5y^2 = 0$$

to the origin in the plane.

8. Find the extreme values of the function $\cos^2 x + \cos^2 y$ subject to the constraint $x - y = \pi/4$ and $0 \leq x \leq \pi$.

9. Find the points on the surface $z^2 - xy = 1$ nearest to the origin.

10. Find the extreme values of the function xy subject to the condition $x + y = 1$.

11. Find the shortest distance between the point $(1, 0)$ and the curve $y^2 = 4x$.

12. Find the maximum and minimum points of the function

$$f(x, y, z) = x + y + z$$

in the region $x^2 + y^2 + z^2 \leq 1$.

13. Find the extremum values of the function $f(x, y, z) = x - 2y + 2z$ on the sphere $x^2 + y^2 + z^2 = 1$.

14. Find the maximum of the function $f(x, y, z) = x + y + z$ on the sphere $x^2 + y^2 + z^2 = 4$.

15. Find the extreme values of the function f given by $f(x, y, z) = xyz$ subject to the condition $x + y + z = 1$.

16. Find the extreme values of the function given by $f(x, y, z) = (x + y + z)^2$ subject to the condition $x^2 + 2y^2 + 3z^2 = 1$.

17. Find the minimum of the function $f(x, y, z) = x^2 + y^2 + z^2$ subject to the condition $3x + 2y - 7z = 5$.

18. In general, if a, b, c, d are numbers with not all of a, b, c equal to 0, find the minimum of the function $x^2 + y^2 + z^2$ subject to the condition

$$ax + by + cz = d.$$

19. Find the maximum and minimum value of the function

$$f(x, y) = x^2 + 2y^2 - x$$

on the disc of radius 1 centered at the origin.

20. Find the shortest distance from a point on the ellipse $x^2 + 4y^2 = 4$ to the line $x + y = 4$. [*Hint:* At a minimum, grad $f(x, y)$ is parallel to grad $g(x, y)$.]

21. A business has \$1 million to spend on three products, each costing an equal amount per unit. How much should be spent on each to maximize the utility, if the utility function is $f(x, y, z) = xyz$?

xy^2

$0 \quad (1, 0)$

$0 \quad (-1, 0)$

$x^2 + y^2 = 1 \quad \frac{2}{3\sqrt{3}} \quad (\frac{4}{\sqrt{3}}, \sqrt{\frac{5}{3}})$

$0 \quad (1, 0)$

$0 \quad (-1, 0)$

$(-\sqrt{\frac{1}{3}}, \sqrt{\frac{2}{3}}) \quad -\frac{2}{3\sqrt{3}}$

$-\frac{2}{3\sqrt{3}} \quad (-\frac{1}{\sqrt{3}}, \sqrt{\frac{2}{3}})$

$\frac{2}{3\sqrt{3}} \quad (\sqrt{\frac{1}{3}} - \sqrt{\frac{2}{3}})$

22. In working x hours at job A and y hours at job B, it can be determined that the satisfaction derived can be roughly expressed in terms of the function

$$f(x, y) = 2\sqrt{x} + \sqrt{y}.$$

How many hours should the person work on each job to maximize this function if the person works a total of 10 hours?

23. Suppose product A costs \$11 per unit and product B costs \$3 per unit. Both are needed to produce product C. When x units of A and y units of B are used, the total number of units of C produced by the production process is:

$$f(x, y) = -3x^2 + 10xy - 3y^2.$$

How many units of A and B should be used to produce 80 units of product C and minimize the costs?

24. A child has \$24 to spend. He can buy candy at \$2 per box and cakes at \$4 per cake. Assuming that his enjoyment as a result of buying x boxes of candy and y cakes is determined by the function

$$f(x, y) = \sqrt{x} + \sqrt{y},$$

find the numbers (x, y) which he should buy to maximize his enjoyment.

2. 条件极值 — 拉格朗日乘数法则.

为了要求函数 $z = f(x, y)$ 带有附加条件 $\varphi(x, y) = 0$ 时的条件极值, 先作一辅助函数 $F(x, y) = f(x, y) + \lambda \varphi(x, y)$

然后写出 (x, y) 取得无条件极值的必要条件 并将必要条件方程与附加条件方程联立, 得方程组

$$\begin{cases} f'_x(x, y) + \lambda \varphi'_x(x, y) = 0 \\ f'_y(x, y) + \lambda \varphi'_y(x, y) = 0 \\ \varphi(x, y) = 0 \end{cases}$$

例 求表面积为 S (限定的) 而体积最大的长方体的长、宽和高.

解: 设所求长方体的长宽高分别是 x, y, z 体积为

$$V = f(x, y, z) = xyz \quad (x > 0, y > 0, z > 0)$$

而附加条件是 $\varphi(x, y, z) = 2xy + 2yz + 2zx - S = 0$

作辅助函数 $F(x, y, z) = xyz + \lambda(2xy + 2yz + 2zx - S)$

令函数的下 (x, y, z) 的一阶偏微分为 0. 再加附加条件. 得方程组

$$\begin{cases} yz + 2\lambda(y + z) = 0 \cdots ⑩ \\ xz + 2\lambda(x + z) = 0 \cdots ⑪ \\ xy + 2\lambda(x + y) = 0 \cdots ⑫ \\ 2xy + 2yz + 2zx - S = 0 \cdots ⑬ \end{cases}$$

(⑩)×x − ⑪)·y 并化简得 $x = y$

(⑪)×x − ⑫)z 并化简得 $x = z$

代入 ⑬. 得 $6x^2 - S = 0$ 即 $x = \sqrt{\dfrac{S}{6}}$

于是求得了一个可解的驻值点, $x = y = z = \sqrt{\dfrac{S}{6}}$

因为问题本身存在着最大值 所以 $x = y = z = \sqrt{\dfrac{S}{6}}$

时也就是长方体为立方体时, 体积最大.

CHAPTER XII

Higher Derivatives

§1. THE FIRST TWO TERMS OF TAYLOR'S FORMULA

In the theory of functions of one variable, we derived an expression for the values of a function f near a point a by means of the derivatives of f at a, namely

$$f(a + h) = f(a) + f'(a)h + \frac{f''(a)}{2!}h^2 + R_3$$

where R_3 is a remainder term given by

$$R_3 = \frac{f^{(3)}(c)}{3!}h^3$$

for some number c between a and $a + h$. We shall now derive a similar formula for functions of two variables. The principle applies just as well to several variables, and also to higher order terms, which you can carry out easily if you understand induction. For our purposes, we are mostly interested in the first and second term of the formula.

We let

$$P = (a, b) \qquad \text{and} \qquad H = (h, k).$$

We assume that P is in an open set U and that f is a function on U all of whose partial derivatives up to order 3 exist and are continuous. We are interested in finding an expression

$$f(P + H) = f(P) + \text{???}$$

The idea is to reduce the problem to the one variable case. Thus we define the function

$$g(t) = f(P + tH) = f(a + th, b + tk)$$

277

for $0 \leq t \leq 1$. We assume that U contains all points $P + tH$ for $0 \leq t \leq 1$. Then

$$g(1) = f(P + H) \quad \text{and} \quad g(0) = f(P).$$

We can use Taylor's formula in one variable applied to the function g and we know that

$$g(1) = g(0) + g'(0) + \frac{g''(0)}{2!} + R_3.$$

Observe here that $g'(0)$ and $g''(0)$ should be multiplied by

$$(1 - 0) = 1,$$

so this factor does not show up explicitly in the present case. The remainder term R_3 has the form

$$R_3 = \frac{1}{3!} g^{(3)}(\tau)$$

for some number τ between 0 and 1. We shall now express $g'(t)$, $g''(t)$ and $g''(0)$ in terms of the partial derivatives of f, and thus obtain the first two terms of the Taylor formula for f itself.

First we have

$$g'(t) = \text{grad } f(P + tH) \cdot H$$
$$= D_1 f(P + tH)h + D_2 f(P + tH)k.$$

Hence

$$\boxed{g'(0) = D_1 f(a, b)h + D_2 f(a, b)k.}$$

If we let $x = a + th$ and $y = b + tk$, then we can also write the above in the form

(1) $$g'(t) = \frac{\partial f}{\partial x} \frac{\partial x}{\partial t} + \frac{\partial f}{\partial y} \frac{\partial y}{\partial t} = \frac{\partial f}{\partial x} h + \frac{\partial f}{\partial y} k.$$

For the second derivative, we must find the derivative with respect to t of each one of the functions $\partial f / \partial x$ and $\partial f / \partial y$. By the chain rule applied to each such function, we have:

(2)
$$\frac{d}{dt}\left(\frac{\partial f}{\partial x} \right) = \frac{\partial^2 f}{\partial x^2} \frac{dx}{dt} + \frac{\partial^2 f}{\partial y \, \partial x} \frac{dy}{dt} = \frac{\partial^2 f}{\partial x^2} h + \frac{\partial^2 f}{\partial y \, \partial x} k,$$
$$\frac{d}{dt}\left(\frac{\partial f}{\partial y} \right) = \frac{\partial^2 f}{\partial x \, \partial y} h + \frac{\partial^2 f}{\partial y^2} k.$$

Hence using (2) to take the derivative of (1), we find:

$$g''(t) = h\left[\frac{\partial^2 f}{\partial x^2}h + \frac{\partial^2 f}{\partial y \, \partial x}k\right] + k\left[\frac{\partial^2 f}{\partial x \, \partial y}h + \frac{\partial^2 f}{\partial y^2}k\right]$$

$$= h^2\frac{\partial^2 f}{\partial x^2} + 2hk\frac{\partial^2 f}{\partial x \, \partial y} + k^2\frac{\partial^2 f}{\partial y^2}.$$

Therefore

$$g''(0) = h^2\frac{\partial^2 f}{\partial x^2}(a, b) + 2hk\frac{\partial^2 f}{\partial x \, \partial y}(a, b) + k^2\frac{\partial^2 f}{\partial y^2}(a, b)$$

which I prefer to write in the form

$$\boxed{g''(0) = h^2 D_1^2 f(a, b) + 2hk D_1 D_2 f(a, b) + k^2 D_2^2 f(a, b).}$$

Hence finally, we can write the **Taylor Formula**

$$\boxed{\begin{aligned}f(a + h, b + k) &= f(a, b) + D_1 f(a, b)h + D_2 f(a, b)k \\ &\quad + \tfrac{1}{2}[h^2 D_1^2 f(a, b) + 2hk D_1 D_2 f(a, b) + k^2 D_2^2 f(a, b)] \\ &\quad + R_3.\end{aligned}}$$

The term

$$D_1 f(a, b) \, h + D_2 f(a, b) \, k$$

is called the **term of degree 1**. The second term is called the **term of degree 2** in Taylor's formula.

Example 1. Find the terms of degree ≤ 2 in the Taylor formula for the function $f(x, y) = \log(1 + x + 2y)$ at the point $(2, 1)$.

We compute the partial derivatives. They are:

$$f(2, 1) = \log 5,$$

$$D_1 f(x, y) = \frac{1}{1 + x + 2y}, \qquad D_1 f(2, 1) = \frac{1}{5} = \frac{\partial f}{\partial x}(2, 1),$$

$$D_2 f(x, y) = \frac{2}{1 + x + 2y}, \qquad D_2 f(2, 1) = \frac{2}{5} = \frac{\partial f}{\partial y}(2, 1),$$

$$D_1^2 f(x, y) = -\frac{1}{(1 + x + 2y)^2}, \qquad D_1^2 f(2, 1) = -\frac{1}{25} = \frac{\partial^2 f}{\partial x^2}(2, 1),$$

$$D_2^2 f(x, y) = -\frac{4}{(1 + x + 2y)^2}, \qquad D_2^2 f(2, 1) = -\frac{4}{25} = \frac{\partial^2 f}{\partial y^2}(2, 1),$$

$$D_1 D_2 f(x, y) = -\frac{2}{(1 + x + 2y)^2}, \qquad D_1 D_2 f(2, 1) = -\frac{2}{25} = \frac{\partial^2 f}{\partial x \, \partial y}(2, 1).$$

Hence

$$f(2 + h, 1 + k) = \log 5 + \left(\frac{1}{5}h + \frac{2}{5}k\right)$$
$$+ \frac{1}{2!}\left[-\frac{1}{25}h^2 - \frac{4}{25}hk - \frac{4}{25}k^2\right] + R_3.$$

When h, k are small, then R_3 is very small compared to the terms of degree 1 and 2 in the middle, so these terms give a good approximation to the function.

Just as we did in one variable, when we work with the point $P = (0, 0)$, and expand a function near the origin, then we write x, y instead of h, k, and in that case we may rewrite the **Taylor formula** as follows:

$$f(x, y) = f(0, 0) + D_1 f(0, 0)x + D_2 f(0, 0)y$$
$$+ \tfrac{1}{2}\left[D_1^2 f(0, 0)x^2 + 2D_1 D_2 f(0, 0)xy + D_2^2 f(0, 0)y^2\right]$$
$$+ R_3.$$

Example 2. Let $P = (0, 0)$. Find the Taylor formula for the function

$$f(x, y) = \log(1 + x + 2y).$$

We had computed the partial derivatives in general in Example 1. Here we substitute $(0, 0)$ to find:

$$D_1 f(0, 0) = 1 \qquad D_2 f(0, 0) = 2$$

and so forth. Then

$$f(x, y) = x + 2y - \tfrac{1}{2}\left[x^2 + 4xy + 4y^2\right] + R_3.$$

EXERCISES

Find the terms up to order 2 in the Taylor formula of the following functions (taking $P = O$).

1. $\sin(xy)$ 2. $\cos(xy)$ 3. $\log(1 + xy)$

4. $\sin(x^2 + y^2)$ 5. e^{x+y} 6. $\cos(x^2 + y)$

7. $(\sin x)(\cos y)$ 8. $e^x \sin y$ 9. $x + xy + 2y^2$

10. In each one of Exercises 1 through 9, find the terms of degree ≤ 2 in the Taylor expansion of the function at the indicated point.

1. $P = (1, \pi)$ 2. $P = (1, \pi)$ 3. $P = (2, 3)$

4. $P = (\sqrt{\pi}, \sqrt{\pi})$ 5. $P = (1, 2)$ 6. $P = (0, \pi)$

7. $P = (\pi/2, \pi)$ 8. $P = (2, \pi/4)$ 9. $P = (1, 1)$

§2. THE QUADRATIC TERM AT CRITICAL POINTS

If the point (a, b) is a critical point of f, that is,

$$D_1 f(a, b) = 0 \quad \text{and} \quad D_2 f(a, b) = 0,$$

then the terms involving the first power of h and k vanish, and the Taylor expansion involves only the terms having the second power of h, k, so that it reads:

$$\boxed{f(a + h, b + k) = f(a, b) + q(h, k) + R_3}$$

where

$$\boxed{q(h, k) = \tfrac{1}{2}\left[D_1^2 f(a, b)h^2 + 2D_1 D_2 f(a, b)hk + D_2^2 f(a, b)k^2 \right].}$$

At a critical point, this expression $q(h, k)$ is called the **quadratic form** associated with the function at the point (a, b). It is often convenient to rewrite $q(h, k)$ with the variables (x, y), so that

$$q(x, y) = \tfrac{1}{2}\left[D_1^2 f(a, b)x^2 + 2D_1 D_2 f(a, b)xy + D_2^2 f(a, b)y^2 \right].$$

Remark on notation. We have used (a, b) to denote the coordinates of the point P. On the other hand, in writing a quadratic expression in x, y it is also standard to use the letters a, b in the context

$$ax^2 + bxy + cy^2,$$

just when one derives the quadratic formula in elementary algebra, giving the roots of the expression

$$ax^2 + bx + c.$$

Clearly, we cannot use the letters a, b, c simultaneously for both purposes. Hence in a discussion when we want to discuss both a point and a quadratic form, we have to make different choices of letters. For instance, we could use the letters

$$P = (p_1, p_2)$$

for the coordinates of the point. In that case, the quadratic form becomes

$$q(x, y) = \tfrac{1}{2}\left[D_1^2 f(p_1, p_2)x^2 + 2D_1 D_2 f(p_1, p_2)xy + D_2^2 f(p_1, p_2)y^2 \right].$$

This merely amounts to changing the letters (a, b) to (p_1, p_2).

It is often the case that the origin itself is a critical point. Furthermore, we can always achieve this by a change of coordinates, e.g. by using the new coordinates

$$x' = x - p_1 \quad \text{and} \quad y' = y - p_2.$$

If $P = (0, 0)$ is the origin itself which is a critical point, then we have

$$\boxed{f(x, y) = f(0, 0) + q(x, y) + R_3}$$

where

$$\boxed{q(x, y) = \tfrac{1}{2}\left[D_1^2 f(O)x^2 + 2D_1 D_2 f(O)xy + D_2^2 f(O)y^2 \right].}$$

This function $q(x, y)$ is called the **quadratic form associated with f at the point** O, whenever O is a critical point of f.

Example 1. Let $f(x, y) = e^{-(x^2 + y^2)}$. Then it is a simple matter to verify that

$$\text{grad } f(0, 0) = O.$$

We let $P = (0, 0)$ be the origin. Standard computations show that

$$D_1^2 f(O) = -2, \qquad D_1 D_2 f(O) = 0, \qquad D_2^2 f(O) = -2.$$

Substituting these values in the general formula gives the expression for the quadratic form, namely

$$q(x, y) = -(x^2 + y^2).$$

In general, at a critical point P, from the expression

$$f(x, y) = f(P) + q(h, k) + R_3,$$

we see that the term R_3 is much smaller than $q(h, k)$ when h, k are small in absolute value.

Application to Local Maxima and Minima.

We say that the point P is a **local maximum** for the function if there exists some open disc U centered at P such that we have

$$f(P) \geq f(X) \qquad \text{for all} \quad X \text{ in } U.$$

Similarly we define a **local minimum** when $f(P) \leq f(X)$ for all X in U. Taking a small open disc U centered at P amounts to considering the value

$$f(x + h, y + k)$$

for small numbers h, k.

Let P be a critical point of f. Let the associated quadratic form be

$$q(x, y) = ax^2 + bxy + cy^2.$$

We say that q is **nonsingular** if $b^2 - 4ac \neq 0$. From the approximate expression $f(P) + q(h, k)$ with much smaller error term R_3, one can then deduce:

Provisional Theorem 1. *Let P be a critical point of the function f, and let q be the associated quadratic form. Assume that q is nonsingular. Then P is a local maximum for f if and only if $O = (0,0)$ is a local maximum for the quadratic form, and similarly for a local minimum.*

We shall not go into the details of the proof.

Note that the quadratic form has the property that for any number t we have

$$q(tx, ty) = t^2 q(x, y).$$

Consequently, if

$$0 = q(0, 0) \geqq q(x, y)$$

when (x, y) are close to $(0, 0)$, then it also follows that

$$0 \geqq q(x, y)$$

for all points (x, y), since multiplying such a point by a small value of t brings it near $(0, 0)$. Thus we may reformulate the provisional form of Theorem 1 as follows.

Theorem 1. *Let P be a critical point of the function f, and let q be the associated quadratic form. Assume that q is nonsingular. Then P is a local maximum for f if and only if we have*

$$q(x, y) \leqq 0 \quad \text{for all points } (x, y).$$

Similarly, P is a local minimum for f if and only if

$$q(x, y) \geqq 0 \quad \text{for all points } (x, y).$$

This theorem may be interpreted as the second derivative test in two variables, analogous to the second derivative test in one variable from the *First Course in Calculus.*

The conditions on the quadratic form in Theorem 1 occur so frequently that a name is given to them. If a quadratic form q has the property that

$$q(x, y) \geqq 0 \qquad \text{for all } (x, y),$$

it is called **positive**. If it has the property that

$$q(x, y) \leqq 0 \qquad \text{for all } (x, y),$$

it is called **negative**. Thus the second derivative test can be formulated for two variables with essentially the same language as for one variable. Let us view q as the "second derivative" of f. We can then formulate Theorem 1 as follows.

Let P be a critical point for the function f. Assume that the second derivative is nonsingular. If the second derivative is positive, then P is a local minimum. If the second derivative is negative, then P is a local maximum.

In the next examples, we graph the level curves of some standard quadratic forms, which allow us to recognize by inspection whether the origin is a maximum or minimum. In the next section, we analyze algebraically those conditions under which an arbitrary quadratic form is positive or negative.

We shall now describe the level curves for some quadratic forms to get an idea of their behavior near the origin.

Example 2. $q(x, y) = x^2 + y^2$. Then a graph of the function q and the level curves look like those in Figs. 1 and 2.

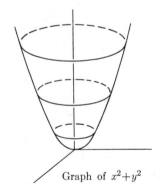

Graph of x^2+y^2

Figure 1

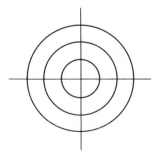

Level curves

Figure 2

In this example, we see that the origin $(0, 0)$ is a local minimum point for the form.

Example 3. $q(x, y) = -(x^2 + y^2)$. The graph and level curves look like Figs. 3 and 4:

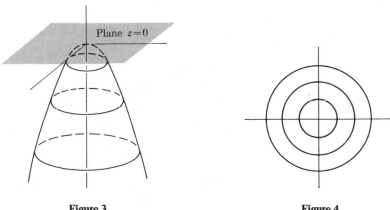

Plane $z=0$

Figure 3   Figure 4

The origin is a local maximum for the form.

Example 4. $q(x, y) = x^2 - y^2$. The level curves are then hyperbolas, determined for each number c by the equation $x^2 - y^2 = c$.

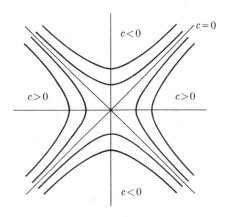

$c<0$ $c=0$

$c>0$ $c>0$

$c<0$

Figure 5

Of course, when $c = 0$, we get the two straight lines as shown (Fig. 5).

Example 5. $q(x, y) = xy$. The level curves look like the following (similar to the preceding example, but turned around):

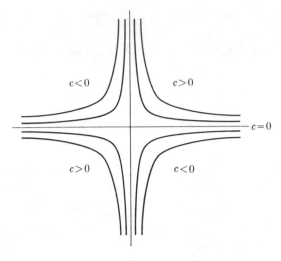

Figure 6

In Examples 4 and 5, we see that the origin, which is a critical point, is neither a local maximum nor local minimum. It is called a **saddle point**, because if you think of the graph of the function, it looks like a saddle.

In the next section, we study more general quadratic forms. The ones above are typical.

EXERCISES

1. Let $f(x, y) = 3x^2 - 4xy + y^2$. Show that the origin is a critical point of f.

2. (a) More generally, let a, b, c be numbers. Show that the function f given by $f(x, y) = ax^2 + bxy + cy^2$ has the origin as a critical point.
   (b) Find the quadratic form $q(x, y)$ associated with $f(x, y)$ at the point $(0, 0)$.

3. Find the quadratic form associated with the function $f(x, y)$ in the following cases, at the critical points P.
   (a) $x^2 + 4xy - y^2 - 8x - 6y$ (b) $x + y \sin x$
   (c) $(x + y) e^{-xy}$ (d) $x^2 y^2$
   (e) $x^4 + y^2$ (f) $(x - y)^4$
   (g) $x \sin y$ (h) $x^2 + 2y^2 - x$

4. Sketch the level curves for the following quadratic forms. Determine whether the origin is a local maximum, minimum, or neither.
   (a) $q(x, y) = 2x^2 - y^2$ (b) $q(x, y) = 3x^2 + 4y^2$
   (c) $q(x, y) = -(4x^2 + 5y^2)$ (d) $q(x, y) = y^2 - x^2$
   (e) $q(x, y) = 2y^2 - x^2$ (f) $q(x, y) = y^2 - 4x^2$
   (g) $q(x, y) = -(3x^2 + 2y^2)$ (h) $q(x, y) = 2xy$

§3. ALGEBRAIC STUDY OF A QUADRATIC FORM

In trying to determine whether a critical point is a maximum or minimum, we are led to study algebraic expressions like

$$q(x, y) = ax^2 + bxy + cy^2,$$

whose coefficients a, b, c are numbers. As we mentioned in the preceding section, such an expression is called a **quadratic form**. Its value at $(0, 0)$ is

$$q(0, 0) = 0.$$

It is easy to see that all the first partial derivatives vanish at the origin $(0, 0)$, i.e.

$$\frac{\partial q}{\partial x} \quad \text{and} \quad \frac{\partial q}{\partial y},$$

evaluated at $(0, 0)$ are equal to 0. Thus the origin is a critical point of $q(x, y)$.

We wish to determine whether the origin is a maximum, minimum, or neither (in which case it would be a saddle point).

First observe that on the line $y = 0$ we have the value

$$q(x, 0) = ax^2.$$

If $a \neq 0$, then $q(x, 0)$ is positive if $a > 0$ and negative if $a < 0$ for all values of $x \neq 0$ because $x^2 > 0$.

Similarly, on the line $x = 0$ we have the value

$$q(0, y) = cy^2.$$

A similar behavior occurs if $c \neq 0$. If both $a = c = 0$, then

$$q(x, y) = bxy$$

which is a hyperbola, and which we know how to graph.

Next we shall analyze the behavior of the quadratic form by pursuing the same idea that we used when we first sketched the level curves of a function, that is by looking at the values of the function on straight lines passing through the origin.

Assume $c \neq 0$. Consider the line

$$y = sx$$

where s is the slope, $s \neq 0$. Then

$$q(x, sx) = ax^2 + bsx^2 + cs^2x^2$$
$$= x^2(a + bs + cs^2).$$

For $x \neq 0$ this last expression differs by the positive factor x^2 from the polynomial

$$a + bs + cs^2.$$

Let

$$u = a + bs + cs^2.$$

In the (s, u)-plane, this is the equation of a parabola, which you should know how to graph from elementary school. The graph may look like any one of these (Fig. 7):

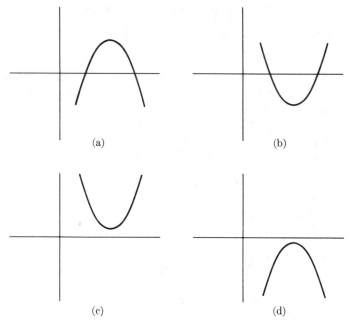

(a)

(b)

(c)

(d)

Figure 7

The roots of the quadratic equation

$$a + bs + cs^2 = 0$$

are given by the quadratic formula

$$s = \frac{-b \pm \sqrt{b^2 - 4ac}}{2c}.$$

If $b^2 - 4ac > 0$, then the graph looks like Fig. 7(a) and (b).

If $b^2 - 4ac < 0$, then the graph looks like Fig. 7(c) and (d).

We will not consider the case $b^2 - 4ac = 0$, where the graph would be tangent to the horizontal axis.

In any case, we see that the sign of the expression $b^2 - 4ac$ determines whether

$$a + bs + cs^2$$

takes on both positive and negative values.

If $b^2 - 4ac > 0$, then we can always find a slope s_1 such that

$$a + bs_1 + cs_1^2 < 0,$$

and we can find a slope s_2 such that

$$a + bs_2 + cs_2^2 > 0.$$

We see that on the lines with slope s_1 we have

$$q(x, s_1 x) < 0,$$

and on the lines with slope s_2 we have

$$q(x, s_2 x) > 0.$$

Therefore we conclude:

Case $c \neq 0$ and $b^2 - 4ac > 0$. The origin is neither a maximum nor a minimum. The quadratic form is neither positive nor negative.

On the other hand, suppose $b^2 - 4ac < 0$. Then for any slope s the sign of

$$q(x, sx)$$

will be always the same because the expression

$$a + bs + cs^2$$

is always negative or always positive (depending on whether $c < 0$ or $c > 0$).

Consequently we conclude:

Case $b^2 - 4ac < 0$.
  If $c < 0$, then

$$q(x, y) < 0 \qquad \text{for all } (x, y) \neq (0, 0)$$

and the origin is a maximum. The quadratic form is negative.
  If $c > 0$, then

$$q(x, y) > 0 \qquad \text{for all } (x, y) \neq (0, 0)$$

and the origin is a minimum. The quadratic form is positive.

Because of the symmetry between a and c in the quadratic form, these two cases also apply to the case when $a \neq 0$, replacing c by a.

Example 1. Let

$$q(x, y) = 3x^2 - 5xy + 7y^2.$$

Determine whether the origin is a maximum, minimum, or neither.
  Here we have

$$b^2 - 4ac = 25 - 4 \cdot 3 \cdot 7 = -59 < 0.$$

Since $c = 7 \neq 0$, we conclude that the origin is a minimum.

Example 2. Let

$$f(x, y) = \log(1 + x^2 + y^2).$$

Find whether the origin is a local maximum or minimum, or neither.
  We compute the first partial derivatives:

$$\frac{\partial f}{\partial x} = \frac{2x}{1 + x^2 + y^2} \qquad \text{and} \qquad \frac{\partial f}{\partial y} = \frac{2y}{1 + x^2 + y^2}.$$

We see that the origin is a critical point because

$$D_1 f(0, 0) = 0 \qquad \text{and} \qquad D_2 f(0, 0) = 0.$$

Now we compute the second partial derivatives:

$$\frac{\partial^2 f}{\partial x^2} = \frac{2(1 + x^2 + y^2) - (2x)(2x)}{(1 + x^2 + y^2)^2}$$

$$\frac{\partial^2 f}{\partial y^2} = \frac{2(1 + x^2 + y^2) - (2y)(2y)}{(1 + x^2 + y^2)^2}$$

$$\frac{\partial^2 f}{\partial x \, \partial y} = \frac{-(2x)(2y)}{(1 + x^2 + y^2)^2}.$$

Hence

$$D_1 D_2 f(0, 0) = 0$$

and

$$D_1^2 f(0, 0) = 2 = D_2^2 f(0, 0).$$

The quadratic form is

$$q(x, y) = \tfrac{1}{2}(2x^2 + 2y^2) = x^2 + y^2.$$

Either by inspection, or by noting that

$$b^2 - 4ac = -4 < 0$$

we conclude that the origin is a local minimum.

EXERCISES

Determine whether the following quadratic forms have a maximum, minimum, or neither at the origin.

1. $3x^2 - 4xy + y^2$
2. $-4x^2 + xy + 5y^2$
3. $6x^2 + xy - 2y^2$
4. $3x^2 + 7xy - y^2$
5. $2x^2 + 3xy + y^2$
6. $x^2 + 3xy + 4y^2$

7. Find all critical points of the function

$$f(x, y) = x^2 y + y^3 - y,$$

and determine whether they are local maxima, local minima, or saddle points.

8. Let $f(x, y) = x^3 + x^2 - y^3 + y^2$. Find all critical points of f and determine whether they are maxima, minima, or saddle points.

§4. PARTIAL DIFFERENTIAL OPERATORS

The main point of this section is to acquaint you with the idea that one can work with differential operators (having constant coefficients) just as one works with polynomials. This will be applied in the next section to Taylor's formula.

We let as usual D_1, D_2, D_3 be the partial derivatives with respect to the 3 variables under consideration. When dealing with two variables, we then just consider D_1, D_2.

In general, suppose that we are given three positive integers m_1, m_2, and m_3. We wish to take the repeated partial derivatives of f by using m_1 times the first partial D_1, using m_2 times the second partial D_2, and using m_3 times the third partial D_3. Then it does not matter in which order we take these partial derivatives, we shall always get the same answer.

To see this, note that by repeated application of Theorem 1 of Chapter III, §4, we can always interchange any occurrence of D_3 with D_2 or D_1 so as to push D_3 towards the right. We can perform such interchanges until all occurrences of D_3 occur furthest to the right, in the same way as we pushed D_3 towards the right going from expression (6) to expression (7). Once this is done, we start interchanging D_2 with D_1 until all occurrences of D_2 pile up just behind D_3. Once this is done, we are left with D_1 repeated a certain number of times on the left.

No matter with what arrangement of D_1, D_2, D_3 we started, we end up with the *same* arrangement, namely

$$\underbrace{D_1 \cdots D_1}_{m_1} \underbrace{D_2 \cdots D_2}_{m_2} \underbrace{D_3 \cdots D_3}_{m_3} f,$$

with D_1 occurring m_1 times, D_2 occurring m_2 times, and D_3 occurring m_3 times.

Exactly the same argument works for functions of more variables.

We shall now describe a notation for iterated derivatives, which generalizes the notation just given for two derivatives.

For simplicity, let us begin with functions of one variable x. We can then take only one type of derivative,

$$D = \frac{d}{dx}.$$

Let f be a function of one variable, and let us assume that all the iterated derivatives of f exist. Let m be a positive integer. Then we can take the m-th derivative of f, which we once denoted by $f^{(m)}$. We now write it

$$DD \cdots Df \quad \text{or} \quad \frac{d}{dx}\left(\frac{d}{dx} \cdots \left(\frac{df}{dx}\right) \cdots \right),$$

the derivative D (or d/dx) being iterated m times. What matters here is the number of times D occurs. We shall use the notation D^m or $(d/dx)^m$ to mean the iteration of D, m times. Thus we write

$$D^m f \qquad \text{or} \qquad \left(\frac{d}{dx}\right)^m f$$

instead of the above expressions. This is shorter. But even better, we have the rule

$$D^m D^n f = D^{m+n} f$$

for any positive integers m, n. So this iteration of derivatives begins to look like a multiplication. Furthermore, if we define $D^0 f$ to be simply f, then the rule above also holds if m, n are ≥ 0.

The expression D^m will be called a **simple differential operator of order** m (in one variable, so far).

Let us now look at the case of two variables, say (x, y). We can then take two partials D_1 and D_2 (or $\partial/\partial x$ and $\partial/\partial y$). Let m_1, m_2 be two integers ≥ 0. Instead of writing

$$\underbrace{D_1 \cdots D_1}_{m_1} \underbrace{D_2 \cdots D_2}_{m_2} f \qquad \text{or} \qquad \underbrace{\frac{\partial}{\partial x} \cdots \left(\frac{\partial}{\partial x}}_{m_1} \underbrace{\left(\frac{\partial}{\partial y} \cdots \left(\frac{\partial f}{\partial y}\right) \cdots \right)\right)}_{m_2},$$

we shall write

$$D_1^{m_1} D_2^{m_2} f \qquad \text{or} \qquad \left(\frac{\partial}{\partial x}\right)^{m_1} \left(\frac{\partial}{\partial y}\right)^{m_2} f.$$

For instance, taking $m_1 = 2$ and $m_2 = 5$ we would write

$$D_1^2 D_2^5 f.$$

This means: take the first partial twice and the second partial five times (in any order). (We assume throughout that all repeated partials exist and are continuous.)

An expression of type

$$D_1^{m_1} D_2^{m_2}$$

will be called a simple differential operator, and we shall say that its **order** is $m_1 + m_2$. In the example we just gave, the order is $5 + 2 = 7$.

It is now clear how to proceed with three or more variables.

If we deal with functions of 3 variables, all of whose repeated partial derivatives exist and are continuous in some open set U, and if D_1, D_2, D_3 denote the partial derivatives with respect to these variables, then we call an expression

$$D_1^{m_1} D_2^{m_2} D_3^{m_3} \qquad \text{or} \qquad \left(\frac{\partial}{\partial x_1}\right)^{m_1} \left(\frac{\partial}{\partial x_2}\right)^{m_2} \left(\frac{\partial}{\partial x_3}\right)^{m_3}$$

a **simple differential operator**, m_1, m_2, m_3 being integers ≥ 0. We say that its **order** is $m_1 + m_2 + m_3$.

Given a function f (satisfying the above stated conditions), and a simple differential operator D, we write Df to mean the function obtained from f by applying repeatedly the partial derivatives D_1, D_2, D_3, the number of times being the number of times each D_i occurs in D.

Example 1. Consider functions of three variables (x, y, z). Then

$$D = \left(\frac{\partial}{\partial x}\right)^3 \left(\frac{\partial}{\partial y}\right)^5 \left(\frac{\partial}{\partial z}\right)^2$$

is a simple differential operator of order $3 + 5 + 2 = 10$. Let f be a function of three variables satisfying the usual hypotheses. To take Df means that we take the partial derivative with respect to z twice, the partial with respect to y five times, and the partial with respect to x three times.

We observe that a simple differential operator gives us a rule which to each function f associates another function Df.

As a matter of notation, referring to Example 1, one would also write the differential operator D in the form

$$D = \frac{\partial^{10}}{\partial x^3\, \partial y^5\, \partial z^2}.$$

We shall show how one can add simple differential operators and multiply them by constants.

Let D, D' be two simple differential operators. For any function f we define $(D + D')f$ to be $Df + D'f$. If c is a number, then we define $(cD)f$ to be $c(Df)$. In this manner, taking iterated sums, and products with constants, we obtain what we shall call **differential operators**. Thus a **differential operator** D is a sum of terms of type

$$cD_1^{m_1}D_2^{m_2}D_3^{m_3},$$

where c is a number and m_1, m_2, m_3 are integers ≥ 0.

Example 2. Dealing with two variables, we see that

$$D = 3\frac{\partial}{\partial x} + 5\left(\frac{\partial}{\partial x}\right)^2 - \pi \frac{\partial}{\partial x}\frac{\partial}{\partial y}$$

is a differential operator. Let $f(x, y) = \sin(xy)$. We wish to find Df. We compute separately:

$$\frac{\partial f}{\partial x} = y\,\cos(xy) \qquad \frac{\partial^2 f}{\partial x^2} = y^2(-\sin(xy))$$

$$\frac{\partial}{\partial y}\frac{\partial f}{\partial x} = y(-\sin(xy))x + \cos xy.$$

Adding these with the appropriate numbers, we get:

$$Df(x, y) = 3\frac{\partial f}{\partial x} + 5\left(\frac{\partial}{\partial x}\right)^2 f - \pi \frac{\partial}{\partial x}\frac{\partial f}{\partial y}$$

$$= 3y\cos(xy) + 5(-y^2\sin(xy))$$
$$- \pi\left[y(-\sin(xy))x + \cos(xy)\right].$$

We see that a differential operator associates with each function f (satisfying the usual conditions) another function Df.

Let c be a number and f a function. Let D_i be any partial derivative. Then

$$D_i(cf) = cD_i f.$$

This is simply the old property that the derivative of a constant times a function is equal to the constant times the derivative of the function. Iterating partial derivatives, we see that this same property applies to differential operators. For any differential operator D, and any number c, we have

$$D(cf) = cDf.$$

Furthermore, if f, g are two functions (defined on the same open set, and having continuous partial derivatives of all orders), then for any partial derivative D_i, we have

$$D_i(f + g) = D_i f + D_i g.$$

Iterating the partial derivatives, we find that for any differential operator D, we have

$$D(f + g) = Df + Dg.$$

Having learned how to add differential operators, we now learn how to multiply them.

Let D, D' be two differential operators. Then we define the differential operator DD' to be the one obtained by taking first D' and then D. In other words, if f is a function, then

$$(DD')f = D(D'f).$$

Example 3. Let

$$D = 3\frac{\partial}{\partial x} + 2\frac{\partial}{\partial y} \qquad \text{and} \qquad D' = \frac{\partial}{\partial x} + 4\frac{\partial}{\partial y}.$$

Then

$$DD' = \left(3\frac{\partial}{\partial x} + 2\frac{\partial}{\partial y}\right)\left(\frac{\partial}{\partial x} + 4\frac{\partial}{\partial y}\right)$$

$$= 3\left(\frac{\partial}{\partial x}\right)^2 + 14\frac{\partial}{\partial x}\frac{\partial}{\partial y} + 8\left(\frac{\partial}{\partial y}\right)^2.$$

Differential operators multiply just like polynomials and numbers, and their addition and multiplication satisfy all the rules of addition and multiplication of polynomials. For instance:

If D, D' are two differential operators, then

$$DD' = D'D.$$

If D, D', D'' are three differential operators, then

$$D(D' + D'') = DD' + DD''.$$

It would be tedious to list all the properties here and to give in detail all the proofs (even though they are quite simple). We shall therefore omit these proofs. The main purpose of this section is to insure that you develop as great a facility in adding and multiplying differential operators as you have in adding and multiplying numbers or polynomials.

When a differential operator is written as a sum of terms of type

$$cD_1^{m_1}D_2^{m_2}D_3^{m_3},$$

then we shall say that it is in **standard form.**

For example,

$$3\left(\frac{\partial}{\partial x}\right)^2 + 14\frac{\partial}{\partial x}\frac{\partial}{\partial y} + 8\left(\frac{\partial}{\partial y}\right)^2$$

is in standard form, but

$$\left(3\frac{\partial}{\partial x} + 2\frac{\partial}{\partial y}\right)\left(\frac{\partial}{\partial x} + 4\frac{\partial}{\partial y}\right)$$

is not.

Each term

$$cD_1^{m_1}D_2^{m_2}D_3^{m_3}$$

is said to have degree $m_1 + m_2 + m_3$. If a differential operator is expressed as a sum of simple differential operators which all have the same degree, say m, then we say that it is **homogeneous** of degree m.

The differential operator of Example 2 is not homogeneous. The differential operator DD' of Example 3 is homogeneous of degree 2.

An important case of differential operators being applied to functions is that of monomials.

Example 4. Let $f(x, y) = x^3y^2$. Then

$$D_1 f(x, y) = 3x^2y^2, \qquad D_1^2 f(x, y) = 2 \cdot 3xy^2$$

$$D_1^3 f(x, y) = 6y^2, \qquad D_1^4 f(x, y) = 0.$$

Also observe that

$$D_1^3 D_2^2 f(x, y) = 3!2!.$$

Example 5. The generalization of the above example is as follows, and will be important for Taylor's formula. Let

$$f(x, y) = x^i y^j$$

be a monomial, with exponents $i, j \geq 0$. Then

$$D_1^i D_2^j f(x, y) = i! j!.$$

This is immediately verified, by differentiating x^i with respect to x, i times, thus getting rid of all powers of x; and differentiating y^j with respect to y, j times, thus getting rid of all powers of y.

On the other hand, let r, s be integers ≥ 0 such that $i \neq r$ or $j \neq s$. Then

$$D_1^r D_2^s f(0, 0) = 0.$$

To see this, suppose that $r \neq i$. If $r > i$, then differentiating r times the power x^i yields 0. If $r < i$, then differentiating r times the power x^i yields

$$i(i - 1) \cdots (i - r + 1) x^{i-r},$$

and $i - r > 0$. Substituting $x = 0$ yields 0. The same argument works if $j \neq s$.

EXERCISES

Put the following differential operators in standard form.

1. $(3D_1 + 2D_2)^2$

2. $(D_1 + D_2 + D_3)^2$

3. $(D_1 - D_2)(D_1 + D_2)$

4. $(D_1 + D_2)^2$

5. $(D_1 + D_2)^3$

6. $(D_1 + D_2)^4$

7. $(2D_1 - 3D_2)(D_1 + D_2)$

8. $(D_1 - D_3)(D_2 + 5D_3)$

9. $\left(\dfrac{\partial}{\partial x} + 4 \dfrac{\partial}{\partial y} \right)^3$

10. $\left(2 \dfrac{\partial}{\partial x} + \dfrac{\partial}{\partial y} \right)^2$

11. $\left(h \dfrac{\partial}{\partial x} + k \dfrac{\partial}{\partial y} \right)^2$

12. $\left(h \dfrac{\partial}{\partial x} + k \dfrac{\partial}{\partial y} \right)^3$

Find the values of the differential operator of Exercise 10 applied to the following functions at the given point.

13. $x^2 y$ at $(0, 1)$

14. xy at $(1, 1)$

15. $\sin(xy)$ at $(0, \pi)$

16. e^{xy} at $(0, 0)$

17. Compute $D_1^4 D_2^3 f(x, y)$ if $f(x, y)$ is
    (a) $x^5 y^4$ (b) $x^4 y^2$
    (c) $x^4 y^3$ (d) $10x^4 y^3$

18. Compute $D_1^7 D_2^9 f(0, 0)$ if $f(x, y)$ is
    (a) $x^8 y^7$ (b) $3x^7 y^9$
    (c) $11x^7 y^9$ (d) $25x^6 y^{11}$

19. Let $f(x, y) = 3x^2 y + 4x^3 y^4 - 7x^9 y^4$. Find
    (a) $D_1^3 D_2^4 f(0, 0)$ (b) $D_1^9 D_2^4 f(0, 0)$
    (c) $D_1^2 D_2 f(0, 0)$ (d) $D_1^3 D_2 f(0, 0)$

20. Let $f(x, y, z) = 4x^2 y z^3 - 5x^3 y^4 z + 7x^6 y^{10} z^7$. Find
    (a) $D_1^2 D_2 D_3^2 f(0, 0, 0)$ (b) $D_1^2 D_2 D_3^3 f(0, 0, 0)$
    (c) $D_1^6 D_2^{10} D_3^7 f(0, 0, 0)$ (d) $D_1^5 D_2 D_3 f(0, 0, 0)$

§5. THE GENERAL EXPRESSION FOR TAYLOR'S FORMULA

(This section is more theoretical than the others and should usually be omitted. We include it because it gives a neat expression for the Taylor formula, which avoids the orgy of indices which are usually present in the treatment for several variables. It may thus be used as a convenient reference for those who want to see how to handle the higher-order terms in the formula.)

Go back to §1, where we let

$$g(t) = f(P + tH) = f(a + th, b + tk).$$

We had found

(1) $$g'(t) = D_1 f(P + tH)h + D_2 f(P + tH)k.$$

If we keep on differentiating as we did a second time in §1, the expressions we shall get look more and more like a horrible mess. We just barely made it through the second derivative because we ended up with only three terms, involving the partial derivatives

$$D_1^2 f(a, b), \qquad D_1 D_2 f(a, b), \qquad D_2^2 f(a, b).$$

Thus to find higher-order derivatives, we have to figure out some notation which will eliminate the mess. This is done as follows.

We rewrite (1) in the form

$$g'(t) = h D_1 f(P + tH) + k D_2 f(P + tH).$$

The expression $h D_1 + k D_2$ looks like a dot product, and thus it is useful to abbreviate the notation and write

$$h D_1 + k D_2 = H \cdot \nabla .$$

With this abbreviation, our first derivative for g can then be written [from (1)]:

$$g'(t) = (H \cdot \nabla)f(P + tH).$$

The higher derivatives of g are determined similarly by induction.

Theorem 2. *Let r be a positive integer. Let f be a function defined on an open set U, and having continuous partial derivatives of orders $\leqq r$. Let P be a point of U, and H a vector. Let $g(t) = f(P + tH)$. Then*

$$g^{(r)}(t) = ((H \cdot \nabla)^r f)(P + tH)$$

for all values of t such that $P + tH$ lies in U.

Proof. The case $r = 1$ has already been verified. Suppose our formula proved for some integer r. Let $\psi = (H \cdot \nabla)^r f$. Then

$$g^{(r)}(t) = \psi(P + tH).$$

Hence by the case for $r = 1$ we get

$$g^{(r+1)}(t) = ((H \cdot \nabla)\psi)(P + tH).$$

Substituting the value for ψ yields

$$g^{(r+1)}(t) = ((H \cdot \nabla)^{r+1}f)(P + tH),$$

thus proving our theorem by induction.

In terms of the $\partial/\partial x$ and $\partial/\partial y$ notation, we see that

$$\boxed{g^{(r)}(t) = \left(h\frac{\partial}{\partial x} + k\frac{\partial}{\partial y}\right)^r f(P + tH).}$$

We repeat that this is equal to

$$\left(h\frac{\partial}{\partial x} + k\frac{\partial}{\partial y}\right)^r f$$

evaluated at the point $P + tH$.

Taylor's Formula. *Let f be a function defined on an open set U, and having continuous partial derivatives up to order r. Let P be a point of U, and H a vector. Assume that the line segment*

$$P + tH, \qquad 0 \leqq t \leqq 1,$$

is contained in U. Then there exists a number τ between 0 and 1 such that

$$f(P + H) = f(P) + \frac{(H \cdot \nabla)f(P)}{1!} + \cdots + \frac{(H \cdot \nabla)^{r-1}f(P)}{(r-1)!}$$

$$+ \frac{(H \cdot \nabla)^r f(P + \tau H)}{r!}.$$

Proof. This is obtained by plugging the expression for the derivatives of the function $g(t) = f(P + tH)$ into the Taylor formula for one variable. We see that

$$g^{(s)}(0) = (H \cdot \nabla)^s f(P)$$

and

$$g^{(r)}(\tau) = (H \cdot \nabla)^r f(P + \tau H).$$

This proves Taylor's formula as stated.

Rewritten in terms of the $\partial/\partial x$ and $\partial/\partial y$ notation, we have

$$f(a + h, b + k) = f(a, b) + \frac{1}{1!}\left(h\frac{\partial}{\partial x} + k\frac{\partial}{\partial y}\right)f(a, b) + \cdots$$

$$+ \frac{1}{(r-1)!}\left(h\frac{\partial}{\partial x} + k\frac{\partial}{\partial y}\right)^{r-1}f(a, b)$$

$$+ \frac{1}{r!}\left(h\frac{\partial}{\partial x} + k\frac{\partial}{\partial y}\right)^r f(a + \tau h, b + \tau k).$$

The powers of the differential operators

$$\left(h\frac{\partial}{\partial x} + k\frac{\partial}{\partial y}\right)^s$$

are found by the usual binomial expansion. For instance:

$$\left(h\frac{\partial}{\partial x} + k\frac{\partial}{\partial y}\right)^2 = h^2\frac{\partial^2}{\partial x^2} + 2hk\frac{\partial^2}{\partial x\,\partial y} + k^2\frac{\partial^2}{\partial y^2},$$

$$\left(h\frac{\partial}{\partial x} + k\frac{\partial}{\partial y}\right)^3 = h^3\left(\frac{\partial}{\partial x}\right)^3 + 3h^2k\left(\frac{\partial}{\partial x}\right)^2\left(\frac{\partial}{\partial y}\right)$$

$$+ 3hk^2\left(\frac{\partial}{\partial x}\right)\left(\frac{\partial}{\partial y}\right)^2 + k^3\left(\frac{\partial}{\partial y}\right)^3.$$

In many cases, we take $P = O$ and we wish to approximate $f(x, y)$ by a polynomial in x, y. Thus we let $H = (x, y)$. In that case, the notation $\partial/\partial x$ and $\partial/\partial y$ becomes even worse than usual since it is not entirely clear in taking the square

$$\left(x\frac{\partial}{\partial x} + y\frac{\partial}{\partial y}\right)^2$$

what is to be treated as a constant and what is not. Thus it is better to write

$$(xD_1 + yD_2)^2,$$

and similarly for higher powers. We then obtain a polynomial expression for f, with a remainder term. The terms of degree ≤ 3 are as follows:

$f(x, y) =$

$f(0, 0) + D_1 f(0, 0)x + D_2 f(0, 0)y$

$+ \dfrac{1}{2!}\left[D_1^2 f(0, 0)x^2 + 2D_1 D_2 f(0, 0)xy + D_2^2 f(0, 0)y^2\right]$

$+ \dfrac{1}{3!}\left[D_1^3 f(0, 0)x^3 + 3D_1^2 D_2 f(0, 0)x^2 y + 3D_1 D_2^2 f(0, 0)xy^2 + D_2^3 f(0, 0)y^3\right]$

$+ R_4.$

In general, the Taylor formula gives us an expression

$$f(x, y) = f(0, 0) + G_1(x, y) + \cdots + G_{r-1}(x, y) + R_r,$$

where $G_d(x, y)$ is a homogeneous polynomial in x, y of degree d, and R_r is the remainder term. We call

$$f(0, 0) + G_1(x, y) + \cdots + G_s(x, y)$$

the **polynomial approximation of** f, of degree $\leq s$.
    We write polynomials in one variable as sums

$$\sum_{i=0}^{n} c_i x^i = c_0 + c_1 x + \cdots + c_n x^n.$$

In a similar way, we can write polynomials in several variables,

$$G(x, y) = \sum_{i=0}^{n} \sum_{j=0}^{m} c_{ij} x^i y^j.$$

Let r, s be a pair of integers ≥ 0. Then

$$D_1^r D_2^s G(0, 0) = r!s!c_{rs},$$

by the example at the end of §2. Hence we have a simple expression for the coefficients of the polynomial,

$$c_{ij} = \frac{D_1^i D_2^j G(0, 0)}{i! j!}.$$

On the other hand, from the binomial expansion

$$(xD_1 + yD_2)^m = \sum_{i=0}^{m} \binom{m}{i} x^i y^{m-i} D_1^i D_2^{m-i},$$

and the value of the binomial coefficient,

$$\binom{m}{i} = \frac{m!}{i! \, (m - i)!},$$

we find that

$$\frac{(xD_1 + yD_2)^m}{m!} = \sum_{i=0}^{m} \frac{x^i y^{m-i}}{i! \, (m - i)!} D_1^i D_2^{m-i}.$$

Consequently,

$$\frac{(xD_1 + yD_2)^m f(0, 0)}{m!} = \sum_{i=0}^{m} c_{ij} x^i y^j = G_m(x, y)$$

is a polynomial in x, y, such that $i + j = m$, so all its monomials have the same degree, and the coefficients are given by

(∗)

$$c_{ij} = \frac{D_1^i D_2^j f(0, 0)}{i! j!}.$$

The general Taylor polynomial of degree $\leq s$ is therefore of the form

$$G(x, y) = \sum_{i+j \leq s} c_{ij} x^i y^j,$$

where the coefficients c_{ij} are given by the above formula (∗). Again, Example 4 at the end of §4 shows that the partial derivatives up to total order s of this polynomial coincide with the derivatives of f, when evaluated at $(0, 0)$. Thus we may say:

> *The Taylor polynomial of a function f up to order s is that polynomial having the same partial derivatives as the function up to order s, when evaluated at* $(0, 0)$.

EXERCISE

Let f be a function of two variables. Assume that $f(O) = 0$ and also that $f(ta, tb) = t^2 f(a, b)$ for all numbers t and all vectors (a, b). Show that for all points $P = (a, b)$ we have

$$f(P) = \frac{(P \cdot \nabla)^2 f(O)}{2!}.$$

CHAPTER XIII

Matrices

§1. MATRICES

We consider a new kind of object, matrices.

Let n, m be two integers ≥ 1. An array of numbers

$$\begin{pmatrix} a_{11} & a_{12} & a_{13} & \cdots & a_{1n} \\ a_{21} & a_{22} & a_{23} & \cdots & a_{2n} \\ \vdots & \vdots & \vdots & & \vdots \\ a_{m1} & a_{m2} & a_{m3} & \cdots & a_{mn} \end{pmatrix}$$

is called a **matrix.** We can abbreviate the notation for this matrix by writing it (a_{ij}), $i = 1, \ldots, m$ and $j = 1, \ldots, n$. We say that it is an m by n matrix, or an $m \times n$ matrix. The matrix has m **rows** and n **columns.** For instance, the first column is

$$\begin{pmatrix} a_{11} \\ a_{21} \\ \vdots \\ a_{m1} \end{pmatrix}$$

and the second row is $(a_{21}, a_{22}, \ldots, a_{2n})$. We call a_{ij} the ij-**entry** or ij-**component** of the matrix.

If you look back at Chapter I, §1, the example of 7-space taken from economics gives rise to a 7×7 matrix (a_{ij}) $(i, j = 1, \ldots, 7)$, where a_{ij} is the amount spent by the i-th industry on the j-th industry. Thus keeping the notation of that example, if $a_{25} = 50$, this means that the auto industry bought 50 million dollars' worth of stuff from the chemical industry during the given year.

Example 1. The following is a 2×3 matrix:

$$\begin{pmatrix} 1 & 1 & -2 \\ -1 & 4 & -5 \end{pmatrix}.$$

It has two rows and three columns.

The rows are $(1, 1, -2)$ and $(-1, 4, -5)$. The columns are

$$\begin{pmatrix} 1 \\ -1 \end{pmatrix}, \quad \begin{pmatrix} 1 \\ 4 \end{pmatrix}, \quad \begin{pmatrix} -2 \\ -5 \end{pmatrix}.$$

Thus the rows of a matrix may be viewed as n-tuples, and the columns may be viewed as vertical m-tuples. A vertical m-tuple is also called a **column vector**.

A vector $(x_1, \ldots, x_n)$ is a $1 \times n$ matrix. A column vector

$$\begin{pmatrix} x_1 \\ \vdots \\ x_n \end{pmatrix}$$

is an $n \times 1$ matrix.

When we write a matrix in the form (a_{ij}), then i denotes the row and j denotes the column. In Example 1, we have for instance $a_{11} = 1$, $a_{23} = -5$.

A single number (a) may be viewed as a 1×1 matrix.

Let (a_{ij}), $i = 1, \ldots, m$ and $j = 1, \ldots, n$ be a matrix. If $m = n$, then we say that it is a **square** matrix. Thus

$$\begin{pmatrix} 1 & 2 \\ -1 & 0 \end{pmatrix} \quad \text{and} \quad \begin{pmatrix} 1 & -1 & 5 \\ 2 & 1 & -1 \\ 3 & 1 & -1 \end{pmatrix}$$

are both square matrices.

We have a **zero matrix**, in which $a_{ij} = 0$ for all i, j. It looks like this:

$$\begin{pmatrix} 0 & 0 & 0 & \cdots & 0 \\ 0 & 0 & 0 & \cdots & 0 \\ \vdots & \vdots & \vdots & & \vdots \\ 0 & 0 & 0 & \cdots & 0 \end{pmatrix}.$$

We shall write it O. We note that we have met so far with the zero number, zero vector, and zero matrix.

We shall now define addition of matrices and multiplication of matrices by numbers.

We define addition of matrices only when they have the same size. Thus let m, n be fixed integers ≥ 1. Let $A = (a_{ij})$ and $B = (b_{ij})$ be two $m \times n$ matrices. We define $A + B$ to be the matrix whose entry in the i-th row and j-th column is $a_{ij} + b_{ij}$. In other words, we add matrices of the same size componentwise.

Example 2. Let

$$A = \begin{pmatrix} 1 & -1 & 0 \\ 2 & 3 & 4 \end{pmatrix} \quad \text{and} \quad B = \begin{pmatrix} 5 & 1 & -1 \\ 2 & 1 & -1 \end{pmatrix}$$

Then

$$A + B = \begin{pmatrix} 6 & 0 & -1 \\ 4 & 4 & 3 \end{pmatrix}.$$

If A, B are both $1 \times n$ matrices, i.e. n-tuples, then we note that our addition of matrices coincides with the addition which we defined in Chapter I for n-tuples.

If O is the zero matrix, then for any matrix A (of the same size, of course), we have $O + A = A + O = A$. This is trivially verified.

We shall now define the multiplication of a matrix by a number. Let c be a number, and $A = (a_{ij})$ be a matrix. We define cA to be the matrix whose ij-component is ca_{ij}. We write $cA = (ca_{ij})$. Thus we multiply each component of A by c.

Example 3. Let A, B be as in Example 2. Let $c = 2$. Then

$$2A = \begin{pmatrix} 2 & -2 & 0 \\ 4 & 6 & 8 \end{pmatrix} \quad \text{and} \quad 2B = \begin{pmatrix} 10 & 2 & -2 \\ 4 & 2 & -2 \end{pmatrix}.$$

For any matrix A we let $-A$ be the matrix obtained by multiplying each component of A with -1. If $A = (a_{ij})$, then

$$-A = (-1)A = (-a_{ij}).$$

For instance, if

$$A = \begin{pmatrix} 1 & -1 & 0 \\ 2 & 3 & 4 \end{pmatrix}$$

is the matrix of Example 2, then

$$-A = (-1)A = \begin{pmatrix} -1 & 1 & 0 \\ -2 & -3 & -4 \end{pmatrix}.$$

Observe that for any matrix A we have

$$A + (-A) = A - A = O.$$

We define one more notion related to a matrix. Let $A = (a_{ij})$ be an $m \times n$ matrix. The $n \times m$ matrix $B = (b_{ji})$ such that $b_{ji} = a_{ij}$ is called the **transpose** of A, and is also denoted by ^tA. Taking the transpose of a matrix amounts to changing rows into columns and vice versa. If A is the matrix which we wrote down at the beginning of this section, then ^tA is the matrix

$$\begin{pmatrix} a_{11} & a_{21} & a_{31} & \cdots & a_{m1} \\ a_{12} & a_{22} & a_{32} & \cdots & a_{m2} \\ \vdots & \vdots & \vdots & & \vdots \\ a_{1n} & a_{2n} & a_{3n} & \cdots & a_{mn} \end{pmatrix}.$$

To take a special case:

$$\text{If } A = \begin{pmatrix} 2 & 1 & 0 \\ 1 & 3 & 5 \end{pmatrix}, \quad \text{then } {}^tA = \begin{pmatrix} 2 & 1 \\ 1 & 3 \\ 0 & 5 \end{pmatrix}.$$

If $A = (2, 1, -4)$ is a **row vector**, then

$$^tA = \begin{pmatrix} 2 \\ 1 \\ -4 \end{pmatrix}$$

is a **column vector**.

EXERCISES

1. Let
$$A = \begin{pmatrix} 1 & 2 & 3 \\ -1 & 0 & 2 \end{pmatrix} \quad \text{and} \quad B = \begin{pmatrix} -1 & 5 & -2 \\ 1 & 1 & -1 \end{pmatrix}.$$
Find $A + B, 3B, -2B, A + 2B, 2A + B, A - B, A - 2B, B - A$.

2. Let
$$A = \begin{pmatrix} 1 & -1 \\ 2 & 1 \end{pmatrix} \quad \text{and} \quad B = \begin{pmatrix} -1 & 1 \\ 0 & -3 \end{pmatrix}.$$
Find $A + B, 3B, -2B, A + 2B, A - B, B - A$.

3. In Exercise 1, find ^tA and ^tB.

4. In Exercise 2, find ^tA and ^tB.

5. If A, B are arbitrary $m \times n$ matrices, show that

$$^t(A + B) = {}^tA + {}^tB.$$

6. If c is a number, show that $'(cA) = c'A$.

7. If $A = (a_{ij})$ is a square matrix, then the elements a_{ii} are called the **diagonal** elements. How do the diagonal elements of A and $'A$ differ?

8. Find $'(A + B)$ and $'A + 'B$ in Exercise 2.

9. Find $A + 'A$ and $B + 'B$ in Exercise 2.

10. A matrix A is said to be **symmetric** if $A = 'A$. Show that for any square matrix A, the matrix $A + 'A$ is symmetric.

11. Write down the row vectors and column vectors of the matrices A, B in Exercise 1.

12. Write down the row vectors and column vectors of the matrices A, B in Exercise 2.

§2. MULTIPLICATION OF MATRICES

We shall now define the product of matrices. Let $A = (a_{ij})$, $i = 1, \ldots, m$ and $j = 1, \ldots, n$ be an $m \times n$ matrix. Let $B = (b_{jk})$, $j = 1, \ldots, n$ and $k = 1, \ldots, s$ be an $n \times s$ matrix.

$$A = \begin{pmatrix} a_{11} & \cdots & a_{1n} \\ \vdots & & \vdots \\ a_{m1} & \cdots & a_{mn} \end{pmatrix}, \quad B = \begin{pmatrix} b_{11} & \cdots & b_{1s} \\ \vdots & & \vdots \\ b_{n1} & \cdots & b_{ns} \end{pmatrix}.$$

We define the product AB to be the $m \times s$ matrix whose ik-coordinate is

$$\sum_{j=1}^{n} a_{ij} b_{jk} = a_{i1} b_{1k} + a_{i2} b_{2k} + \cdots + a_{in} b_{nk}.$$

If $A_1, \ldots, A_m$ are the row vectors of the matrix A, and if $B^1, \ldots, B^s$ are the column vectors of the matrix B, then the ik-coordinate of the product AB is equal to $A_i \cdot B^k$. Thus

$$AB = \begin{pmatrix} A_1 \cdot B^1 & \cdots & A_1 \cdot B^s \\ \vdots & & \vdots \\ A_m \cdot B^1 & \cdots & A_m \cdot B^s \end{pmatrix}.$$

Multiplication of matrices is therefore a generalization of the dot product.

Example 1. Let

$$A = \begin{pmatrix} 2 & 1 & 5 \\ 1 & 3 & 2 \end{pmatrix}, \quad B = \begin{pmatrix} 3 & 4 \\ -1 & 2 \\ 2 & 1 \end{pmatrix}.$$

Then AB is a 2×2 matrix, and computations show that

$$AB = \begin{pmatrix} 2 & 1 & 5 \\ 1 & 3 & 2 \end{pmatrix} \begin{pmatrix} 3 & 4 \\ -1 & 2 \\ 2 & 1 \end{pmatrix} = \begin{pmatrix} 15 & 15 \\ 4 & 12 \end{pmatrix}.$$

Example 2. Let

$$C = \begin{pmatrix} 1 & 3 \\ -1 & -1 \end{pmatrix}.$$

Let A, B be as in Example 1. Then

$$BC = \begin{pmatrix} 3 & 4 \\ -1 & 2 \\ 2 & 1 \end{pmatrix} \begin{pmatrix} 1 & 3 \\ -1 & -1 \end{pmatrix} = \begin{pmatrix} -1 & 5 \\ -3 & -5 \\ 1 & 5 \end{pmatrix}$$

and

$$A(BC) = \begin{pmatrix} 2 & 1 & 5 \\ 1 & 3 & 2 \end{pmatrix} \begin{pmatrix} -1 & 5 \\ -3 & -5 \\ 1 & 5 \end{pmatrix} = \begin{pmatrix} 0 & 30 \\ -8 & 0 \end{pmatrix}.$$

Compute $(AB)C$. What do you find?

Let A be an $m \times n$ matrix and let B be an $n \times 1$ matrix, i.e. a column vector. Then AB is again a column vector. The product looks like this:

$$\begin{pmatrix} a_{11} & \cdots & a_{1n} \\ \vdots & & \vdots \\ a_{m1} & \cdots & a_{mn} \end{pmatrix} \begin{pmatrix} b_1 \\ \vdots \\ b_n \end{pmatrix} = \begin{pmatrix} c_1 \\ \vdots \\ c_m \end{pmatrix}$$

where

$$c_i = \sum_{j=1}^{n} a_{ij}b_j = a_{i1}b_1 + \cdots + a_{in}b_n.$$

If $X = (x_1, \ldots, x_m)$ is a row vector, i.e. a $1 \times m$ matrix, then we can form the product XA, which looks like this:

$$(x_1, \ldots, x_m) \begin{pmatrix} a_{11} & \cdots & a_{1n} \\ \vdots & & \vdots \\ a_{m1} & \cdots & a_{mn} \end{pmatrix} = (y_1, \ldots, y_n),$$

where

$$y_k = x_1 a_{1k} + \cdots + x_m a_{mk}.$$

In this case, XA is a $1 \times n$ matrix, i.e. a row vector.

If A is a square matrix, then we can form the product AA, which will be a square matrix of the same size as A. It is denoted by A^2. Similarly, we can form A^3, A^4, and in general, A^n for any positive integer n.

We define the unit $n \times n$ matrix to be the matrix having diagonal components all equal to 1, and all other components equal to 0. Thus the unit $n \times n$ matrix, denoted by I_n, looks like this:

$$\begin{pmatrix} 1 & 0 & 0 & \cdots & 0 \\ 0 & 1 & 0 & \cdots & 0 \\ 0 & 0 & 1 & \cdots & 0 \\ \vdots & \vdots & \vdots & \ddots & \vdots \\ 0 & 0 & 0 & & 1 & 0 \\ 0 & 0 & 0 & \cdots & 1 \end{pmatrix}.$$

We can then define $A^0 = I$ (the unit matrix of the same size as A).

Theorem 1. *Let A, B, C be matrices. Assume that A, B can be multiplied, and A, C can be multiplied, and B, C can be added. Then A, $B + C$ can be multiplied, and we have*

$$A(B + C) = AB + AC.$$

If x is a number, then

$$A(xB) = x(AB).$$

Proof. Let A_i be the i-th row of A, and let B^k, C^k be the k-th column of B and C respectively. Then $B^k + C^k$ is the k-th column of $B + C$. By definition, the ik-component of AB is $A_i \cdot B^k$, the ik-component of AC is $A_i \cdot C^k$, and the ik-component of $A(B + C)$ is $A_i \cdot (B^k + C^k)$. Since

$$A_i \cdot (B^k + C^k) = A_i \cdot B^k + A_i \cdot C^k,$$

our first assertion follows. As for the second, observe that the k-th column of xB is xB^k. Since

$$A_i \cdot xB^k = x(A_i \cdot B^k),$$

our second assertion follows.

Theorem 2. *Let A, B, C be matrices such that A, B can be multiplied and B, C can be multiplied. Then A, BC can be multiplied, so can AB, C, and we have*

$$(AB)C = A(BC).$$

Proof. Let $A = (a_{ij})$ be an $m \times n$ matrix, let $B = (b_{jk})$ be an $n \times r$ matrix, and let $C = (c_{kl})$ be an $r \times s$ matrix. The product AB is an $m \times r$ matrix, whose

ik-component is equal to the sum

$$a_{i1}b_{1k} + a_{i2}b_{2k} + \cdots + a_{in}b_{nk}.$$

We shall abbreviate this sum using our Σ notation by writing

$$\sum_{j=1}^{n} a_{ij}b_{jk}.$$

By definition, the il-component of $(AB)C$ is equal to

$$\sum_{k=1}^{r}\left[\sum_{j=1}^{n} a_{ij}b_{jk}\right] c_{kl} = \sum_{k=1}^{r}\left[\sum_{j=1}^{n} a_{ij}b_{jk}c_{kl}\right].$$

The sum on the right can also be described as the sum of all terms

$$a_{ij}b_{jk}c_{kl},$$

where j, k range over all integers $1 \leq j \leq n$ and $1 \leq k \leq r$ respectively.

If we had started with the jl-component of BC and then computed the il-component of $A(BC)$ we would have found exactly the same sum, thereby proving the theorem.

EXERCISES

1. Let I be the unit $n \times n$ matrix. Let A be an $n \times r$ matrix. What is IA? If A is an $m \times n$ matrix, what is AI?

2. Let O be the matrix all of whose coordinates are 0. Let A be a matrix of a size such that the product AO is defined. What is AO?

3. In each one of the following cases, find $(AB)C$ and $A(BC)$.

(a) $A = \begin{pmatrix} 2 & 1 \\ 3 & 1 \end{pmatrix}$, $B = \begin{pmatrix} -1 & 1 \\ 1 & 0 \end{pmatrix}$, $C = \begin{pmatrix} 1 & 4 \\ 2 & 3 \end{pmatrix}$

(b) $A = \begin{pmatrix} 2 & 1 & -1 \\ 3 & 1 & 2 \end{pmatrix}$, $B = \begin{pmatrix} 1 & 1 \\ 2 & 0 \\ 3 & -1 \end{pmatrix}$, $C = \begin{pmatrix} 1 \\ 3 \end{pmatrix}$

(c) $A = \begin{pmatrix} 2 & 4 & 1 \\ 3 & 0 & -1 \end{pmatrix}$, $B = \begin{pmatrix} 1 & 1 & 0 \\ 2 & 1 & -1 \\ 3 & 1 & 5 \end{pmatrix}$, $C = \begin{pmatrix} 1 & 2 \\ 3 & 1 \\ -1 & 4 \end{pmatrix}$

4. Let A, B be square matrices of the same size, and assume that $AB = BA$. Show that $(A + B)^2 = A^2 + 2AB + B^2$, and

$$(A + B)(A - B) = A^2 - B^2,$$

using the properties of matrices stated in Theorem 1.

5. Let

$$A = \begin{pmatrix} 1 & 2 \\ 3 & -1 \end{pmatrix}, \qquad B = \begin{pmatrix} 2 & 0 \\ 1 & 1 \end{pmatrix}.$$

Find AB and BA.

6. Let

$$C = \begin{pmatrix} 7 & 0 \\ 0 & 7 \end{pmatrix}.$$

Let A, B be as in Exercise 5. Find CA, AC, CB, and BC. State the general rule including this exercise as a special case.

7. Let $X = (1, 0, 0)$ and let

$$A = \begin{pmatrix} 3 & 1 & 5 \\ 2 & 0 & 1 \\ 1 & 1 & 7 \end{pmatrix}.$$

What is XA?

8. Let $X = (0, 1, 0)$, and let A be an arbitrary 3×3 matrix. How would you describe XA? What if $X = (0, 0, 1)$? Generalize to similar statements concerning $n \times n$ matrices, and their products with unit vectors.

9. Let $A = \begin{pmatrix} 0 & 2 \\ 0 & 0 \end{pmatrix}$. What is A^2?

10. Let $A = \begin{pmatrix} 0 & 0 \\ -5 & 0 \end{pmatrix}$. What is A^2?

11. (a) Let A be the matrix

$$\begin{pmatrix} 0 & 1 & 1 \\ 0 & 0 & 1 \\ 0 & 0 & 0 \end{pmatrix}.$$

Find A^2, A^3. Generalize to 4×4 matrices.

(b) Let A be the matrix

$$\begin{pmatrix} 1 & 1 & 1 \\ 0 & 1 & 1 \\ 0 & 0 & 1 \end{pmatrix}.$$

Compute A^2, A^3, and A^4.

12. Let X be the indicated column vector, and A the indicated matrix. Find AX as a column vector.

(a) $X = \begin{pmatrix} 3 \\ 2 \\ 1 \end{pmatrix}$, $A = \begin{pmatrix} 1 & 0 & 1 \\ 2 & 1 & 1 \\ 2 & 0 & -1 \end{pmatrix}$ (b) $X = \begin{pmatrix} 1 \\ 1 \\ 0 \end{pmatrix}$, $A = \begin{pmatrix} 2 & 1 & 5 \\ 0 & 1 & 1 \end{pmatrix}$

(c) $X = \begin{pmatrix} x_1 \\ x_2 \\ x_3 \end{pmatrix}$, $A = \begin{pmatrix} 0 & 1 & 0 \\ 0 & 0 & 0 \end{pmatrix}$ (d) $X = \begin{pmatrix} x_1 \\ x_2 \\ x_3 \end{pmatrix}$, $A = \begin{pmatrix} 0 & 0 & 0 \\ 1 & 0 & 0 \end{pmatrix}$

13. Let

$$A = \begin{pmatrix} 2 & 1 & 3 \\ 4 & 1 & 5 \end{pmatrix}.$$

Find AX for each of the following values of X.

(a) $X = \begin{pmatrix} 1 \\ 0 \\ 0 \end{pmatrix}$ (b) $X = \begin{pmatrix} 0 \\ 1 \\ 0 \end{pmatrix}$ (c) $X = \begin{pmatrix} 0 \\ 0 \\ 1 \end{pmatrix}$ (d) $X = \begin{pmatrix} 0 \\ 1 \\ 1 \end{pmatrix}$.

14. Let
$$A = \begin{pmatrix} 3 & 7 & 5 \\ 1 & -1 & 4 \\ 2 & 1 & 8 \end{pmatrix}.$$

Find AX for each of the values of X given in Exercise 13.

15. Let
$$X = \begin{pmatrix} 0 \\ 1 \\ 0 \\ 0 \end{pmatrix} \quad \text{and} \quad A \quad \begin{pmatrix} a_{11} & \cdots & a_{14} \\ \vdots & & \vdots \\ a_{m1} & \cdots & a_{m4} \end{pmatrix}.$$

What is AX?

16. Let X be a column vector having all its components equal to 0 except the i-th component which is equal to 1. Let A be an arbitrary matrix, whose size is such that we can form the product AX. What is AX?

17. Let X be a row vector having all its components equal to 0 except the j-th component which is equal to 1. Let A be an arbitrary matrix, whose size is such that we can form the product XA. What is XA? Work out special cases when X has 2 components, then when X has 3 components.

18. Let a, b be numbers, and let
$$A = \begin{pmatrix} 1 & a \\ 0 & 1 \end{pmatrix} \quad \text{and} \quad B = \begin{pmatrix} 1 & b \\ 0 & 1 \end{pmatrix}.$$

What is AB? What is A^n where n is a positive integer?

19. If A is a square $n \times n$ matrix, we call a square matrix B an **inverse** for A if $AB = BA = I_n$. Show that if B, C are inverses for A, then $B = C$.

20. Show that the matrix A in Exercise 18 has an inverse. What is this inverse?

21. Show that if A, B are $n \times n$ matrices which have inverses, then AB has an inverse.

22. Determine all 2×2 matrices A such that $A^2 = O$.

23. Let $A = \begin{pmatrix} \cos\theta & -\sin\theta \\ \sin\theta & \cos\theta \end{pmatrix}$. Show that $A^2 = \begin{pmatrix} \cos 2\theta & -\sin 2\theta \\ \sin 2\theta & \cos 2\theta \end{pmatrix}$.

Determine A^n by induction for any positive integer n.

24. Find a 2×2 matrix A such that $A^2 = -I = \begin{pmatrix} -1 & 0 \\ 0 & -1 \end{pmatrix}$.

25. Let
$$A = \begin{pmatrix} 1 & 0 & 0 \\ 0 & 2 & 0 \\ 0 & 0 & 3 \end{pmatrix}.$$

Find A^2, A^3, A^4.

26. Let A be a diagonal matrix, with diagonal elements $a_1, \ldots, a_n$. What is A^2, A^3, A^k for any positive integer k?

27. Let

$$A = \begin{pmatrix} 0 & 1 & 6 \\ 0 & 0 & 4 \\ 0 & 0 & 0 \end{pmatrix}.$$

Find A^3.

CHAPTER XIV

Linear Mappings

We shall first define the general notion of a mapping, which generalizes the notion of a function. Among mappings, the linear mappings are the most important. A good deal of mathematics is devoted to reducing questions concerning arbitrary mappings to linear mappings. For one thing, they are interesting in themselves, and many mappings are linear. On the other hand, it is often possible to approximate an arbitrary mapping by a linear one, whose study is much easier than the study of the original mapping. This is done in the calculus of several variables.

§1. MAPPINGS

As usual, a collection of objects will be called a **set**. A member of the collection is also called an **element** of the set. It is useful in practice to use short symbols to denote certain sets. For instance we denote by **R** the set of all numbers. To say that "x is a number" or that "x is an element of **R**" amounts to the same thing. The set of n-tuples of numbers will be denoted by $\mathbf{R}^n$. Thus "X is an element of $\mathbf{R}^n$" and "X is an n-tuple" mean the same thing. Instead of saying that u is an element of a set S, we shall also frequently say that u **lies in** S and we sometimes write $u \in S$. If S and S' are two sets, and if every element of S' is an element of S, then we say that S' is a **subset** of S. Thus the set of rational numbers is a subset of the set of (real) numbers. To say that S is a subset of S' is to say that S is part of S'. To denote the fact that S is a subset of S', we write $S \subset S'$.

If S_1, S_2 are sets, then the **intersection** of S_1 and S_2, denoted by $S_1 \cap S_2$, is the set of elements which lie in both S_1 and S_2. The **union** of S_1 and S_2, denoted by $S_1 \cup S_2$, is the set of elements which lie in S_1 or S_2.

Let S, S' be two sets. A **mapping** from S to S' is an association which to every element of S associates an element of S'. Instead of saying that F is a mapping from S into S', we shall often write the symbols $F : S \rightarrow S'$. A mapping will also be called a **map,** for the sake of brevity.

A function is a special type of mapping, namely it is a mapping from a set into the set of numbers, i.e. into **R**.

We extend to mappings some of the terminology we have used for functions. For instance, if $T : S \rightarrow S'$ is a mapping, and if u is an element of S, then we denote by $T(u)$, or Tu, the element of S' associated to u by T. We call $T(u)$ the **value** of T at u, or also the **image** of u under T. The symbols $T(u)$ are read "T of u". The set of all elements $T(u)$, when u ranges over all elements of S, is called the **image** of T. If W is a subset of S, then the set of elements $T(w)$, when w ranges over all elements of W, is called the **image** of W under T, and is denoted by $T(W)$.

Let $F: S \rightarrow S'$ be a map from a set S into a set S'. If x is an element of S, we often write

$$x \longmapsto F(x)$$

with a special arrow $\longmapsto$ to denote the image of x under F. Thus, for instance, we would speak of the map F such that $F(x) = x^2$ as the map $x \longmapsto x^2$.

Example 1. Let S and S' be both equal to **R**. Let $f : \mathbf{R} \rightarrow \mathbf{R}$ be the function $f(x) = x^2$ (i.e. the function whose value at a number x is x^2). Then f is a mapping from **R** into **R**. Its image is the set of numbers ≥ 0.

Example 2. Let S be the set of numbers ≥ 0, and let $S' = \mathbf{R}$. Let

$$g : S \rightarrow S'$$

be the function such that $g(x) = x^{1/2}$. Then g is a mapping from S into **R**.

Example 3. Let S be the set $\mathbf{R}^3$, i.e. the set of 3-tuples. Let $A = (2, 3, -1)$. Let $L: \mathbf{R}^3 \rightarrow \mathbf{R}$ be the mapping whose value at a vector $X = (x, y, z)$ is $A \cdot X$. Then $L(X) = A \cdot X$. If $X = (1, 1, -1)$, then the value of L at X is 6.

Just as we did with functions, we describe a mapping by giving its values. Thus, instead of making the statement in Example 3 describing the mapping L, we would also say: Let $L: \mathbf{R}^3 \rightarrow \mathbf{R}$ be the mapping $L(X) = A \cdot X$. This is somewhat incorrect, but is briefer, and does not usually give rise to confusion. More correctly, we can write $X \longmapsto L(X)$ or $X \longmapsto A \cdot X$ with the special arrow $\longmapsto$ to denote the effect of the map L on the element X.

Example 4. Let $F: \mathbf{R}^2 \rightarrow \mathbf{R}^2$ be the mapping given by

$$F(x, y) = (2x, 2y).$$

Describe the image under F of the points lying on the circle $x^2 + y^2 = 1$.

Let (x, y) be a point on the circle of radius 1.

Let $u = 2x$ and $v = 2y$. Then u, v satisfy the relation

$$(u/2)^2 + (v/2)^2 = 1$$

or in other words,

$$\frac{u^2}{4} + \frac{v^2}{4} = 1.$$

Hence (u, v) is a point on the circle of radius 2. Therefore the image under F of the circle of radius 1 is a subset of the circle of radius 2. Conversely, given a point (u, v) such that

$$u^2 + v^2 = 4,$$

let $x = u/2$ and $y = v/2$. Then the point (x, y) satisfies the equation

$$x^2 + y^2 = 1,$$

and hence is a point on the circle of radius 1. Furthermore, $F(x, y) = (u, v)$. Hence every point on the circle of radius 2 is the image of some point on the circle of radius 1. We conclude finally that the image of the circle of radius 1 under F is precisely the circle of radius 2.

Note. In general, let S, S' be two sets. To prove that $S = S'$, one frequently proves that S is a subset of S' and that S' is a subset of S. This is what we did in the preceding argument.

Observe that the association

$$(x, y) \mapsto (2x, 2y)$$

is a dilation, i.e. a stretching by a factor of 2. Each point (x, y) is set on the point $(2x, 2y)$ which lies on the same ray from the origin, at twice the distance from the origin, as illustrated on Fig. 1.

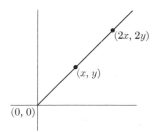

Figure 1

Example 5. In general, let r be a positive number. The association

$$(x, y) \mapsto (rx, ry)$$

is called **dilation** by the factor of r. We can also define it in 3-space, by

$$(x, y, z) \mapsto (rx, ry, rz).$$

We shall study such dilations later when we take up area and volume, and we shall see how these change under dilations.

Example 6. A curve in space as we studied in Chapter II was a mapping. For instance, we can define a map

$$F: \mathbf{R} \longmapsto \mathbf{R}^3$$

by the association

$$t \longmapsto (2t, 10^t, t^3).$$

Thus $F(t) = (2t, 10^t, t^3)$, and the value of F at 2 is

$$F(2) = (4, 100, 8).$$

In such a mapping we call

$$f_1(t) = 2t, \qquad f_2(t) = 10^t, \qquad f_3(t) = t^3$$

the **coordinate functions** of the mapping.

In general, a mapping $F: \mathbf{R} \longrightarrow \mathbf{R}^3$ can always be expressed in terms of such functions, and we write

$$F(t) = (f_1(t), f_2(t), f_3(t)).$$

Example 7. Polar Coordinate Mapping. Let $F: \mathbf{R}^2 \longrightarrow \mathbf{R}^2$ be the mapping defined by

$$F(r, \theta) = (r \cos \theta, r \sin \theta).$$

Thus we may put

$$x = r \cos \theta$$
$$y = r \sin \theta.$$

Then F is a mapping, which is called the **polar coordinate mapping.** We see that x and y depend on r, θ, and x, y are the coordinate functions of the mapping. We studied this mapping when we changed coordinates in a double integral. You should get well acquainted with this mapping, and we work out one example of what it does. Let S be the rectangle consisting of all points (r, θ) such that

$$0 \leqq r \leqq 2 \qquad \text{and} \qquad 0 \leqq \theta \leqq \pi/2.$$

We want to describe the image of S under the polar coordinate mapping.

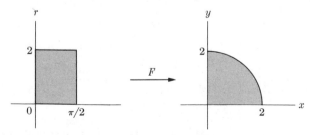

Figure 2

The image of S under the polar coordinate map F consists of all points (x, y) whose polar coordinates (r, θ) satisfy the above inequalities. We see that the image is just the sector of radius 2 in the first quadrant as shown on Fig. 2.

Example 8. Translations. Let A be a vector, say in the plane. We let

$$T_A: \mathbf{R}^2 \to \mathbf{R}^2$$

be the mapping such that

$$T_A(X) = X + A.$$

We call T_A the translation by A. On Fig. 3 we have drawn the translations of various points P, Q, M under translation by A. We may describe the image of a point P under translation by A as the point obtained from P by moving P in the direction of A, for a distance equal to the distance between O and A. Of course, the same notion also works in higher dimensional space. If A is an n-tuple, then

$$T_A: \mathbf{R}^n \to \mathbf{R}^n$$

is the mapping defined by the same equation as above, namely

$$T_A(X) = X + A.$$

You can visualize the picture (at least in $\mathbf{R}^3$) similarly.

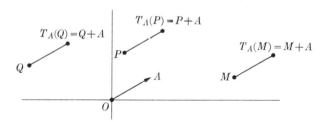

Figure 3

Example 9. You should not forget the identity mapping I, defined on any set S, and such that $I(x) = x$ for all x in S.

EXERCISES

1. Let $L(X) = A \cdot X$, where $A = (2, 3, -1)$. Give $L(X)$ when X is the vector:
   (a) $(1, 2, -3)$ (b) $(-1, 5, 0)$ (c) $(2, 1, 1)$

2. Let $F: \mathbf{R} \to \mathbf{R}^2$ be the mapping such that $F(t) = (e^t, t)$. What is $F(1)$, $F(0)$, $F(-1)$?

3. Let $A = (1, 1, -1, 3)$. Let $F: \mathbf{R}^4 \to \mathbf{R}$ be the mapping such that for any vector $X = (x_1, x_2, x_3, x_4)$ we have $F(X) = X \cdot A + 2$. What is the value of $F(X)$ when (a) $X = (1, 1, 0, -1)$ and (b) $X = (2, 3, -1, 1)$?

In each case, to prove that the image is equal to a certain set S, you must prove that the image is contained in S, and also that every element of S is in the image.

4 Let $F: \mathbf{R}^2 \to \mathbf{R}^2$ be the mapping defined by $F(x, y) = (2x, 3y)$. Describe the image of the points lying on the circle $x^2 + y^2 = 1$.

5. Let $F: \mathbf{R}^2 \to \mathbf{R}^2$ be the mapping defined by $F(x, y) = (xy, y)$. Describe the image under F of the straight line $x = 2$.

6. Let F be the mapping defined by $F(x, y) = (e^x \cos y, e^x \sin y)$. Describe the image under F of the line $x = 1$. Describe more generally the image under F of a line $x = c$, where c is a constant.

7. Let F be the mapping defined by $F(t, u) = (\cos t, \sin t, u)$. Describe geometrically the image of the (t, u)-plane under F.

8. Let F be the mapping defined by $F(x, y) = (x/3, y/4)$. What is the image under F of the ellipse

$$\frac{x^2}{9} + \frac{y^2}{16} = 1?$$

9. Draw the images of the following sets S under the polar coordinate mapping. In each case, the set S consists of all points (r, θ) satisfying the stated inequalities.
   (a) $0 \leq r \leq 1$ and $0 \leq \theta \leq \pi/3$
   (b) $0 \leq r \leq 3$ and $0 \leq \theta \leq 3\pi/4$
   (c) $1 \leq r \leq 2$ and $\pi/4 \leq \theta \leq 3\pi/4$
   (d) $1 \leq r \leq 2$ and $\pi/3 \leq \theta \leq 2\pi/3$
   (e) $2 \leq r \leq 3$ and $\pi/6 \leq \theta \leq \pi/4$
   (f) $2 \leq r \leq 3$ and $\pi/6 \leq \theta \leq \pi/3$
   (g) $3 \leq r \leq 4$ and $\pi/2 \leq \theta \leq 2\pi/3$

10. In general, let S be the rectangle defined by the inequalities

$$0 < r_1 \leq r \leq r_2 \quad \text{and} \quad 0 \leq \theta_1 \leq \theta \leq \theta_2.$$

Describe the image of S under the polar coordinate mapping.

11. Let $A = (-1, 2)$. Draw the image of the point X under translation by A when
   (a) $X = (2, 3)$ (b) $X = (-5, 2)$ (c) $X = (1, 1)$

12. The identity mapping of $\mathbf{R}^n$ is equal to a translation T_A for some vector A. True or false? If true, which vector A?

13. Draw the image of the following figures under translation T_A, where $A = (-1, 2)$.

(a) The circle as shown:

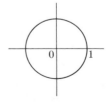

Figure 4

(b)   The square as shown:

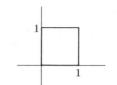

Figure 5

(c)   The circle as shown:

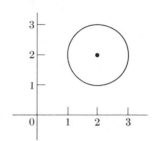

Figure 6

(d)   The square as shown:

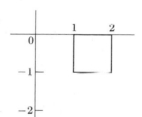

Figure 7

§2. LINEAR MAPPINGS

Consider two Euclidean spaces $\mathbf{R}^n$ and $\mathbf{R}^m$. In the applications, the values for m and n are 1, 2, or 3, but they can all occur, so it is just as easy to leave them indeterminate for what we are about to say.

A mapping

$$L: \mathbf{R}^n \to \mathbf{R}^m$$

is called a **linear mapping** if it satisfies the following properties:

LM 1. *For any elements X, Y in $\mathbf{R}^n$ we have*

$$L(X + Y) = L(X) + L(Y).$$

LM 2. *If c is a number, then*

$$L(cX) = cL(X).$$

These properties should remind you of properties of multiplication of matrices, and also of the dot product of n-tuples. These in fact provide us with the examples which interest us for this course.

Example 1. Let $A = (3, 1, -2)$. Then we have a linear map

$$L_A: \mathbf{R}^3 \to \mathbf{R}$$

defined by the dot product,

$$L_A(X) = A \cdot X,$$

where X is a column vector in $\mathbf{R}^3$. If $X = \begin{pmatrix} x \\ y \\ z \end{pmatrix}$, then

$$L_A(X) = 3x + y - 2z.$$

In general, let

$$A = \begin{pmatrix} a_{11} & \cdots & a_{1n} \\ \vdots & & \vdots \\ a_{m1} & \cdots & a_{mn} \end{pmatrix}$$

be an $m \times n$ matrix. We can then associate with A a map

$$L_A: \mathbf{R}^n \to \mathbf{R}^m$$

by letting

$$L_A(X) = AX$$

for every **column** vector X in $\mathbf{R}^n$. Thus L_A is defined by the association $X \mapsto AX$, the product being the product of matrices. That L_A is linear is simply a special case of Theorem 1, Chapter XIII, §2, namely the theorem concerning properties of multiplication of matrices. Indeed, we have

$$A(X + Y) = AX + AY \qquad \text{and} \qquad A(cX) = cAX$$

for all vectors X, Y in $\mathbf{R}^n$ and all numbers c. We call L_A the linear map **associated** with the matrix A. We also say that A is the **matrix representing** the linear map L_A.

Example 2. If

$$A = \begin{pmatrix} 2 & 1 \\ -1 & 5 \end{pmatrix} \qquad \text{and} \qquad X = \begin{pmatrix} 3 \\ 7 \end{pmatrix},$$

then

$$L_A(X) = \begin{pmatrix} 2 & 1 \\ -1 & 5 \end{pmatrix}\begin{pmatrix} 3 \\ 7 \end{pmatrix} = \begin{pmatrix} 6 + 7 \\ -3 + 35 \end{pmatrix} = \begin{pmatrix} 13 \\ 32 \end{pmatrix}.$$

Theorem 1. *If A, B are $m \times n$ matrices and if $L_A = L_B$, then $A = B$. In other words, if matrices A, B give rise to the same linear map, then they are equal.*

Proof. By definition, we have $A_i \cdot X = B_i \cdot X$ for all i, if A_i is the i-th row of A and B_i is the i-th row of B. Hence $(A_i - B_i) \cdot X = 0$ for all i and all X. Hence $A_i - B_i = O$, and $A_i = B_i$ for all i. Hence $A = B$.

It can easily be shown that every linear map from $\mathbf{R}^n$ into $\mathbf{R}^m$ is of the form L_A for some matrix A, in other words, the above example is the most general type of linear map from $\mathbf{R}^n$ into $\mathbf{R}^m$. The matrix A is called the **matrix associated with the linear map.** We shall give the proof when $n = 2$.

Let $E^1 = \begin{pmatrix} 1 \\ 0 \end{pmatrix}$ and $E^2 = \begin{pmatrix} 0 \\ 1 \end{pmatrix}$ be the standard unit vectors. Let $L: \mathbf{R}^2 \to \mathbf{R}^2$ be a linear map such that

$$L(E^1) = \begin{pmatrix} a \\ c \end{pmatrix} \quad \text{and} \quad L(E^2) = \begin{pmatrix} b \\ d \end{pmatrix}.$$

We shall prove that the matrix associated with L is precisely

$$A = \begin{pmatrix} a & b \\ c & d \end{pmatrix}.$$

First note that

$$\begin{pmatrix} a & b \\ c & d \end{pmatrix} \begin{pmatrix} 1 \\ 0 \end{pmatrix} = \begin{pmatrix} a \\ c \end{pmatrix} = L(E^1)$$

and

$$\begin{pmatrix} a & b \\ c & d \end{pmatrix} \begin{pmatrix} 0 \\ 1 \end{pmatrix} = \begin{pmatrix} b \\ d \end{pmatrix} = L(E^2).$$

Let $X = \begin{pmatrix} x \\ y \end{pmatrix}$, so that $X = xE^1 + yE^2$. Then

$$L(X) = L(xE^1) + L(yE^2) = xL(E^1) + yL(E^2)$$
$$= xAE^1 + yAE^2$$
$$= A(xE^1 + yE^2)$$
$$= AX.$$

This proves that $L(X) = AX$, and therefore that A is the matrix representing L. A similar proof can be given for $\mathbf{R}^3$, or $\mathbf{R}^n$.

Example 3. Let $L: \mathbf{R}^2 \to \mathbf{R}^2$ be a linear map such that

$$L(E^1) = \begin{pmatrix} 3 \\ 5 \end{pmatrix} \quad \text{and} \quad L(E^2) = \begin{pmatrix} -2 \\ 9 \end{pmatrix}.$$

Then the matrix associated with L is the matrix

$$A = \begin{pmatrix} 3 & -2 \\ 5 & 9 \end{pmatrix}.$$

You can check that it has the desired effect on the unit vectors, namely:

$$\begin{pmatrix} 3 & -2 \\ 5 & 9 \end{pmatrix} \begin{pmatrix} 1 \\ 0 \end{pmatrix} = \begin{pmatrix} 3 \\ 5 \end{pmatrix}$$

and

$$\begin{pmatrix} 3 & -2 \\ 5 & 9 \end{pmatrix} \begin{pmatrix} 0 \\ 1 \end{pmatrix} = \begin{pmatrix} -2 \\ 9 \end{pmatrix}.$$

EXERCISES

1. In each case, find the vector $L_A(X)$.

   (a) $A = \begin{pmatrix} 2 & 1 \\ 1 & 0 \end{pmatrix}$, $X = \begin{pmatrix} 3 \\ -1 \end{pmatrix}$

   (b) $A = \begin{pmatrix} 1 & 0 \\ 0 & 0 \end{pmatrix}$, $X = \begin{pmatrix} 5 \\ 1 \end{pmatrix}$

   (c) $A = \begin{pmatrix} 1 & 1 \\ 0 & 1 \end{pmatrix}$, $X = \begin{pmatrix} 4 \\ 1 \end{pmatrix}$

   (d) $A = \begin{pmatrix} 0 & 0 \\ 0 & 1 \end{pmatrix}$, $X = \begin{pmatrix} 7 \\ -3 \end{pmatrix}$

2. Let r be a number. Let $F_r: \mathbf{R}^n \to \mathbf{R}^n$ be the dilation mapping, defined by the formula

   $$F_r(X) = rX.$$

   Exhibit a matrix A such that $F_r(X) = AX$.

3. Let a, b be numbers. Let $F_{a,b}: \mathbf{R}^2 \to \mathbf{R}^2$ be the mapping such that

   $$F_{a,b}\begin{pmatrix} x \\ y \end{pmatrix} = \begin{pmatrix} ax \\ by \end{pmatrix}.$$

   Exhibit a matrix A such that $F_{a,b}(X) = AX$.

4. Let a_1, a_2, a_3 be numbers. If $X = (x, y, z)$ let $F(X) = (a_1x, a_2y, a_3z)$. Writing X as a column vector, exhibit a matrix A such that $F(X) = AX$.

5. Let $F(x, y, z) = (x, y)$. Writing X as a column vector, exhibit a matrix A such that $F(X) = AX$.

6. Let $F(x, y, z) = x$. Writing X as a column vector, exhibit a matrix A such that $F(X) = AX$.

7. Let $F(x, y, z) = (x, z)$. Writing X as a column vector, exhibit a matrix A such that $F(X) = AX$.

8. Same question if $F(x, y, z) = (y, z)$.

9. Let $F: \mathbf{R}^4 \to \mathbf{R}^2$ be the mapping such that

$$F(x_1, x_2, x_3, x_4) = (x_1, x_2).$$

Writing X as a column vector, exhibit a matrix A such that $F(X) = AX$.

10. Let $F: \mathbf{R}^4 \to \mathbf{R}^3$ be the mapping such that

$$F(x_1, x_2, x_3, x_4) = (x_1, x_2, x_3).$$

Writing X as a column vector, exhibit a matrix A such that $F(X) = AX$.

11. Let A be an element of $\mathbf{R}^3$. Suppose that the translation by A is a linear map. What is the only possibility for A? If $A \neq O$, can T_A be a linear map? Proof?

12. Let $L: \mathbf{R}^2 \mapsto \mathbf{R}^2$ be the linear map such that

$$L(E^1) = \begin{pmatrix} -5 \\ 7 \end{pmatrix} \quad \text{and} \quad L(E^2) = \begin{pmatrix} 3 \\ 1 \end{pmatrix}.$$

What is the matrix associated with L?

13. Same question if

$$L(E^1) = \begin{pmatrix} -1 \\ 4 \end{pmatrix} \quad \text{and} \quad L(E^2) = \begin{pmatrix} 2 \\ 6 \end{pmatrix}.$$

14. Let $L: \mathbf{R}^3 \mapsto \mathbf{R}^3$ be a linear map such that

$$L(E^1) = \begin{pmatrix} 1 \\ 3 \\ 4 \end{pmatrix}, \quad L(E^2) = \begin{pmatrix} -2 \\ 7 \\ 9 \end{pmatrix}, \quad L(E^3) = \begin{pmatrix} 8 \\ -5 \\ 2 \end{pmatrix}.$$

Here,

$$E^1 = \begin{pmatrix} 1 \\ 0 \\ 0 \end{pmatrix}, E^2 = \begin{pmatrix} 0 \\ 1 \\ 0 \end{pmatrix}, \text{ and } E^3 = \begin{pmatrix} 0 \\ 0 \\ 1 \end{pmatrix}.$$

What is the matrix associated with L? Verify that it has the desired effect on the unit vectors.

15. Write out the proof that if E^1, E^2, E^3 are the standard unit vectors in $\mathbf{R}^3$, and if $L: \mathbf{R}^3 \mapsto \mathbf{R}^3$ is the linear map such that

$$L(E^1) = \begin{pmatrix} a_{11} \\ a_{21} \\ a_{31} \end{pmatrix}, \quad L(E^2) = \begin{pmatrix} a_{12} \\ a_{22} \\ a_{32} \end{pmatrix}, \quad L(E^3) = \begin{pmatrix} a_{13} \\ a_{23} \\ a_{33} \end{pmatrix},$$

then the matrix A associated with L is the matrix (a_{ij}), that is

$$\begin{pmatrix} a_{11} & a_{12} & a_{13} \\ a_{21} & a_{22} & a_{23} \\ a_{31} & a_{32} & a_{33} \end{pmatrix}.$$

16. Let $L: \mathbf{R}^3 \mapsto \mathbf{R}^3$ be the linear map such that

$$L(E^1) = \begin{pmatrix} -3 \\ 5 \\ 0 \end{pmatrix}, \quad L(E^2) = \begin{pmatrix} 4 \\ 1 \\ -7 \end{pmatrix}, \quad L(E^3) = \begin{pmatrix} 5 \\ -2 \\ 8 \end{pmatrix}.$$

What is the matrix associated with L? Verify directly that it has the desired effect on the unit vectors.

17. Let $L: \mathbf{R} \mapsto \mathbf{R}^n$ be a linear map. Prove that there exists a vector A in $\mathbf{R}^n$ such that for all t in $\mathbf{R}$ we have

$$L(t) = tA.$$

18. Let $L: \mathbf{R}^2 \mapsto \mathbf{R}^3$ be a linear map. Let

$$E^1 = \begin{pmatrix} 1 \\ 0 \end{pmatrix} \quad \text{and} \quad E^2 = \begin{pmatrix} 0 \\ 1 \end{pmatrix}$$

be the unit vectors in $\mathbf{R}^2$. Suppose that

$$L(E^1) = \begin{pmatrix} a_{11} \\ a_{21} \\ a_{31} \end{pmatrix}, \quad L(E^2) = \begin{pmatrix} a_{12} \\ a_{22} \\ a_{32} \end{pmatrix}.$$

In terms of the a_{ij}, what is the matrix A associated with L?

19. Let $L: \mathbf{R}^2 \mapsto \mathbf{R}^3$ be a linear map, and suppose that E^1, E^2 are the unit vectors in $\mathbf{R}^2$. Let

$$L(E^1) = \begin{pmatrix} 3 \\ 1 \\ -4 \end{pmatrix} \quad \text{and} \quad L(E^2) = \begin{pmatrix} -5 \\ 7 \\ -8 \end{pmatrix}.$$

What is the matrix A associated with L?

§3. GEOMETRIC APPLICATIONS

Let P, A be elements of $\mathbf{R}^n$. We define the **line segment** between P and $P + A$ to be the set of all points

$$P + tA, \qquad\qquad 0 \leqq t \leqq 1.$$

This line segment is illustrated in Fig. 8.

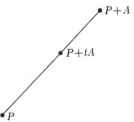

Figure 8

For instance, if $t = \frac{1}{2}$, then $P + \frac{1}{2}A$ is the point midway between P and $P + A$. Similarly, if $t = \frac{1}{3}$, then $P + \frac{1}{3}A$ is the point one-third of the way between P and $P + A$ (Fig. 9).

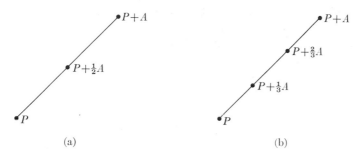

(a)   (b)

Figure 9

If P, Q are elements of $\mathbf{R}^n$, let $A = Q - P$. Then the line segment between P and Q is the set of all points $P + tA$, or

$$P + t(Q - P), \qquad\qquad 0 \leq t \leq 1.$$

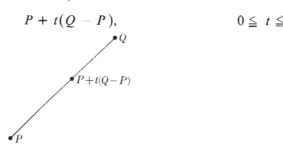

Figure 10

Observe that we can rewrite the expression for these points in the form

(1) $$(1 - t)P + tQ, \qquad\qquad 0 \leq t \leq 1,$$

and letting $s = 1 - t, t = 1 - s$, we can also write it as

$$sP + (1 - s)Q, \qquad\qquad 0 \leq s \leq 1.$$

Finally, we can write the points of our line segment in the form

(2) $$t_1 P + t_2 Q$$

with $t_1, t_2 \geq 0$ and $t_1 + t_2 = 1$. Indeed, letting $t = t_2$, we see that every point which can be written in the form (2) satisfies (1). Conversely, we let $t_1 = 1 - t$ and $t_2 = t$ and see that every point of the form (1) can be written in the form (2).

Let $L: \mathbf{R}^n \to \mathbf{R}^m$ be a linear map. Let S be the line segment in $\mathbf{R}^n$ between two points P, Q. Then the image $L(S)$ of this line segment is the line segment in

$\mathbf{R}^m$ between the points $L(P)$ and $L(Q)$. This is obvious from (2), because

$$L(t_1 P + t_2 Q) = t_1 L(P) + t_2 L(Q).$$

We shall now generalize this discussion to higher dimensional figures. Let P, Q be elements of $\mathbf{R}^n$, and assume that they are $\neq O$, and Q is not a scalar multiple of P. We define the **parallelogram spanned** by P and Q to be the set of all points

$$t_1 P + t_2 Q,$$

with

$$0 \leq t_i \leq 1 \quad \text{for} \quad i = 1, 2.$$

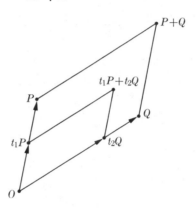

Figure 11

This definition is clearly justified since $t_1 P$ is a point of the segment between O and P (Fig. 11), and $t_2 Q$ is a point of the segment between O and Q. For all values of t_1, t_2 ranging independently between 0 and 1, we see geometrically that $t_1 v + t_2 w$ describes all points of the parallelogram.

At the end of §1 we defined **translations.** We obtain the most general parallelogram (Fig. 12) by taking the translation of the parallelogram just described. Thus if A is an element of $\mathbf{R}^n$, the translation by A of the parallelogram spanned by P and Q consists of all points

$$A + t_1 P + t_2 Q,$$

with

$$0 \leq t_i \leq 1 \quad \text{for} \quad i = 1, 2.$$

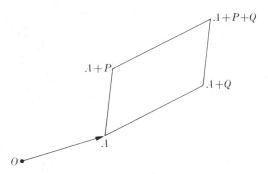

Figure 12

As with line segments, we see that if

$$L: \mathbf{R}^n \to \mathbf{R}^m$$

is a linear map, and if S is a parallelogram as described above, then the image of S is again a parallelogram, provided that $L(P)$ and $L(Q)$ do not lie on the same line through the origin (i.e. $L(P)$ is not a scalar multiple of $L(Q)$). This is immediately seen, because the image of S under L consists of all points

$$L(A + t_1 P + t_2 Q) = L(A) + t_1 L(P) + t_2 L(Q),$$

with

$$0 \leq t_i \leq 1 \qquad \text{for} \qquad i = 1, 2.$$

We see again the usefulness of the conditions for linearity **LM 1** and **LM 2**.

Example. Let S be the parallelogram spanned by the vectors $P = (1, 2)$ and $Q = (-1, 5)$. Let $L: \mathbf{R}^2 \to \mathbf{R}^2$ be the linear map L_A, where A is the matrix

$$\begin{pmatrix} 3 & 1 \\ -1 & 5 \end{pmatrix}.$$

Then, writing P, Q as vertical vectors, we obtain

$$L(P) = AP = \begin{pmatrix} 3 & 1 \\ -1 & 5 \end{pmatrix}\begin{pmatrix} 1 \\ 2 \end{pmatrix} = \begin{pmatrix} 5 \\ 9 \end{pmatrix}$$

$$L(Q) = AQ = \begin{pmatrix} 3 & 1 \\ -1 & 5 \end{pmatrix}\begin{pmatrix} -1 \\ 5 \end{pmatrix} = \begin{pmatrix} 2 \\ 26 \end{pmatrix}.$$

Hence the image of S under L is the parallelogram spanned by the vectors $(5, 9)$ and $(2, 26)$.

On the next figure, we have drawn a typical situation of the image of a parallelogram under a linear map.

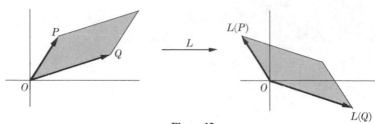

Figure 13

A similar discussion can be carried out in 3-space. It is good practice for you to write it up yourself. Do Exercise 5.

EXERCISES

1. Let L be the linear map represented by the matrix

$$\begin{pmatrix} 1 & -1 \\ 2 & 3 \end{pmatrix}.$$

Let S be the line segment between P and Q. Draw the image of S under L, indicating $L(P)$ and $L(Q)$ in each of the following cases.
   (a) $P = (2, 1)$ and $Q = (-1, 1)$
   (b) $P = (3, -1)$ and $Q = (1, 2)$
   (c) $P = (1, 1)$ and $Q = (1, -1)$
   (d) $P = (2, -1)$ and $Q = (1, 2)$

2. In cases (a), (b), (c), and (d) of Exercise 1, let T be the parallelogram spanned by P and Q. Draw the image of T by the linear map L of Exercise 1, indicating in each case $L(P)$ and $L(Q)$.

3. Let $E^1 = \begin{pmatrix} 1 \\ 0 \end{pmatrix}$ and $E^2 = \begin{pmatrix} 0 \\ 1 \end{pmatrix}$ be the standard unit vectors. Write down their images under the linear map L represented by the matrix

$$\begin{pmatrix} 3 & -1 \\ 5 & 2 \end{pmatrix}.$$

Let S be the square spanned by E^1 and E^2. Draw the image of this square under L, indicating $L(E^1)$ and $L(E^2)$.

4. Let E^1, E^2 again be the standard unit vectors, drawn vertically. Let L be the linear map represented by the matrix

$$\begin{pmatrix} -2 & 3 \\ 1 & 5 \end{pmatrix}.$$

Let S be the square spanned by E^1, E^2. Draw the image $L(S)$, again indicating $L(E^1)$ and $L(E^2)$.

5. (a) Give a definition of the **box (parallelepiped)** spanned by three vectors A, B, C in $\mathbf{R}^3$.

(b) Let $L: \mathbf{R}^3 \rightarrow \mathbf{R}^3$ be a linear map. Prove that the image of such a box under L is again a box, spanned by $L(A)$, $L(B)$, $L(C)$ (provided that the segments from O to $L(A)$, $L(B)$, $L(C)$, respectively, do not all lie in a plane, otherwise you get a "degenerate" box).

(c) Draw a picture for this in 3-dimensional space.

6. Let L be the linear map of $\mathbf{R}^3$ into itself represented by the matrix

$$\begin{pmatrix} -3 & 1 & 4 \\ 2 & 2 & 1 \\ 1 & -2 & 5 \end{pmatrix}.$$

Let S be the cube spanned by the three unit vectors E^1, E^2, E^3. Give explicitly three vectors spanning $L(S)$.

7. Same questions as in Exercise 6, if L is represented by the matrix

$$\begin{pmatrix} 2 & 4 & -6 \\ 3 & 7 & 5 \\ -1 & 2 & -8 \end{pmatrix}.$$

8. Let $X(t) = P + tA$, with t in $\mathbf{R}$, be the parametrization of a straight line in $\mathbf{R}^n$. Let $L: \mathbf{R}^n \rightarrow \mathbf{R}^m$ be a linear map. Suppose that $L(A) \neq O$. Prove that the image of the straight line is a straight line.

9. Let S be a line passing through two distinct points P and Q, in $\mathbf{R}^n$. Let $L: \mathbf{R}^n \rightarrow \mathbf{R}^m$ be a linear map, such that $L(P) \neq L(Q)$.
(a) Give a parametric representation of the line S.
(b) Give a parametric representation of the line $L(S)$.

10. Let A, B be non-zero vectors in $\mathbf{R}^n$ and assume that neither is a scalar multiple of the other. Such vectors are called **independent**. We define the **plane spanned by A and B** to be the set of all points

$$tA + sB,$$

for all real numbers t, s. Observe that this is the 2-dimensional analogue of the parametrization of a line. Let $L: \mathbf{R}^n \rightarrow \mathbf{R}^m$ be a linear map. Assume that $L(A)$ and $L(B)$ are independent. Prove that the image of the plane spanned by A and B is a plane (spanned by which vectors?).

11. Let A, B be independent vectors in $\mathbf{R}^n$, and let P be a point. We define the **plane through P parallel to A, B** to be the set of all points

$$P + tA + sB,$$

where t, s range over all real numbers. Let $L: \mathbf{R}^n \rightarrow \mathbf{R}^m$ be a linear map such that $L(A)$ and $L(B)$ are independent. Prove that the image of the preceding plane is also a plane.

The plane of Exercise 11 looks like this.

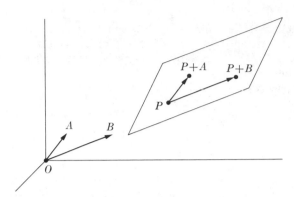

Figure 14

It is the translation by P of the plane in Exercise 10.

§4. COMPOSITION AND INVERSE OF MAPPINGS

This section will be useful for Chapter XVI, §2, §3 and Chapter XVII.

Before we discuss linear mappings, we have to make some more remarks on mappings in general. You recall that in studying functions of one variable, you met composite functions and the chain rule for differentiation. We shall meet a similar situation in several variables.

In one variable, let

$$f: \mathbf{R} \to \mathbf{R} \quad \text{and} \quad g: \mathbf{R} \to \mathbf{R}$$

be functions. Then we can form the **composite function** $g \circ f$, defined by

$$(g \circ f)(x) = g(f(x)).$$

Let U, V, W be sets. Let

$$F: U \to V \quad \text{and} \quad G: V \to W$$

be mappings. Then we can form the **composite mapping** from U into W, denoted by $G \circ F$. It is by definition the mapping defined by

$$(G \circ F)(u) = G(F(u))$$

for all u in U.

Example 1. Let $G: \mathbf{R}^2 \to \mathbf{R}^2$ be the mapping such that

$$G(Y) = 3Y.$$

Let $F: \mathbf{R}^2 \to \mathbf{R}^2$ be the mapping such that $F(X) = X + A$, where

$$A = (1, -2).$$

Then

$$G(F(X)) = G(X + A) = 3(X + A) = 3X + 3A.$$

Our mapping $G \circ F$ is the composite of a translation and a dilation.

Example 2. Let $G: \mathbf{R}^2 \to \mathbf{R}^3$ be the mapping such that

$$G(x, y) = (x^2, xy, \sin y).$$

If (u, v, w) are the coordinates of $\mathbf{R}^3$, we may set

$$u = x^2, \qquad v = xy, \qquad w = \sin y.$$

Let $F: \mathbf{R}^3 \to \mathbf{R}^3$ be the mapping such that

$$F(u, v, w) = (u^3, uv, vw).$$

Then

$$F(G(x, y)) = (x^6, x^3 y, xy \sin y).$$

The composition of mappings is associative. More precisely, let U, V, W, S be sets. Let

$$F: U \to V, \qquad G: V \to W, \qquad \text{and} \qquad H: W \to S$$

be mappings. Then

$$H \circ (G \circ F) = (H \circ G) \circ F.$$

Proof. Here again, the proof is very simple. By definition, we have, for any element u of U:

$$(H \circ (G \circ F))(u) = H((G \circ F)(u)) = H(G(F(u))).$$

On the other hand,

$$((H \circ G) \circ F)(u) = (H \circ G)(F(u)) = H(G(F(u))).$$

By definition, this means that $(H \circ G) \circ F = H \circ (G \circ F)$.

If S is any set, the **identity mapping** I_S is defined to be the map such that $I_S(x) = x$ for all $x \in S$. If we do not need to specify the reference to S (because it is made clear by the context), then we write I instead of I_S. Thus we have $I(x) = x$ for all $x \in S$.

Finally, we define inverse mappings. Let $F: S \to S'$ be a mapping from one set into another set. We say that F has an **inverse** if there exists a mapping

$$G: S' \to S$$

such that

$$G \circ F = I_S \qquad \text{and} \qquad F \circ G = I_{S'}.$$

By this we mean that the composite maps $G \circ F$ and $F \circ G$ are the identity mappings of S and S' respectively.

Example 3. Let $S = S'$ be the set of all numbers ≥ 0. Let

$$f: S \longrightarrow S'$$

be the map such that $f(x) = x^2$. Then f has an inverse mapping, namely the map $g: S \longrightarrow S$ such that $g(x) = \sqrt{x}$.

Example 4. Let $\mathbf{R}^+$ be the set of numbers > 0 and let $f: \mathbf{R} \longrightarrow \mathbf{R}^+$ be the map such that $f(x) = e^x$. Then f has an inverse mapping which is nothing but the logarithm. .

Example 5. Let A be a vector in $\mathbf{R}^3$ and let

$$T_A: \mathbf{R}^3 \longrightarrow \mathbf{R}^3$$

be the translation by A. By definition, we recall that this means

$$T_A(X) = X + A.$$

If B is another vector in $\mathbf{R}^3$, then the composite mapping $T_B \circ T_A$ has the value

$$(T_B \circ T_A)(X) = T_B(T_A(X))$$
$$= T_B(X + A)$$
$$= X + A + B.$$

If $B = -A$, we see that

$$T_{-A}(T_A(X)) = X + A - A = X,$$

and similarly that $T_A(T_{-A}(X)) = X$. Hence T_{-A} is the inverse mapping of T_A. In words, we may say that the inverse mapping of translation by A is translation by $-A$. Of course, the same holds in $\mathbf{R}^n$.

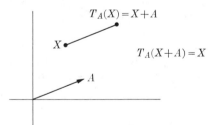

Figure 15

Let

$$f: S \longrightarrow S'$$

be a map. We say that f is **injective** if whenever $x, y \in S$ and $x \neq y$, then $f(x) \neq f(y)$. In other words, f is injective means that f takes on distinct values at distinct elements of S. For example, the map

$$f: \mathbf{R} \to \mathbf{R}$$

such that $f(x) = x^2$, is not injective, because $f(1) = f(-1) = 1$. Also the function $x \mapsto \sin x$ is not injective, because $\sin x = \sin(x + 2\pi)$. However, the map $f: \mathbf{R} \mapsto \mathbf{R}$ such that $f(x) = x + 1$ is injective, because if $x + 1 = y + 1$, then $x = y$.

Again, let $f: S \to S'$ be a mapping. We shall say that f is **surjective** if the image of f is all of S'. Again, the map

$$f: \mathbf{R} \to \mathbf{R}$$

such that $f(x) = x^2$, is not surjective, because its image consists of all numbers ≥ 0, and this image is not equal to all of $\mathbf{R}$. On the other hand, the map of $\mathbf{R}$ into $\mathbf{R}$ given by $x \mapsto x^3$ is surjective, because given a number y there exists a number x such that $y = x^3$ (the cube root of y). Thus every number is in the image of our map.

Let $\mathbf{R}^+$ be the set of real numbers ≥ 0. As a matter of convention, we agree to distinguish between the maps

$$\mathbf{R} \to \mathbf{R} \quad \text{and} \quad \mathbf{R}^+ \to \mathbf{R}^+$$

given by the same formula $x \mapsto x^2$. The point is that when we view the association $x \mapsto x^2$ as a map of $\mathbf{R}$ into $\mathbf{R}$, then it is not surjective, and it is not injective. But when we view this formula as defining a map from $\mathbf{R}^+$ into $\mathbf{R}^+$, then it gives both an injective and surjective map of $\mathbf{R}^+$ into itself, because every positive number has a positive square root, and such a positive square root is uniquely determined.

In general, when dealing with a map $f: S \to S'$, we must therefore always specify the sets S and S', to be able to say that f is injective, or surjective, or neither. To have a completely accurate notation, we should write

$$f_{S,\,S'}$$

or some such symbol which specifies S and S' into the notation, but this becomes too clumsy, and we prefer to use the context to make our meaning clear.

Let

$$f: S \to S'$$

be a map which has an inverse mapping g. Then f is both injective and surjective.

Proof. Let $x, y \in S$ and $x \neq y$. Let $g: S' \to S$ be the inverse mapping of f. If $f(x) = f(y)$, then we must have

$$x = g(f(x)) = g(f(y)) = y,$$

which is impossible. Hence $f(x) \neq f(y)$, and therefore f is injective. To prove that f is surjective, let $z \in S'$. Then

$$f(g(z)) = z$$

by definition of the inverse mapping, and hence $z = f(x)$, where $x = g(z)$. This proves that f is surjective.

The converse of the statement we just proved is also true, namely:

Let $f: S \rightarrow S'$ be a map which is both injective and surjective. Then f has an inverse mapping.

Proof. Given $z \in S'$, since f is surjective, there exists $x \in S$ such that $f(x) = z$. Since f is injective, this element x is uniquely determined by z, and we can therefore define

$$g(z) = x.$$

By definition of g, we find that $f(g(z)) = z$, and $g(f(x)) = x$, so that g is an inverse mapping for f.

Thus we can say that a map $f: S \rightarrow S'$ has an inverse mapping if and only if f is both injective and surjective.

Using another terminology, we can also say that a map

$$f: S \rightarrow S'$$

which has an inverse mapping establishes a one-one correspondence between the elements of S and the elements of S'.

We shall be mostly concerned with linear mappings.

Let $F: \mathbf{R}^n \rightarrow \mathbf{R}^m$ and $G: \mathbf{R}^m \rightarrow R^s$ be linear maps. Then the composite map $G \circ F$ is also a linear map.

Proof. This is very easy to prove. Let u, v be elements of U. Since F is linear, we have $F(u + v) = F(u) + F(v)$. Hence

$$(G \circ F)(u + v) = G(F(u + v)) = G(F(u) + F(v)).$$

Since G is linear, we obtain

$$G(F(u) + F(v)) = G(F(u)) + G(F(v)).$$

Hence

$$(G \circ F)(u + v) = (G \circ F)(u) + (G \circ F)(v).$$

Next, let c be a number. Then

$$(G \circ F)(cu) = G(F(cu))$$
$$= G(cF(u)) \qquad \text{(because } F \text{ is linear)}$$
$$= cG(F(u)) \qquad \text{(because } G \text{ is linear)}.$$

This proves that $G \circ F$ is a linear mapping.

We can also see this with matrices. Suppose that A is the matrix associated with F, and B is the matrix associated with G. Then by definition, we have

$$F(X) = AX, \qquad\qquad \text{for } X \text{ in } \mathbf{R}^n,$$

and

$$G(Y) = BY, \qquad\qquad \text{for } Y \text{ in } \mathbf{R}^m.$$

Hence

$$G(F(X)) = B(AX) = (BA)X,$$

and we see that the product BA is the matrix associated with the linear map $G \circ F$. In other words, the product of the matrices associated with G and F, respectively, is the matrix associated with $G \circ F$.

Let $F \colon \mathbf{R}^n \to \mathbf{R}^n$ be a linear mapping. We shall say that F is **invertible** if there exists a linear mapping

$$G \colon \mathbf{R}^n \to \mathbf{R}^n$$

such that $G \circ F = I$ and $F \circ G = I$. [It can be shown that if an inverse for F exists as a mapping, then this inverse is necessarily linear, but we don't give the proof. It is an easy exercise.] Similarly, let A be an $n \times n$ matrix. We say that A is **invertible** if there exists an $n \times n$ matrix B such that $AB = BA = I_n$ is the unit $n \times n$ matrix. We denote B by A^{-1}.

If F is a linear mapping as above, then we know that it has an associated matrix A, such that

$$F(X) = AX, \qquad\qquad \text{all } X \text{ in } \mathbf{R}^n.$$

Suppose that F is invertible, and that G is its inverse linear mapping. Then G also has an associated matrix B, and since $G(F(X)) = X$, we must have

$$BAX = X,$$

for all X in $\mathbf{R}^n$. Similarly, we must also have $ABX = X$ for all X in $\mathbf{R}^n$. In particular, this must be true if X is any one of the standard unit vectors, and from this we see that $AB = BA = I_n$ is the unit $n \times n$ matrix. Thus $B = A^{-1}$. In other words:

If A is the matrix associated with an invertible linear mapping $L \colon \mathbf{R}^n \to \mathbf{R}^n$, then A^{-1} is the matrix associated with the inverse of L.

It is usually a tedious process to find the inverse of a matrix, and this process involves linear equations. For 2×2 matrices, however, the process is short. We shall discuss it in connection with determinants.

EXERCISES

1. Let $F: \mathbf{R}^3 \to \mathbf{R}^3$ be the map such that $F(X) = 7X$. Prove that F has an inverse mapping, and that this inverse is linear. Do the same if $F: \mathbf{R}^n \to \mathbf{R}^n$ is defined by the same formula.

2. Let $F: \mathbf{R}^n \to \mathbf{R}^n$ be the map such that $F(X) = -8X$. Prove that F is invertible, and write down its inverse explicitly.

3. Let c be a number $\neq 0$ and let $L: \mathbf{R}^n \to \mathbf{R}^n$ be the map such that $F(X) = cX$. Prove that L has an inverse linear map, and write it down explicitly.

4. Let A, B, C be square matrices of the same size and assume that they are invertible. Prove that AB is invertible, and express its inverse in terms of A^{-1} and B^{-1}. Also show that ABC is invertible.

5. Let A be a square matrix such that $A^2 = O$. Show that $I - A$ is invertible. (I is the unit matrix of the same size as A.)

6. Let A be a square matrix such that $A^2 + 2A + I = O$. Show that A is invertible.

7. Let A be a square matrix such that $A^3 = O$. Show that $I - A$ is invertible.

CHAPTER XV

Determinants

In this chapter we carry out the theory of determinants for the case of 2×2 and 3×3 matrices. Those interested in the general case of $n \times n$ matrices can look it up in my Introduction to Linear Algebra.

§1. DETERMINANTS OF ORDER 2

Let

$$A = \begin{pmatrix} a & b \\ c & d \end{pmatrix}$$

be a 2×2 matrix. We define its **determinant** to be $ad - cb$. Thus the determinant is a number. We denote it by

$$\begin{vmatrix} a & b \\ c & d \end{vmatrix} = ad - bc.$$

For example, the determinant of the matrix

$$\begin{pmatrix} 2 & 1 \\ 1 & 4 \end{pmatrix}$$

is equal to $2 \cdot 4 - 1 \cdot 1 = 7$. The determinant of

$$\begin{pmatrix} -2 & -3 \\ 4 & 5 \end{pmatrix}$$

is equal to $(-2) \cdot 5 - (-3) \cdot 4 = -10 + 12 = 2$.

Theorem 1. *If A is a 2×2 matrix, then the determinant of A is equal to the determinant of the transpose of A. In other words,*

$$D(A) = D({}^t A).$$

Proof. This is immediate from the definition of the determinant. We have

$$|A| = \begin{vmatrix} a & b \\ c & d \end{vmatrix} \quad \text{and} \quad |{}^t A| = \begin{vmatrix} a & c \\ b & d \end{vmatrix},$$

and

$$ad - bc = ad - cb.$$

Of course, the property expressed in Theorem 1 is very simple. We give it here because it is satisfied by 3×3 determinants which will be studied later.

Consider a 2×2 matrix A with columns A^1, A^2. The determinant $D(A)$ has interesting properties with respect to these columns, which we shall describe. Thus it is useful to use the notation

$$D(A) = D(A^1, A^2)$$

to emphasize the dependence of the determinant on its columns. If the two columns are denoted by

$$B = \begin{pmatrix} b_1 \\ b_2 \end{pmatrix} \quad \text{and} \quad C = \begin{pmatrix} c_1 \\ c_2 \end{pmatrix},$$

then we would write

$$D(B, C) = \begin{vmatrix} b_1 & c_1 \\ b_2 & c_2 \end{vmatrix} = b_1 c_2 - c_1 b_2.$$

We may view the determinant as a certain type of "product" between the columns B and C. To what extent does this product satisfy the same rules as the product of numbers? Answer: To some extent, which we now determine precisely.

To begin with, this "product" satisfies distributivity. In the determinant notation, this means:

D1. *If $B = B' + B''$, i.e.*

$$\begin{pmatrix} b_1 \\ b_2 \end{pmatrix} = \begin{pmatrix} b_1' \\ b_2' \end{pmatrix} + \begin{pmatrix} b_1'' \\ b_2'' \end{pmatrix},$$

then

$$D(B' + B'', C) = D(B', C) + D(B'', C).$$

Similarly, if $C = C' + C''$, then

$$D(B, C' + C'') = D(B, C') + D(B, C'').$$

Proof. Of course, the proof is quite simple using the definition of the determinant. We have

$$D(B' + B'', C) = \begin{vmatrix} b_1' + b_1'' & c_1 \\ b_2' + b_2'' & c_2 \end{vmatrix}$$

$$= (b_1' + b_1'')c_2 - (b_2' + b_2'')c_1$$

$$= b_1'c_2 + b_1''c_2 - b_2'c_1 - b_2''c_1$$

$$= D(B', C) + D(B'', C).$$

Distributivity on the other side is proved similarly.

D2. *If x is a number, then*

$$D(xB, C) = x \cdot D(B, C) = D(B, xC).$$

Proof. We have

$$D(xB, C) = \begin{vmatrix} xb_1 & c_1 \\ xb_2 & c_2 \end{vmatrix} = xb_1c_2 - xb_2c_1 = x(b_1c_2 - b_2c_1)$$

$$= xD(B, C).$$

Again, the other equality is proved similarly.

Properties **D1** and **D2** may be expressed by saying that *the determinant is linear as a function of each column.*

D3. *If the two columns of the matrix are equal, then the determinant is equal to 0. In other words,*

$$D(B, B) = 0.$$

Proof. This is obvious, because

$$\begin{vmatrix} b_1 & b_1 \\ b_2 & b_2 \end{vmatrix} = b_1b_2 - b_2b_1 = 0.$$

The two vectors

$$E^1 = \begin{pmatrix} 1 \\ 0 \end{pmatrix} \quad \text{and} \quad E^2 = \begin{pmatrix} 0 \\ 1 \end{pmatrix}$$

are the standard **unit vectors**. The matrix formed by them, namely

$$E = \begin{pmatrix} 1 & 0 \\ 0 & 1 \end{pmatrix},$$

is the **unit matrix.** We have:

D4. *If E is the unit matrix, then* $D(E) = D(E^1, E^2) = 1.$

This is obvious.

These four basic properties are fundamental, and other properties can be deduced from them, without going back to the definition of the determinant in terms of the components of the matrix.

D5. *If we add a multiple of one column to the other, then the value of the determinant does not change. In other words, let x be a number. Then*

$$D(B + xC, C) = D(B, C) \quad and \quad D(B, C + xB) = D(B, C).$$

Written out in terms of components, the first relation reads

$$\begin{vmatrix} b_1 + xc_1 & c_1 \\ b_2 + xc_2 & c_2 \end{vmatrix} = \begin{vmatrix} b_1 & c_1 \\ b_2 & c_2 \end{vmatrix}.$$

Proof. Using **D1, D2, D3** in succession, we find that

$$D(B + xC, C) = D(B, C) + D(xC, C)$$
$$= D(B, C) + xD(C, C) = D(B, C).$$

A similar proof applies to $D(B, C + xB)$.

D6. *If the two columns are interchanged, then the value of the determinant changes by a sign. In other words, we have*

$$D(B, C) = -D(C, B).$$

Proof. Again, we use **D1, D2, D4** successively, and get

$$0 = D(B + C, B + C) = D(B, B + C) + D(C, B + C)$$
$$= D(B, B) + D(B, C) + D(C, B) + D(C, C)$$
$$= D(B, C) + D(C, B).$$

This proves that $D(B, C) = -D(C, B)$, as desired.

Of course, you can also give a proof using the components of the matrix. Do this as an exercise. However, there is some point in doing it as above, because in the study of determinants in the higher-dimensional case later, a proof with components becomes much messier, while the proof following the same pattern as the one we have given remains neat.

EXERCISES

1. Compute the following determinants.

(a) $\begin{vmatrix} 3 & -5 \\ 4 & 2 \end{vmatrix}$ (b) $\begin{vmatrix} 2 & -1 \\ -3 & 4 \end{vmatrix}$ (c) $\begin{vmatrix} -3 & 4 \\ 2 & -1 \end{vmatrix}$

(d) $\begin{vmatrix} -5 & 3 \\ 4 & 6 \end{vmatrix}$ (e) $\begin{vmatrix} 3 & 3 \\ -7 & -8 \end{vmatrix}$ (f) $\begin{vmatrix} -5 & -4 \\ 6 & 3 \end{vmatrix}$

2. Compute the determinant

$$\begin{vmatrix} \cos \theta & -\sin \theta \\ \sin \theta & \cos \theta \end{vmatrix}$$

for any real number θ.

3. Compute the determinant

$$\begin{vmatrix} \cos \theta & \sin \theta \\ \sin \theta & \cos \theta \end{vmatrix}$$

when

(a) $\theta = \pi$,   (b) $\theta = \pi/2$,   (c) $\theta = \pi/3$,   (d) $\theta = \pi/4$.

4. Prove:
   (a) The other half of **D1**.
   (b) The other half of **D2**.
   (c) The other half of **D5**.

5. Let c be a number, and let A be a 2×2 matrix. Define cA to be the matrix obtained by multiplying all components of A by c. How does $D(cA)$ differ from $D(A)$?

§2. DETERMINANTS OF ORDER 3

We shall define the determinant for 3×3 matrices, and we shall see that it satisfies properties analogous to those of the 2×2 case.

Let

$$A = (a_{ij}) = \begin{pmatrix} a_{11} & a_{12} & a_{13} \\ a_{21} & a_{22} & a_{23} \\ a_{31} & a_{32} & a_{33} \end{pmatrix}$$

be a 3×3 matrix. We define its **determinant** according to the formula known as the **expansion by a row,** say the first row. That is, we define

(1) $$D(A) = a_{11} \begin{vmatrix} a_{22} & a_{23} \\ a_{32} & a_{33} \end{vmatrix} - a_{12} \begin{vmatrix} a_{21} & a_{23} \\ a_{31} & a_{33} \end{vmatrix} + a_{13} \begin{vmatrix} a_{21} & a_{22} \\ a_{31} & a_{32} \end{vmatrix}$$

and we denote $D(A)$ also with the two vertical bars

$$D(A) = \begin{vmatrix} a_{11} & a_{12} & a_{13} \\ a_{21} & a_{22} & a_{23} \\ a_{31} & a_{32} & a_{33} \end{vmatrix}.$$

We may describe the sum in (1) as follows. Let A_{ij} be the matrix obtained from A by deleting the i-th row and the j-th column. Then the sum for $D(A)$ can be written as

$$a_{11}D(A_{11}) - a_{12}D(A_{12}) + a_{13}D(A_{13}).$$

In other words, each term consists of the product of an element of the first row and the determinant of the 2×2 matrix obtained by deleting the first row and the j-th column, and putting the appropriate sign to this term as shown.

Example 1. Let

$$A = \begin{pmatrix} 2 & 1 & 0 \\ 1 & 1 & 4 \\ -3 & 2 & 5 \end{pmatrix}.$$

Then

$$A_{11} = \begin{pmatrix} 1 & 4 \\ 2 & 5 \end{pmatrix}, \quad A_{12} = \begin{pmatrix} 1 & 4 \\ -3 & 5 \end{pmatrix}, \quad A_{13} = \begin{pmatrix} 1 & 1 \\ -3 & 2 \end{pmatrix}$$

and our formula for the determinant of A yields

$$D(A) = 2\begin{vmatrix} 1 & 4 \\ 2 & 5 \end{vmatrix} - 1\begin{vmatrix} 1 & 4 \\ -3 & 5 \end{vmatrix} + 0\begin{vmatrix} 1 & 1 \\ -3 & 2 \end{vmatrix}$$

$$= 2(5 - 8) - 1(5 + 12) + 0$$

$$= -23.$$

Thus the determinant is a number. To compute this number in the above example, we computed the determinants of the 2×2 matrices explicitly. We can also expand these in the general definition, and thus we find a six-term expression for the determinant of a general 3×3 matrix $A = (a_{ij})$, namely:

(2)
$$D(A) = a_{11}a_{22}a_{33} - a_{11}a_{32}a_{23} - a_{12}a_{21}a_{33}$$
$$+ a_{12}a_{23}a_{31} + a_{13}a_{21}a_{32} - a_{13}a_{22}a_{31}.$$

Do not memorize (2). Remember only (1), and write down (2) only when needed for specific purposes.

We could have used the other rows to expand the determinant, instead of the first row. For instance, the expansion according to the second row is given by

$$- a_{21} \begin{vmatrix} a_{12} & a_{13} \\ a_{32} & a_{33} \end{vmatrix} + a_{22} \begin{vmatrix} a_{11} & a_{13} \\ a_{31} & a_{33} \end{vmatrix} - a_{23} \begin{vmatrix} a_{11} & a_{12} \\ a_{31} & a_{32} \end{vmatrix}$$

$$= - a_{21}D(A_{21}) + a_{22}D(A_{22}) - a_{23}D(A_{23}).$$

Again, each term is the product of a_{2j} with the determinant of the 2×2 matrix obtained by deleting the second row and j-th column, together with the appropriate sign in front of each term. This sign is determined according to the pattern:

$$\begin{pmatrix} + & - & + \\ - & + & - \\ + & - & + \end{pmatrix}.$$

If you write down the two terms for each one of the 2×2 determinants in the expansion according to the second row, you will obtain six terms, and you will find immediately that they give you the same value which we wrote down in formula (2). Thus expanding according to the second row gives the same value for the determinant as expanding according to the first row.

Furthermore, we can also expand according to any one of the columns. For instance, expanding according to the first column, we find that

$$a_{11} \begin{vmatrix} a_{22} & a_{23} \\ a_{32} & a_{33} \end{vmatrix} - a_{21} \begin{vmatrix} a_{12} & a_{13} \\ a_{32} & a_{33} \end{vmatrix} + a_{31} \begin{vmatrix} a_{12} & a_{13} \\ a_{22} & a_{23} \end{vmatrix}$$

yields precisely the same six terms as in (2), if you write down each one of the two terms corresponding to each one of the 2×2 determinants in the above expression.

Example 2. We compute the determinant

$$\begin{vmatrix} 3 & 0 & 1 \\ 1 & 2 & 5 \\ -1 & 4 & 2 \end{vmatrix}$$

by expanding according to the second column. The determinant is equal to

$$2 \begin{vmatrix} 3 & 1 \\ -1 & 2 \end{vmatrix} - 4 \begin{vmatrix} 3 & 1 \\ 1 & 5 \end{vmatrix} = 2(6 - (-1)) - 4(15 - 1) = -42.$$

Note that the presence of 0 in the first row and second column eliminates one term in the expansion, since this term is equal to 0.

If we expand the above determinant according to the third column, we find the same value, namely

$$+1\begin{vmatrix} 1 & 2 \\ -1 & 4 \end{vmatrix} - 5\begin{vmatrix} 3 & 0 \\ -1 & 4 \end{vmatrix} + 2\begin{vmatrix} 3 & 0 \\ 1 & 2 \end{vmatrix} = -42.$$

Theorem 2. *If A is a 3 × 3 matrix, then D(A) = D(^{t}A). In other words, the determinant of A is equal to the determinant of the transpose of A.*

Proof This is true because expanding D(A) according to rows or columns gives the same value, namely the expression in (2).

EXERCISES

1. Write down the expansion of a 3 × 3 determinant according to the third row, the second column, and the third column, and verify in each case that you get the same six terms as in (2).

2. Compute the following determinants by expanding according to the second row, and also according to the third column, as a check for your computation. Of course, you should find the same value.

(a) $\begin{vmatrix} 2 & 1 & 2 \\ 0 & 3 & -1 \\ 4 & 1 & 1 \end{vmatrix}$
(b) $\begin{vmatrix} 3 & -1 & 5 \\ -1 & 2 & 1 \\ -2 & 4 & 3 \end{vmatrix}$
(c) $\begin{vmatrix} 2 & 4 & 3 \\ -1 & 3 & 0 \\ 0 & 2 & 1 \end{vmatrix}$

(d) $\begin{vmatrix} 1 & 2 & -1 \\ 0 & 1 & 1 \\ 0 & 2 & 7 \end{vmatrix}$
(e) $\begin{vmatrix} -1 & 5 & 3 \\ 4 & 0 & 0 \\ 2 & 7 & 8 \end{vmatrix}$
(f) $\begin{vmatrix} 3 & 1 & 2 \\ 4 & 5 & 1 \\ -1 & 2 & -3 \end{vmatrix}$

3. Compute the following determinants.

(a) $\begin{vmatrix} 4 & 0 & 0 \\ 0 & 5 & 0 \\ 0 & 0 & 7 \end{vmatrix}$
(b) $\begin{vmatrix} -3 & 0 & 0 \\ 0 & 5 & 0 \\ 0 & 0 & -8 \end{vmatrix}$
(c) $\begin{vmatrix} 6 & 0 & 0 \\ 0 & 5 & 0 \\ 0 & 0 & -2 \end{vmatrix}$

4. Let a, b, c be numbers. In terms of a, b, c, what is the value of the determinant

$$\begin{vmatrix} a & 0 & 0 \\ 0 & b & 0 \\ 0 & 0 & c \end{vmatrix}?$$

5. Find the determinants of the following matrices.

(a) $\begin{pmatrix} 1 & 2 & 5 \\ 0 & 1 & 7 \\ 0 & 0 & 3 \end{pmatrix}$
(b) $\begin{pmatrix} -1 & 5 & 20 \\ 0 & 4 & 8 \\ 0 & 0 & 6 \end{pmatrix}$

(c) $\begin{pmatrix} 2 & -6 & 9 \\ 0 & 1 & 4 \\ 0 & 0 & 8 \end{pmatrix}$

(d) $\begin{pmatrix} -7 & 98 & 54 \\ 0 & 2 & 46 \\ 0 & 0 & -1 \end{pmatrix}$

(e) $\begin{pmatrix} 1 & 4 & 6 \\ 0 & 0 & 1 \\ 0 & 0 & 8 \end{pmatrix}$

(f) $\begin{pmatrix} 4 & 0 & 0 \\ -5 & 2 & 0 \\ 79 & 54 & 1 \end{pmatrix}$

(g) $\begin{pmatrix} 1 & 5 & 2 \\ 0 & 2 & 7 \\ 0 & 0 & 4 \end{pmatrix}$

(h) $\begin{pmatrix} -5 & 0 & 0 \\ 7 & 2 & 0 \\ -9 & 4 & 1 \end{pmatrix}$

6. In terms of the components of the matrix, what is the value of the determinant:

(a) $\begin{vmatrix} a_{11} & a_{12} & a_{13} \\ 0 & a_{22} & a_{23} \\ 0 & 0 & a_{33} \end{vmatrix}$?

(b) $\begin{vmatrix} a_{11} & 0 & 0 \\ a_{21} & a_{22} & 0 \\ a_{31} & a_{32} & a_{33} \end{vmatrix}$?

§3. ADDITIONAL PROPERTIES OF DETERMINANTS

We shall now see that 3×3 determinants satisfy the properties **D1** through **D6**, listed previously for 2×2 determinants. These properties are concerned with the columns of the matrix, and hence it is useful to use the same notation which we used before. If A^1, A^2, A^3 are the columns of the 3×3 matrix A, then we write

$$D(A) = D(A^1, A^2, A^3).$$

For the rest of this section, we assume that our column and row vectors have dimension 3; that is, that they have three components. Thus any column vector B in this section can be written in the form

$$B = \begin{pmatrix} b_1 \\ b_2 \\ b_3 \end{pmatrix}.$$

D1. *Suppose that the first column can be written as a sum,*

$$A^1 = B + C,$$

that is,

$$\begin{pmatrix} a_{11} \\ a_{21} \\ a_{31} \end{pmatrix} = \begin{pmatrix} b_1 \\ b_2 \\ b_3 \end{pmatrix} + \begin{pmatrix} c_1 \\ c_2 \\ c_3 \end{pmatrix}.$$

Then

$$D(B + C, A^2, A^3) = D(B, A^2, A^3) + D(C, A^2, A^3).$$

and the analogous rule holds with respect to the second and third columns.

Proof. We use the definition of the determinant, namely the expansion according to the first row. We see that each term splits into a sum of two terms corresponding to B and C. For instance,

$$a_{11}\begin{vmatrix} a_{22} & a_{23} \\ a_{31} & a_{33} \end{vmatrix} = b_1 \begin{vmatrix} a_{22} & a_{23} \\ a_{31} & a_{33} \end{vmatrix} + c_1 \begin{vmatrix} a_{22} & a_{23} \\ a_{31} & a_{33} \end{vmatrix},$$

$$a_{12}\begin{vmatrix} b_2 + c_2 & a_{23} \\ b_3 + c_3 & a_{33} \end{vmatrix} = a_{12}\begin{vmatrix} b_2 & a_{23} \\ b_3 & a_{33} \end{vmatrix} + a_{12}\begin{vmatrix} c_2 & a_{23} \\ c_3 & a_{33} \end{vmatrix},$$

$$a_{13}\begin{vmatrix} b_2 + c_2 & a_{22} \\ b_3 + c_3 & a_{32} \end{vmatrix} = a_{13}\begin{vmatrix} b_2 & a_{22} \\ b_3 & a_{32} \end{vmatrix} + a_{13}\begin{vmatrix} c_2 & a_{22} \\ c_3 & a_{32} \end{vmatrix}.$$

Summing with the appropriate sign yields the desired relation.

D2. *If x is a number, then*

$$D(xA^1, A^2, A^3) = x \cdot D(A^1, A^2, A^3),$$

and similarly for the other columns.

Proof. We have

$$D(xA^1, A^2, A^3) = xa_{11}\begin{vmatrix} a_{22} & a_{23} \\ a_{32} & a_{33} \end{vmatrix} - a_{12}\begin{vmatrix} xa_{21} & a_{23} \\ xa_{31} & a_{33} \end{vmatrix} + a_{13}\begin{vmatrix} xa_{21} & a_{22} \\ xa_{31} & a_{32} \end{vmatrix}$$

$$= x \cdot D(A^1, A^2, A^3).$$

The proof is similar for the other columns.

D3. *If two columns of the matrix are equal, then the determinant is equal to 0.*

Proof. Suppose that $A^1 = A^2$, and look at the expansion of the determinant according to the first row. Then $a_{11} = a_{12}$, and the first two terms cancel. The third term is equal to 0 because it involves a 2×2 determinant whose two columns are equal. The proof for the other cases is similar. (Other cases: $A^2 = A^3$ and $A^1 = A^3$.)

In the 3×3 case, we also have the **unit vectors,** namely

$$E^1 = \begin{pmatrix} 1 \\ 0 \\ 0 \end{pmatrix}, \qquad E^2 = \begin{pmatrix} 0 \\ 1 \\ 0 \end{pmatrix}, \qquad E^3 = \begin{pmatrix} 0 \\ 0 \\ 1 \end{pmatrix}.$$

and the unit **3 × 3 matrix,** namely

$$E = \begin{pmatrix} 1 & 0 & 0 \\ 0 & 1 & 0 \\ 0 & 0 & 1 \end{pmatrix}.$$

D4. *If E is the unit matrix, then $D(E) = D(E^1, E^2, E^3) = 1$.*

Proof. This is obvious from the expansion according to the first row.

Observe that to prove our basic four properties, we needed to use the definition of the determinant, i.e. its expansion according to the first row. For the remaining properties, we can give a proof which is not based directly on this expansion, but only on the formalism of **D1** through **D4**. This has the advantage of making the arguments easier, and in fact of making them completely analogous to those used in the 2 × 2 case. We carry them out.

D5. *If we add a multiple of one column to another, then the value of the determinant does not change. In other words, let x be a number. Then for instance,*

$$D(A^1, A^2 + xA^1, A^3) = D(A^1, A^2, A^3),$$

and similarly in all other cases.

Proof. We have

$$
\begin{aligned}
D(A^1, A^2 + xA^1, A^3) &= D(A^1, A^2, A^3) + D(A^1, xA^1, A^3) && \text{(by \textbf{D1})} \\
&= D(A^1, A^2, A^3) + x \cdot D(A^1, A^1, A^3) && \text{(by \textbf{D2})} \\
&= D(A^1, A^2, A^3) && \text{(by \textbf{D3})}.
\end{aligned}
$$

This proves what we wanted. The proofs of the other cases are similar.

D6. *If two adjacent columns are interchanged, then the determinant changes by a sign. In other words, we have*

$$D(A^1, A^3, A^2) = -D(A^1, A^2, A^3),$$

and similarly in the other case.

Proof. We use the same method as before. We find

$$
\begin{aligned}
0 &= D(A^1, A^2 + A^3, A^2 + A^3) \\
&= D(A^1, A^2, A^2 + A^3) + D(A^1, A^3, A^2 + A^3) \\
&= D(A^1, A^2, A^2) + D(A^1, A^2, A^3) + D(A^1, A^3, A^2) + D(A^1, A^3, A^3) \\
&= D(A^1, A^2, A^3) + D(A^1, A^3, A^2),
\end{aligned}
$$

using **D1** and **D3**. This proves **D6** in this case, and the other cases are proved similarly.

Using these rules, especially **D5**, we can compute determinants a little more efficiently. For instance, we have already noticed that when a 0 occurs in the given matrix, we can expand according to the row (or column) in which this 0 occurs, and it eliminates one term. Using **D5** repeatedly, we can change the matrix so as to get as many zeros as possible, and then reduce the computation to one term.

Furthermore, knowing that the determinant of A is equal to the determinant of its tranpose, we can also conclude that properties **D1** through **D6** hold for rows instead of columns. For instance, we can state **D6** for rows:

If two adjacent rows are interchanged, then the determinant changes by a sign.

As an exercise, state all the other properties for rows.

Example 1. Compute the determinant

$$\begin{vmatrix} 3 & 0 & 1 \\ 1 & 2 & 5 \\ -1 & 4 & 2 \end{vmatrix}.$$

We already have 0 in the first row. We subtract two times the second row from the third row. Our determinant is then equal to

$$\begin{vmatrix} 3 & 0 & 1 \\ 1 & 2 & 5 \\ -3 & 0 & -8 \end{vmatrix}.$$

We expand according to the second column. The expansion has only one term $\neq 0$, with a $+$ sign, and that is:

$$2\begin{vmatrix} 3 & 1 \\ -3 & -8 \end{vmatrix}.$$

The 2×2 determinant can be evaluated by our definition of $ad - bc$, and we find the value

$$2(-24 - (-3)) = -42.$$

Example 2. We compute the determinant

$$\begin{vmatrix} 4 & 7 & 10 \\ 3 & 7 & 5 \\ 5 & -1 & 10 \end{vmatrix}.$$

We subtract two times the second row from the first row, and then from the third row, yielding

$$\begin{vmatrix} -2 & -7 & 0 \\ 3 & 7 & 5 \\ -1 & -15 & 0 \end{vmatrix},$$

which we expand according to the third column, and get

$$-5(30 - 7) = -5(23)$$
$$= -115.$$

Note that the term has a minus sign, determined by our usual pattern of signs.

Determinants can also be defined for $n \times n$ matrices, satisfying analogous properties to **D1** through **D6**. The proofs are similar, but involve sometimes more complicated notation, so we shall not go into them.

EXERCISES

1. (a) Write out in full and prove property **D1** with respect to the second column and the third column.
   (b) Same thing for property **D2**.

2. Prove the two cases not treated in the text for property **D3**.

3. Prove **D5** in the case
   (a) you add a multiple of the third column to the first;
   (b) you add a multiple of the second column to the first;
   (c) you add a multiple of the third column to the second.

4. If you interchange the first and third columns of the given matrix, how does its determinant change? What about interchanging the first and third row?

5. Compute the following determinants.

(a) $\begin{vmatrix} 2 & 1 & 2 \\ 0 & 3 & -1 \\ 4 & 1 & 1 \end{vmatrix}$ (b) $\begin{vmatrix} 3 & -1 & 5 \\ -1 & 2 & 1 \\ -2 & 4 & 3 \end{vmatrix}$ (c) $\begin{vmatrix} 2 & 4 & 3 \\ -1 & 3 & 0 \\ 0 & 2 & 1 \end{vmatrix}$

(d) $\begin{vmatrix} 1 & 2 & -1 \\ 0 & 1 & 1 \\ 0 & 2 & 7 \end{vmatrix}$ (e) $\begin{vmatrix} -1 & 5 & 3 \\ 4 & 0 & 0 \\ 2 & 7 & 8 \end{vmatrix}$ (f) $\begin{vmatrix} 3 & 1 & 2 \\ 4 & 5 & 1 \\ -1 & 2 & -3 \end{vmatrix}$

6. Compute the following determinants.

(a) $\begin{vmatrix} 1 & 1 & 3 \\ -1 & 1 & 0 \\ 1 & 2 & 5 \end{vmatrix}$ (b) $\begin{vmatrix} 3 & 2 & 1 \\ 4 & 1 & 2 \\ 1 & 5 & 7 \end{vmatrix}$ (c) $\begin{vmatrix} 3 & 1 & 1 \\ 2 & 5 & 5 \\ 8 & 7 & 7 \end{vmatrix}$

(d) $\begin{vmatrix} 4 & -9 & 2 \\ 4 & -9 & 2 \\ 3 & 1 & 0 \end{vmatrix}$ (e) $\begin{vmatrix} 4 & -1 & 1 \\ 2 & 0 & 0 \\ 1 & 5 & 7 \end{vmatrix}$ (f) $\begin{vmatrix} 2 & 0 & 0 \\ 1 & 1 & 0 \\ 8 & 5 & 7 \end{vmatrix}$

(g) $\begin{vmatrix} 4 & 0 & 0 \\ 0 & 1 & 0 \\ 0 & 0 & 27 \end{vmatrix}$ (h) $\begin{vmatrix} 5 & 0 & 0 \\ 0 & 3 & 0 \\ 0 & 0 & 9 \end{vmatrix}$ (i) $\begin{vmatrix} 2 & -1 & 4 \\ 3 & 1 & 5 \\ 1 & 2 & 3 \end{vmatrix}$

7. In general, what is the determinant of a diagonal matrix

$$\begin{vmatrix} a_{11} & 0 & 0 \\ 0 & a_{22} & 0 \\ 0 & 0 & a_{33} \end{vmatrix}?$$

8. Compute the following determinants, making the computation as easy as you can.

(a) $\begin{vmatrix} 4 & -9 & 2 \\ 4 & -9 & 2 \\ 3 & 1 & 5 \end{vmatrix}$ (b) $\begin{vmatrix} 4 & -1 & 1 \\ 2 & 0 & 0 \\ 1 & 5 & 7 \end{vmatrix}$ (c) $\begin{vmatrix} 2 & -1 & 4 \\ 1 & 1 & 5 \\ 1 & 2 & 3 \end{vmatrix}$

(d) $\begin{vmatrix} 3 & 1 & 1 \\ 2 & 5 & 5 \\ 8 & 7 & 7 \end{vmatrix}$ (e) $\begin{vmatrix} 2 & 1 & 1 \\ 3 & 1 & 5 \\ 4 & -2 & 3 \end{vmatrix}$ (f) $\begin{vmatrix} -4 & 4 & 2 \\ 5 & 1 & 3 \\ 2 & 1 & 4 \end{vmatrix}$

(g) $\begin{vmatrix} 7 & 3 & 2 \\ 1 & -1 & 1 \\ 2 & 1 & 3 \end{vmatrix}$ (h) $\begin{vmatrix} 3 & 2 & 1 \\ 1 & 1 & 1 \\ -1 & 3 & 4 \end{vmatrix}$ (i) $\begin{vmatrix} -2 & -1 & 1 \\ 3 & 1 & -1 \\ -1 & 2 & 3 \end{vmatrix}$

(j) $\begin{vmatrix} 2 & 1 & 1 \\ 1 & 1 & 1 \\ 2 & 2 & 2 \end{vmatrix}$ (k) $\begin{vmatrix} -4 & 1 & 2 \\ 3 & 2 & 1 \\ -1 & -1 & 1 \end{vmatrix}$ (l) $\begin{vmatrix} -1 & 3 & 2 \\ 3 & -1 & 1 \\ 6 & -2 & 2 \end{vmatrix}$

9. Let c be a number and multiply each component a_{ij} of a 3×3 matrix A by c, thus obtaining a new matrix which we denote by cA. How does $D(A)$ differ from $D(cA)$?

10. Let x_1, x_2, x_3 be numbers. Show that

$$\begin{vmatrix} 1 & x_1 & x_1^2 \\ 1 & x_2 & x_2^2 \\ 1 & x_3 & x_3^2 \end{vmatrix} = (x_2 - x_1)(x_3 - x_2)(x_3 - x_1).$$

11. Suppose that A^1 is a sum of three columns, say

$$A^1 = B^1 + B^2 + B^3.$$

Using **D1** twice, prove that

$$D(B^1 + B^2 + B^3, A^2, A^3) = D(B^1, A^2, A^3) + D(B^2, A^2, A^3) + D(B^3, A^2, A^3).$$

Using summation notation, we can write this in the form

$$D(B^1 + B^2 + B^3, A^2, A^3) = \sum_{j=1}^{3} D(B^j, A^2, A^3),$$

which is shorter. In general, suppose that

$$A^1 = \sum_{j=1}^{n} B^j$$

is a sum of n columns. Using the summation notation, express similarly

$$D(A^1, A^2, A^3)$$

as a sum of (how many?) terms.

12. Let x_j ($j = 1, 2, 3$) be numbers. Let

$$A^1 = x_1 C^1 + x_2 C^2 + x_3 C^3.$$

Prove that

$$D(A^1, A^2, A^3) = \sum_{j=1}^{3} x_j D(C^j, A^2, A^3).$$

State and prove the analogous statement when

$$A^1 = \sum_{j=1}^{n} x_j C^j.$$

13. State the analogous property to that of Exercise 12 with respect to the second column. Then with respect to the third column.

14. If $a(t), b(t), c(t), d(t)$ are functions of t, one can form the determinant

$$\begin{vmatrix} a(t) & b(t) \\ c(t) & d(t) \end{vmatrix},$$

just as with numbers. Write out in full the determinant

$$\begin{vmatrix} \sin t & \cos t \\ -\cos t & \sin t \end{vmatrix}.$$

15. Write out in full the determinant

$$\begin{vmatrix} t+1 & t-1 \\ t & 2t+5 \end{vmatrix}.$$

16. Let $f(t), g(t)$ be two functions having derivatives of all orders. Let $\varphi(t)$ be the function obtained by taking the determinant

$$\varphi(t) = \begin{vmatrix} f(t) & g(t) \\ f'(t) & g'(t) \end{vmatrix}.$$

Show that

$$\varphi'(t) = \begin{vmatrix} f(t) & g(t) \\ f''(t) & g''(t) \end{vmatrix},$$

i.e. the derivative is obtained by taking the derivative of the bottom row.

17. Let

$$A(t) = \begin{pmatrix} b_1(t) & c_1(t) \\ b_2(t) & c_2(t) \end{pmatrix}$$

be a 2×2 matrix of differentiable functions. Let $B(t)$ and $C(t)$ be its column vectors. Let

$$\varphi(t) = \mathrm{Det}(A(t)).$$

Show that

$$\varphi'(t) = D(B'(t), C(t)) + D(B(t), C'(t)).$$

§4. INDEPENDENCE OF VECTORS

In the geometric applications of Chapter XIV, we studied parallelograms and parallelotopes spanned by vectors. Let us look at the situation in 3-space. Let A, B, C be vectors in $\mathbf{R}^3$, and suppose that A, B are independent. We define the **plane spanned by** A **and** B to be the set of all points

$$xA + yB,$$

with all real numbers x, y. When $x = y = 0$ we obtain the origin, so the plane passes through the origin and looks like Fig. 1.

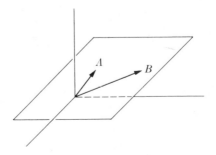

Figure 1

We say that C is **independent** of A and B if C does not lie in the above plane, i.e. if C cannot be written in the form

$$C = xA + yB$$

with some numbers x and y. Geometrically, this means that C points in a direction outside the plane, as shown on Fig. 2.

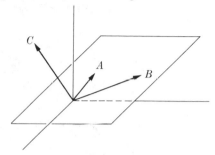

Figure 2

We shall now see that the determinant gives us a criterion when C is independent of A and B.

Theorem 3. *Let A, B, C be in $\mathbf{R}^3$. If $D(A, B, C) \neq 0$, then C is independent of A and B.*

Proof. Suppose that $C = xA + yB$ with some numbers x, y. Then

$$
\begin{aligned}
D(A, B, C) &= D(A, B, xA + yB) \\
&= D(A, B, xA) + D(A, B, yB) \\
&= xD(A, B, A) + yD(A, B, B) \\
&= 0 \qquad\qquad\qquad\qquad\qquad\text{(why?).}
\end{aligned}
$$

This is against our hypothesis, and thus proves our theorem.

EXERCISES

In the following exercises, let A, B, C be in $\mathbf{R}^3$ and assume that the determinant $D(A, B, C)$ is $\neq 0$. Prove

1. There is no number x such that $B = xA$.

2. There is no number x such that $B = xC$.

3. A is independent of B and C.

4. B is independent of A and C.

5. Let x, y, z be numbers such that $xA + yB + zC = 0$. Then $x = y = z = 0$.

6. Draw a picture of the set of all points

$$xA + yB + zC,$$

with $0 \leq x \leq 1, 0 \leq y \leq 1$, and $0 \leq z \leq 1$, in 3-space. This set is called the **box** (or **parallelotope**) spanned by A, B, C.

§5. DETERMINANT OF A PRODUCT

Theorem 4. *Let A, B be 3×3 matrices. Then*

$$D(AB) = D(A)D(B).$$

In other words, the determinant of a product is the product of the determinants.

Proof. Let $AB = C$ and let C^m be the m-th column of C. From the definition of the product of matrices, one sees that if X is a column vector, then

$$AX = x_1 A^1 + x_2 A^2 + x_3 A^3.$$

Apply this remark to each one of the columns of B successively, that is, $X = B^1$, $X = B^2$, and $X = B^3$ to find the respective columns of C. We conclude that

$$C^m = b_{1m} A^1 + b_{2m} A^2 + b_{3m} A^3.$$

Therefore

$$D(AB) = D(C) = D\left(\sum_{i=1}^{3} b_{i1} A^i, \sum_{j=1}^{3} b_{j2} A^j, \sum_{k=1}^{3} b_{k3} A^k \right)$$

$$= \sum_{i=1}^{3} \sum_{j=1}^{3} \sum_{k=1}^{3} b_{i1} b_{j2} b_{k3} D(A^i, A^j, A^k).$$

Here we have used repeatedly linearity with respect to each column. Any term on the right in the sum will be 0 if $i = j$, or $i = k$, or $j = k$. The other terms will correspond to a permutation of A^1, A^2, A^3, and there will be six such terms. If you write them out, and interchange columns making the appropriate sign change, you will find that the sum is equal to the six-term expansion for the determinant of B times the determinant of A, in other words

$$D(AB) = D(B)D(A).$$

This proves our theorem.

Observe that if A is invertible and $AB = I$, then we necessarily have $D(A) \neq 0$, because according to Theorem 4,

$$1 = D(I) = D(A)D(B).$$

The converse is also true, that is: If $D(A) \neq 0$, then A is invertible. We shall discuss it in the next section.

§6. INVERSE OF A MATRIX

Theorem 5. *Let A be a square matrix such that $D(A) \neq 0$. Then A is invertible.*

Let us consider the 2×2 case. Let

$$A = \begin{pmatrix} a & b \\ c & d \end{pmatrix}$$

be a 2×2 matrix, and assume that its determinant $ad - bc \neq 0$. We wish to find an inverse for A, that is a 2×2 matrix

$$X = \begin{pmatrix} x & y \\ z & w \end{pmatrix}$$

such that

$$AX = XA = I.$$

Let us look at the first requirement, $AX = I$, which, written out in full, looks like this:

$$\begin{pmatrix} a & b \\ c & d \end{pmatrix}\begin{pmatrix} x & y \\ z & w \end{pmatrix} = \begin{pmatrix} 1 & 0 \\ 0 & 1 \end{pmatrix}.$$

Let us look at the first column of AX. We must solve the equations

(*) $ax + bz = 1, \qquad cx + dz = 0.$

This is a system of two equations in two unknowns, x and z, which we know how to solve. Similarly, looking at the second column, we see that we must solve a system of two equations in the unknowns y, w, namely

(**) $ay + bw = 0, \qquad cy + dw = 1.$

Example. Let

$$A = \begin{pmatrix} 2 & 1 \\ 4 & 3 \end{pmatrix}.$$

We seek a matrix X such that $AX = I$. We must therefore solve the systems of linear equations

$$2x + z = 1, \qquad\qquad 2y + w = 0,$$
$$\text{and}$$
$$4x + 3z = 0, \qquad\qquad 4y + 3w = 1.$$

By the ordinary method of solving two equations in two unknowns, we find

$$x = 1, \quad z = -1 \quad \text{and} \quad y = -\tfrac{1}{2}, \quad w = 1.$$

Thus the matrix

$$X = \begin{pmatrix} 1 & -\tfrac{1}{2} \\ -1 & 1 \end{pmatrix}$$

is such that $AX = I$. The reader will also verify by direct multiplication that $XA = I$. This solves for the desired inverse.

The same procedure, of course, works for the general systems (*) and (**). Consider (*). Multiply the first equation by d, multiply the second equation by b, and subtract. We get

$$(ad - bc)x = d,$$

whence

$$x = \frac{d}{ad - bc}.$$

We see that the determinant of A occurs in the denominator. You can solve similarly for y, z, w and you will find similar expressions with only $D(A)$ in the denominator. This proves Theorem 5 in the 2×2 case.

The proof in the 3×3 case is also done by solving linear equations, but we shall omit it.

EXERCISES

1. Find the inverses of the following matrices.

(a) $\begin{pmatrix} 2 & -1 \\ 5 & 2 \end{pmatrix}$ (b) $\begin{pmatrix} 3 & 4 \\ -2 & 1 \end{pmatrix}$ (c) $\begin{pmatrix} 5 & 1 \\ 1 & 2 \end{pmatrix}$ (d) $\begin{pmatrix} -2 & -1 \\ -3 & -4 \end{pmatrix}$

2. Write down the general formula for the inverse of a 2×2 matrix

$$\begin{pmatrix} a & b \\ c & d \end{pmatrix}.$$

CHAPTER XVI

Applications to Functions of Several Variables

§1. THE JACOBIAN MATRIX

Throughout this section, all our vectors will be vertical vectors. We let $D_1, \ldots, D_n$ be the usual partial derivatives. Thus $D_i = \partial/\partial x_i$.

Let $F: \mathbf{R}^n \longrightarrow \mathbf{R}^m$ be a mapping. We can represent F by coordinate functions. In other words, there exist functions $f_1, \ldots, f_m$ such that

$$F(X) = \begin{pmatrix} f_1(X) \\ f_2(X) \\ \vdots \\ f_m(X) \end{pmatrix} = {}^t(f_1(X), \ldots, f_m(X)).$$

To simplify the typography, we shall sometimes write a vertical vector as the transpose of a horizontal vector, as we have just done.

We view X as a column vector, $X = {}^t(x_1, \ldots, x_n)$.

Let us assume that the partial derivatives of each function f_i $(i = 1, \ldots, m)$ exist. We can then form the matrix of partial derivatives:

$$\left(\frac{\partial f_i}{\partial x_j} \right) = \begin{pmatrix} \dfrac{\partial f_1}{\partial x_1} & \dfrac{\partial f_1}{\partial x_2} & \cdots & \dfrac{\partial f_1}{\partial x_n} \\[2ex] \dfrac{\partial f_2}{\partial x_1} & \dfrac{\partial f_2}{\partial x_2} & \cdots & \dfrac{\partial f_2}{\partial x_n} \\[2ex] \vdots & \vdots & & \vdots \\[2ex] \dfrac{\partial f_m}{\partial x_1} & \dfrac{\partial f_m}{\partial x_2} & \cdots & \dfrac{\partial f_m}{\partial x_n} \end{pmatrix} = \begin{pmatrix} D_1 f_1(X) & \cdots & D_n f_1(X) \\ \vdots & & \vdots \\ D_1 f_m(X) & \cdots & D_n f_m(X) \end{pmatrix}$$

$i = 1, \ldots, m$ and $j = 1, \ldots, n$. This matrix is called the **Jacobian** matrix of F, and is denoted by $J_F(X)$.

In the case of two variables (x, y), say F is given by functions (f, g), so that

$$F(x, y) = (f(x, y), g(x, y)),$$

then Jacobian matrix is

$$J_F(x, y) = \begin{pmatrix} \dfrac{\partial f}{\partial x} & \dfrac{\partial f}{\partial y} \\ \dfrac{\partial g}{\partial x} & \dfrac{\partial g}{\partial y} \end{pmatrix}.$$

(As we have done just now, we sometimes write the vectors horizontally, although to be strictly correct, they should be written vertically.)

Example 1. Let $F: \mathbf{R}^2 \to \mathbf{R}^2$ be the mapping defined by

$$F(x, y) = \begin{pmatrix} x^2 + y^2 \\ e^{xy} \end{pmatrix} = \begin{pmatrix} f(x, y) \\ g(x, y) \end{pmatrix}.$$

Find the Jacobian matrix $J_F(P)$ for $P = (1, 1)$.

The Jacobian matrix at an arbitrary point (x, y) is

$$\begin{pmatrix} \dfrac{\partial f}{\partial x} & \dfrac{\partial f}{\partial y} \\ \dfrac{\partial g}{\partial x} & \dfrac{\partial g}{\partial y} \end{pmatrix} = \begin{pmatrix} 2x & 2y \\ ye^{xy} & xe^{yx} \end{pmatrix}.$$

Hence when $x = 1, y = 1$, we find:

$$J_F(1, 1) = \begin{pmatrix} 2 & 2 \\ e & e \end{pmatrix}.$$

Example 2. Let $F: \mathbf{R}^2 \to \mathbf{R}^3$ be the mapping defined by

$$F(x, y) = \begin{pmatrix} xy \\ \sin x \\ x^2 y \end{pmatrix}.$$

Find $J_F(P)$ at the point $P = (\pi, \pi/2)$.

The Jacobian matrix at an arbitrary point (x, y) is

$$J_F(x, y) = \begin{pmatrix} y & x \\ \cos x & 0 \\ 2xy & x^2 \end{pmatrix}.$$

Hence

$$J_F\left(\pi, \frac{\pi}{2}\right) = \begin{pmatrix} \pi/2 & \pi \\ -1 & 0 \\ \pi^2 & \pi^2 \end{pmatrix}.$$

Example 3. Let $f: \mathbf{R}^n \rightarrow \mathbf{R}$ be a function of n variables. Then its Jacobian matrix is simply the row vector

$$\left(\frac{\partial f}{\partial x_1}, \ldots, \frac{\partial f}{\partial x_n}\right),$$

and this is just the gradient grad $f(X)$ studied in the early chapters.

For an arbitrary mapping

$$F: \mathbf{R}^n \rightarrow \mathbf{R}^m$$

we observe that the row vectors of the Jacobian matrix are the gradients of the coordinate functions $f_1, \ldots, f_m$, so we may rewrite the Jacobian matrix as

$$J_F(X) = \begin{pmatrix} \text{grad } f_1(X) \\ \vdots \\ \text{grad } f_m(X) \end{pmatrix}.$$

Thus the Jacobian matrix is a generalization of the gradient.

Let U be open in $\mathbf{R}^n$ and $F: U \rightarrow \mathbf{R}^n$ be a map into the same dimensional space. Then the Jacobian matrix $J_F(X)$ is a square matrix, and its determinant is called the **Jacobian determinant** of F at X. We denote it by

$$\Delta_F(X).$$

Example 4. Let F be as in Example 1, $F(x, y) = (x^2 + y^2, e^{xy})$. Then the Jacobian determinant is equal to

$$\Delta_F(x, y) = \begin{vmatrix} 2x & 2y \\ ye^x & xe^y \end{vmatrix} = 2x^2 e^y - 2y^2 e^x.$$

In particular,

$$\Delta_F(1, 1) = 2e - 2e = 0,$$
$$\Delta_F(1, 2) = 2e^2 - 8e.$$

Example 5. An important map is given by the polar coordinates,

$$F: \mathbf{R}^2 \rightarrow \mathbf{R}^2$$

such that

$$F(r, \theta) = (r \cos \theta, r \sin \theta),$$

We can view the map as defined on all of $\mathbf{R}^2$, although when selecting polar coordinates, we take $r \geq 0$. We see that F maps a rectangle into a circular sector (Fig. 1).

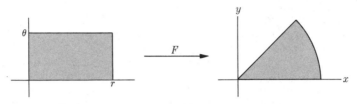

Figure 1

It is easy to compute the Jacobian matrix and determinant. Do this as Exercise 7.

EXERCISES

1. In each of the following cases, compute the Jacobian matrix of F.
   (a) $F(x, y) = (x + y, x^2y)$ (b) $F(x, y) = (\sin x, \cos xy)$
   (c) $F(x, y) = (e^{xy}, \log x)$ (d) $F(x, y, z) = (xz, xy, yz)$
   (e) $F(x, y, z) = (xyz, x^2z)$ (f) $F(x, y, z) = (\sin xyz, xz)$

2. Find the Jacobian matrix of the mappings in Exercise 1 evaluated at the following points.
   (a) $(1, 2)$ (b) $(\pi, \pi/2)$ (c) $(1, 4)$
   (d) $(1, 1, -1)$ (e) $(2, -1, -1)$ (f) $(\pi, 2, 4)$

3. Find the Jacobian matrix of the following maps.
   (a) $F(x, y) = (xy, x^2)$ (b) $F(x, y, z) = (\cos xy, \sin xy, xz)$

4. Find the Jacobian determinant of the map in Exercise 1(a). Determine all points where the Jacobian determinant is equal to 0.

5. Find the Jacobian determinant of the map in Exercise 1(b).

6. Let $F: \mathbf{R}^2 \rightarrow \mathbf{R}^2$ be the map defined by

$$F(r, \theta) = (r \cos \theta, r \sin \theta),$$

in other words the polar coordinates map

$$x = r \cos \theta, \qquad y = r \sin \theta.$$

Find the Jacobian matrix and Jacobian determinant of this mapping. Determine all points (r, θ) where the Jacobian determinant vanishes.

7. Let $F: \mathbf{R}^3 \rightarrow \mathbf{R}^3$ be the mapping defined by

$$F(r, \theta, \varphi) = (r \sin \varphi \cos \theta, r \sin \varphi \sin \theta, r \cos \varphi)$$

or in other words

$$x = r \sin \varphi \cos \theta, \qquad y = r \sin \varphi \sin \theta, \qquad z = r \cos \varphi.$$

Find the Jacobian matrix and Jacobian determinant of this mapping.

8. Find the Jacobian matrix and determinant of the map

$$F(r, \theta) = (e^r \cos \theta, e^r \sin \theta).$$

Show that the Jacobian determinant is never 0. Show that there exist two distinct points (r_1, θ_1) and (r_2, θ_2) such that

$$F(r_1, \theta_1) = F(r_2, \theta_2).$$

§2. DIFFERENTIABILITY

Let U be an open set in $\mathbf{R}^n$. Let

$$F: U \rightarrow \mathbf{R}^m$$

be a mapping. Let X be a point of U. Let

$$F(X) = {}^t(f_1(X), \ldots, f_m(X))$$

be the coordinate functions of F. We shall say that F is **differentiable** at X if all the partial derivatives

$$D_i f_j(X)$$

exist (so the Jacobian matrix $J_F(X)$ exists), and if there exists a mapping G, defined for sufficiently small vectors H such that

$$F(X + H) = F(X) + J_F(X)H + \|H\|G(H)$$

and

$$\lim_{\|H\| \to 0} G(H) = O.$$

Observe that this definition is entirely analogous to the definition of differentiability of a function given in Chapter III. In writing

$$J_F(X)H,$$

we must of course view H as a column vector,

$$H = \begin{pmatrix} h_1 \\ h_2 \\ \vdots \\ h_n \end{pmatrix}.$$

Then we see that

$$J_F(X)H = \begin{pmatrix} \operatorname{grad} f_1(X) \cdot H \\ \vdots \\ \operatorname{grad} f_m(X) \cdot H \end{pmatrix}.$$

Theorem 1. *Let U be an open set in* $\mathbf{R}^n$. *Let* $F: U \to \mathbf{R}^m$ *be a mapping, having coordinate functions* $f_1, \ldots, f_m$. *Assume that each function* f_i *is differentiable at a point X of U. Then F is differentiable at X.*

Proof. For each integer i between 1 and n, there is a function g_i such that

$$\lim_{\|H\| \to 0} g_i(H) = 0,$$

and such that we can write

$$f_i(X + H) = f_i(X) + \operatorname{grad} f_i(X) \cdot H + \|H\| g_i(H).$$

We view X and $F(X)$ as vertical vectors. By definition, we can then write

$$F(X + H) = {}^t(f_1(X + H), \ldots, f_m(X + H)).$$

Hence

$$F(X + H) = \begin{pmatrix} f_1(X) \\ \vdots \\ f_m(X) \end{pmatrix} + \begin{pmatrix} \operatorname{grad} f_1(X) \cdot H \\ \vdots \\ \operatorname{grad} f_m(X) \cdot H \end{pmatrix} + \|H\| \begin{pmatrix} g_1(H) \\ \vdots \\ g_m(H) \end{pmatrix}.$$

The term in the middle, involving the gradients, is precisely equal to the product of the Jacobian matrix, times H, i.e. to

$$J_F(X)H.$$

Let $G(H) = {}^t(g_1(H), \ldots, g_m(H))$ be the vector on the right. Then

$$F(X + H) = F(X) + J_F(X)H + \|H\| G(H).$$

As $\|H\|$ approaches 0, each coordinate of $G(H)$ approaches 0. Hence $G(H)$ approaches O; in other words,

$$\lim_{\|H\| \to 0} G(H) = O.$$

This proves the theorem.

Observe that the Jacobian matrix $J_F(X)$ when applied to H may be viewed as a linear map.

It is convenient to use the standard notation for the derivative in one variable, and write

$$F'(X) \qquad \text{instead of} \qquad J_F(X)$$

when we interpret the Jacobian matrix as a linear map.

§3. THE CHAIN RULE

In the *First Course*, we proved a chain rule for composite functions. Earlier in this book, a chain rule was given for a composite of a function and a map defined for real numbers, but having values in $\mathbf{R}^n$. In this section, we give a general formulation of the chain rule for arbitrary compositions of mappings.

Let U be an open set in $\mathbf{R}^n$, and let V be an open set in $\mathbf{R}^m$. Let $F: U \to \mathbf{R}^m$ be a mapping, and assume that all values of F are contained in V. Let $G: V \to \mathbf{R}^s$ be a mapping. Then we can form the composite mapping $G \circ F$ from U into $\mathbf{R}^s$ (Fig. 2).

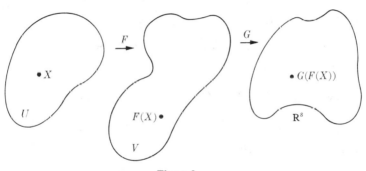

Figure 2

The next theorem tells us what the derivative of $G \circ F$ is in terms of the derivative of F at X, and the derivative of G at $F(X)$.

Theorem 2. *Let U be an open set in $\mathbf{R}^n$, let V be an open set in $\mathbf{R}^m$. Let*

$$F: U \to V \qquad \text{and} \qquad G: V \to \mathbf{R}^s$$

be mappings. Let X be a point of U such that F is differentiable at X. Assume that G is differentiable at $F(X)$. Then the composite mapping $G \circ F$ is differentiable at X, and its derivative is given by

$$(G \circ F)'(X) = G'(F(X)) \circ F'(X).$$

Proof. By definition of differentiability, there exists a mapping Φ_1 such that

$$\lim_{\|H\| \to 0} \Phi_1(H) = O$$

and

$$F(X + H) = F(X) + F'(X)H + \|H\|\Phi_1(H).$$

Similarly, there exists a mapping Φ_2 such that

$$\lim_{\|K\|\to 0} \Phi_2(K) = O,$$

and

$$G(Y + K) = G(Y) + G'(Y)K + \|K\|\Phi_2(K).$$

We let $K = K(H)$ be

$$K = F(X + H) - F(X) = F'(X)H + \|H\|\Phi_1(H).$$

Then

$$G(F(X + H)) = G(F(X) + K)$$

$$= G(F(X)) + G'(F(X))K + \|K\|\Phi_2(K).$$

Using the fact that $G'(F(X))$ is linear, and

$$K = F(X + H) - F(X) = F'(X)H + \|H\|\Phi_1(H),$$

we can write

$$(G \circ F)(X + H) = (G \circ F)(X) + G'(F(X))F'(X)H$$

$$+ \|H\|G'(F(X))\Phi_1(H) + \|K\|\Phi_2(K).$$

Using simple estimates which we do not give in detail, we conclude that

$$(G \circ F)(X + H) = (G \circ F)(X) + G'(F(X))F'(X)H + \|H\|\Phi_3(H).$$

where

$$\lim_{\|H\|\to 0} \Phi_3(H) = O.$$

This proves the theorem.

Observe how the proof follows the same pattern as the old proof for the chain rule in Chapter IV. We used the notation $F'(X)$ and $G'(F(X))$ to be as close as possible to the old notation for the derivative in the calculus of one variable. We could of course write down the Jacobian matrices instead of this notation, and we obtain the formula:

$$\boxed{J_{G \circ F}(X) = J_G(F(X))J_F(X).}$$

Thus the Jacobian matrix of the composite mapping $G \circ F$ is the product of the

Jacobian matrices of G and F respectively, evaluated at the appropriate points, namely

$$J_G(F(X)) \qquad and \qquad J_F(X).$$

The discussion of inverse mappings and implicity functions, which follows in the next two sections, is independent of the discussion of the Hessian in §6. They may thus be covered in any order at the discretion of the instructor. Furthermore §6 is not necessary for the considerations which follow.

§4. INVERSE MAPPINGS

Let U be open in $\mathbf{R}^n$ and let $F: U \to \mathbf{R}^n$ be a map, given by coordinate functions:

$$F(X) = (f_1(X), \ldots, f_n(X)).$$

If all the partial derivatives of all functions f_i exist and are continuous, we say that F is a C^1-**map.** We say that F is C^1-**invertible** on U if the image $F(U)$ is an open set V, and if there exists a C^1-map $G: V \to U$ such that $G \circ F$ and $F \circ G$ are the respective identity mappings on U and V (Fig. 3).

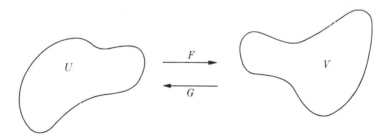

Figure 3

Example 1. Let A be a fixed vector, and let $F: \mathbf{R}^n \to \mathbf{R}^n$ be the translation by A, namely $F(X) = X + A$. Then F is C^1-invertible, its inverse being translation by $-A$.

Example 2. Let U be the subset of $\mathbf{R}^2$ consisting of all pairs (r, θ) with $r > 0$ and $0 < \theta < \pi$. Let

$$F(r, \theta) = (r \cos \theta, r \sin \theta).$$

Let $x = r \cos \theta$ and $y = r \sin \theta$. Then the image of U is the upper half-plane consisting of all (x, y) such that $y > 0$, and arbitrary x (Fig. 4).

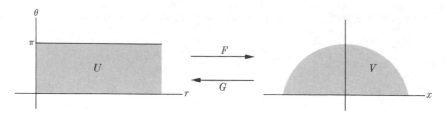

Figure 4

We can solve for the inverse map G, namely:

$$r = \sqrt{x^2 + y^2} \qquad \text{and} \qquad \theta = \arccos \frac{x}{r}$$

so that

$$G(x, y) = \left(\sqrt{x^2 + y^2}, \arccos \frac{x}{r} \right).$$

In many applications, a map is not necessarily invertible, but has still a useful property locally. Let P be a point of U. We say that F is **locally** C^1-**invertible at** P if there exists an open set U_1 contained in U and containing P such that F is C^1-invertible on U_1.

Example 3. If we view $F(r, \theta) = (r \cos \theta, r \sin \theta)$ as defined on all of $\mathbf{R}^2$, then F is not C^1-invertible on all of $\mathbf{R}^2$, but given any point other than the origin, it is locally invertible at that point. One could see this by giving an explicit inverse map as we did in Example 2. At any rate, from Example 2, we see that F is C^1-invertible on the set $r > 0$ and $0 < \theta < \pi$.

In most cases, it is not possible to define an inverse map by explicit formulas. However, there is a very important theorem which allows us to conclude that a map is locally invertible at a point.

Inverse Mapping Theorem. *Let $F: U \to \mathbf{R}^n$ be a C^1-map. Let P be a point of U. If the Jacobian determinant $\Delta_F(P)$ is not equal to 0, then F is locally C^1-invertible at P.*

A proof of this theorem is too involved to be given in this book. However, we make the following comment. The fact that the determinant $\Delta_F(P)$ is not 0 implies (and in fact is equivalent with) the fact that the Jacobian matrix is invertible. Since it is usually very easy to determine whether the Jacobian determinant vanishes or not, we see that the inverse mapping theorem gives us a simple criterion for local invertibility.

Example 4. Consider the case of one variable, $y = f(x)$. In the *First Course*, we proved that if $f'(x_0) \neq 0$ at a point x_0, then there is an inverse function defined near $y_0 = f(x_0)$. Indeed, say $f'(x_0) > 0$. By continuity, assuming that f' is continuous (i.e. f is C^1), we know that $f'(x) > 0$ for x close to x_0. Hence f is strictly increasing, and an inverse function exists near x_0. In fact, we determined the derivative. If g is the inverse function, then we proved that

$$g'(y_0) = f'(x_0)^{-1}.$$

Example 5. The formula for the derivative of the inverse function in the case of one variable can be generalized to the case of the inverse mapping theorem. Suppose that the map $F: U \to V$ has a C^1-inverse $G: V \to U$. Let X be a point of U. Then $G \circ F = I$ is the identity, and since I is linear, we see directly from the definition of the derivative that $I'(X) = I$. Using the chain rule, we find that

$$I = (G \circ F)'(X) = G'(F(X)) \circ F'(X)$$

for all X in U. In particular, this means that if $Y = F(X)$, then

$$\boxed{G'(Y) = F'(X)^{-1}}$$

where the inverse in this last expression is to be understood as the inverse of the linear map $F'(X)$. Thus we have generalized the formula for the derivative of an inverse function.

Example 6. Let $F(x, y) = (e^x \cos y, e^x \sin y)$. Show that F is locally invertible at every point.

We find that

$$J_F(x, y) = \begin{pmatrix} e^x \cos y & -e^x \sin y \\ e^x \sin y & e^x \cos y \end{pmatrix}, \quad \text{whence} \quad \Delta_F(x, y) = e^{2x} \neq 0.$$

Since the Jacobian determinant is not 0, it follows that F is locally invertible at (x, y) for all x, y.

EXERCISES

1. Determine whether the following mappings are locally C^1-invertible at the given point.
   (a) $F(x, y) = (x^2 - y^2, 2xy)$ at $(x, y) \neq (0, 0)$
   (b) $F(x, y) = (x^3 y + 1, x^2 + y^2)$ at $(1, 2)$
   (c) $F(x, y) = (x + y, y^{1/4})$ at $(1, 16)$
   (d) $F(x, y) = \left(\dfrac{x}{x^2 + y^2}, \dfrac{y}{x^2 + y^2} \right)$ at $(x, y) \neq (0, 0)$
   (e) $F(x, y) = (x + x^2 + y, x^2 + y^2)$ at $(x, y) = (5, 8)$

2. Determine whether the following mappings are locally C^1-invertible at the indicated point.
   (a) $F(x, y) = (x + y, x^2 y)$ at $(1, 2)$
   (b) $F(x, y) = (\sin x, \cos xy)$ at $(\pi, \pi/2)$
   (c) $F(x, y) = (e^{xy}, \log x)$ at $(1, 4)$
   (d) $F(x, y, z) = (xz, xy, yz)$ at $(1, 1, -1)$

3. Show that the map defined by $F(x, y) = (e^x \cos y, e^x \sin y)$ is not invertible on all of $\mathbf{R}^2$, even though it is locally invertible everywhere.

§5. IMPLICIT FUNCTIONS

Let U be an open set in 2-space, and let

$$f: U \to \mathbf{R}$$

be a C^1-function. Let (a, b) be a point of U, and let

$$f(a, b) = c.$$

We ask whether there is some differentiable function $y = \varphi(x)$ defined near $x = a$ such that $\varphi(a) = b$ and

$$f(x, \varphi(x)) = c$$

for all x near a. If such a function φ exists, then we say that $y = \varphi(x)$ is the **function determined implicitly by** f.

Implicit Function Theorem. *Let U be open in $\mathbf{R}^2$ and let $f: U \to \mathbf{R}$ be a C^1-function. Let (a, b) be a point of U, and let $f(a, b) = c$. Assume that $D_2 f(a, b) \neq 0$. Then there exists an implicit function $y = \varphi(x)$ which is C^1 in some interval containing a, and such that $\varphi(a) = b$.*

Before giving the proof, we discuss some examples.

Example 1. Let $f(x, y) = x^2 + y^2$ and let $(a, b) = (1, 1)$. Then $c = f(1, 1) = 2$. We have $D_2 f(x, y) = 2y$ so that

$$D_2 f(1, 1) = 2 \neq 0,$$

so the implicit function $y = \varphi(x)$ near $x = 1$ exists. In this case, we can of course solve explicitly for y, namely

$$y = \sqrt{2 - x^2}.$$

Example 2. We take $f(x, y) = x^2 + y^2$ as in Example 1, and

$$(a, b) = (-1, -1).$$

Then again $c = f(-1, -1) = 2$, and

$$D_2 f(-1, -1) = -2 \neq 0.$$

In this case we can still solve for y in terms of x, namely

$$y = -\sqrt{2 - x^2} .$$

In general, the equation $f(x, y) = c$ defines some curve as in the following picture (Fig. 5).

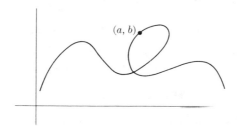

Figure 5

Near the point (a, b) as indicated in the picture, we see that there is an implicit function (Fig. 6):

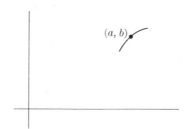

Figure 6

but that one could not define the implicit function for all x, only for those x near a.

Example 3. Let $f(x, y) = x^2 y + 3y^3 x^4 - 4$. Take $(a, b) = (1, 1)$ so that $f(a, b) = 0$. Then $D_2 f(x, y) = x^2 + 9y^2 x^4$ and

$$D_2 f(1, 1) = 10 \neq 0.$$

Hence the implicit function $y = \varphi(x)$ exists, but there is no simple way to solve for it. We can also determine the derivative $\varphi'(1)$. Indeed, differentiating the equation $f(x, y) = 0$, knowing that $y = \varphi(x)$ is a differentiable function, we find

$$2xy + x^2y' + 12y^3x^3 + 9y^2y'x^4 = 0,$$

whence we can solve for $y' = \varphi'(x)$, namely

$$\varphi'(x) = y' = -\frac{2xy + 12y^3x^3}{x^2 + 9y^2x^4}.$$

Hence

$$\varphi'(1) = -\frac{2 + 12}{1 + 9} = -\frac{7}{5}.$$

In Exercise 4 we give the general formula for an arbitrary function f.

Example 4. In general, given any function $f(x, y) = 0$ and $y = \varphi(x)$ we can find $\varphi'(x)$ by differentiating in the usual way. For instance, suppose

$$x^3 + 4y \sin(xy) = 0.$$

Then taking the derivative with respect to x, we find

$$3x^2 + 4y' \sin(xy) + 4y \cos(xy)(y + xy').$$

We then solve for y' as

$$y' = -\frac{4y^2 \cos(xy) + 3x^2}{4 \sin(xy) + 4xy \cos(xy)}$$

whenever $4 \sin(xy) + 4xy \cos(xy) \neq 0$. Similarly, we can solve for y'' by differentiating either of the last two expressions. In the present case, this gets complicated.

Proof of the theorem. We shall first need explicitly a special case of the inverse mapping theorem, stated as a lemma.

Lemma. *Let U be open in $\mathbf{R}^2$ and let $f \colon U \to \mathbf{R}$ be a C^1-function. Let (a, b) be a point of U. Assume that $D_2 f(a, b) \neq 0$. Then the map F given by*

$$(x, y) \mapsto F(x, y) = (x, f(x, y))$$

is locally invertible at (a, b).

Proof. All we have to do is compute the Jacobian matrix and determinant. We have

$$J_F(x, y) = \begin{pmatrix} 1 & 0 \\ \dfrac{\partial f}{\partial x} & \dfrac{\partial f}{\partial y} \end{pmatrix}$$

so that

$$J_F(a, b) = \begin{pmatrix} 1 & 0 \\ D_1 f(a, b) & D_2 f(a, b) \end{pmatrix}$$

and hence

$$\Delta_F(a, b) = D_2 f(a, b).$$

By assumption, this is not 0, and the inverse mapping theorem implies what we want.

We apply the lemma to the situation of the theorem. Thus we let

$$F(x, y) = (x, f(x, y)).$$

We know that $F(a, b) = (a, c)$ and that there exists a C^1-inverse G defined locally near (a, c). The inverse map G has two coordinate functions, and we can write $G(x, z) = (x, g(x, z))$ for some function g. Thus we put $y = g(x, z)$, and $z = f(x, y)$. We define

$$\psi(x) = g(x, c).$$

Then on the one hand,

$$F(x, \varphi(x)) = F(x, g(x, c)) = F(G(x, c)) = (x, c),$$

and on the other hand,

$$F(x, \varphi(x)) = (x, f(x, \varphi(x))).$$

This proves that $f(x, \varphi(x)) = c$. Furthermore, by definition of an inverse map, $G(a, c) = (a, b)$ so that $\varphi(a) = b$. This proves the implicit function theorem.

EXERCISES

1. Let $y = \varphi(x)$ be an implicit function satisfying $f(x, \varphi(x)) = 0$, both f, φ being C^1. Show that

$$\varphi'(x) = -\frac{D_1 f(x, \varphi(x))}{D_2 f(x, \varphi(x))}$$

wherever $D_2 f(x, \varphi(x)) \neq 0$.

2. Find an expression for $\varphi''(x)$ by differentiating the preceding expression for $\varphi'(x)$.

3. Let $f(x, y) = (x - 2)^3 y + xe^{y-1}$. Is $D_2 f(a, b) \neq 0$ at the following points (a, b)?
   (a) $(1, 1)$ (b) $(0, 0)$ (c) $(2, 1)$

4. For each of the following functions f, show that $f(x, y) = 0$ defines an implicit function $y = \varphi(x)$ at the given point (a, b), and find $\varphi'(a)$.
   (a) $f(x, y) = x^2 - xy + y^2 - 3$ at $(1, 2)$
   (b) $f(x, y) = x \cos xy$ at $(1, \pi/2)$

(c) $f(x, y) = 2e^{x+y} - x + y$ at $(1, -1)$
(d) $f(x, y) = xe^y - y + 1$ at $(-1, 0)$
(e) $f(x, y) = x + y + x \sin y$ at $(0, 0)$
(f) $f(x, y) = x^5 + y^5 + xy + 4$ at $(2, -2)$

5. Let f be a C^1-function of 3 variables (x, y, z) defined on an open set U of $\mathbf{R}^3$. Let (a, b, c) be a point of U, and assume $f(a, b, c) = 0$, $D_3 f(a, b, c) \neq 0$. Show that there exists a C^1-function $\varphi(x, y)$ defined near (a, b) such that

$$f(x, y, \varphi(x, y)) = 0 \qquad \text{and} \qquad \varphi(a, b) = c.$$

We call φ the implicit function $z = \varphi(x, y)$ determined by f at (a, b).

6. In Exercise 5, show that

$$D_1\varphi(a, b) = -\frac{D_1 f(a, b, c)}{D_3 f(a, b, c)}.$$

7. For each of the following functions $f(x, y, z)$, show that $f(x, y, z) = 0$ defines an implicit function $z = \varphi(x, y)$ at the given point (a, b, c) and find $D_1\varphi(a, b)$ and $D_2\varphi(a, b)$.
(a) $f(x, y, z) = x + y + z + \cos xyz$ at $(0, 0, -1)$
(b) $f(x, y, z) = z^3 - z - xy \sin z$ at $(0, 0, 0)$
(c) $f(x, y, z) = x^3 + y^3 + z^3 - 3xyz - 4$ at $(1, 1, 2)$
(d) $f(x, y, z) = x + y + z - e^{xyz}$ at $(0, \frac{1}{2}, \frac{1}{2})$

8. Let $f(x, y, z) = x^3 - 2y^2 + z^2$. Show that $f(x, y, z) = 0$ defines an implicit function $x = \varphi(y, z)$ at the point $(1, 1, 1)$. Find $D_1\varphi$ and $D_2\varphi$ at the point $(1, 1)$.

9. If possible, show that $f(x, y, z) = 0$ in Exercise 7 also determines y as an implicit function of (x, z) and x as an implicit function of (y, z). Find the partial derivatives of these functions at the given point.

§6. THE HESSIAN

Let U be an open set in $\mathbf{R}^n$ and let

$$f: U \to R$$

be a function which is twice continuously differentiable. Let P be a point of U. We have already defined P to be a **critical point** if

$$\operatorname{grad} f(P) = O,$$

or in other words,

$$D_i f(P) = 0 \qquad \text{for} \quad i = 1, \dots, n.$$

It is then interesting to look at the analogue of the second derivative, which for functions of several variables is called the Hessian. If f is a function of

$X = (x_1, \ldots, x_n)$, then its **Hessian** $H_f(X)$ is the matrix

$$H_f(X) = \left(\frac{\partial^2 f}{\partial x_i \, \partial x_j} \right).$$

Example. Suppose $n = 2$ and the variables are x, y. Then

$$H_f(X) = \begin{pmatrix} \dfrac{\partial^2 f}{\partial x^2} & \dfrac{\partial^2 f}{\partial x \, \partial y} \\[2mm] \dfrac{\partial^2 f}{\partial y \, \partial x} & \dfrac{\partial^2 f}{\partial y^2} \end{pmatrix} = \begin{pmatrix} D_1^2 f(X) & D_1 D_2 f(X) \\ D_1 D_2 f(X) & D_2^2 f(X) \end{pmatrix}.$$

When we discussed relative maxima and minima in Chapter XII, we encountered quadratic forms. We may now use matrix notation to express quadratic forms. Suppose we have a quadratic form

$$q(x, y) = ax^2 + 2bxy + cy^2.$$

We may write this in terms of matrices as the product

$$(x, y) \begin{pmatrix} a & b \\ b & c \end{pmatrix} \begin{pmatrix} x \\ y \end{pmatrix}.$$

The partial product

$$(x, y) \begin{pmatrix} a & b \\ b & c \end{pmatrix} = (ax + by, \; bx + cy)$$

is a row vector, which, when multiplied by the column vector $\begin{pmatrix} x \\ y \end{pmatrix}$ yields precisely the value $q(x, y)$, which is a *number*. The matrix

$$A = \begin{pmatrix} a & b \\ b & d \end{pmatrix}$$

is called the **matrix associated with the quadratic form.**

Let us apply this to the Hessian. Let $P = (p_1, p_2)$. Then

$$H_f(P) = \begin{pmatrix} a & b \\ b & c \end{pmatrix}$$

where

$$a = D_1^2 f(p_1, p_2), \qquad b = D_1 D_2 f(p_1, p_2), \qquad c = D_2^2 f(p_1, p_2).$$

Thus the quadratic form associated with the Hessian is precisely

$$q(x, y) = D_1^2 f(p_1, p_2)x + 2D_1 D_2 f(p_1, p_2)xy + D_2^2 f(p_1, p_2)y^2.$$

If the reader now looks back at Chapter XII, §2, he will see that this is exactly the same quadratic form considered in that chapter. All we have done here is to show how to express it in terms of a matrix multiplication, and introduced a name for that matrix, namely the Hessian.

Remark on notation. In studying the Hessian, the associated quadratic form has the type

$$ax^2 + 2bxy + cy^2 = ax^2 + b'xy + cy^2$$

where

$$b' = 2b.$$

Of course it does not matter what we write for the coefficient of xy in the quadratic form, we must just be clear which letters denote what. In terms of b', the matrix of the quadratic form can be written as

$$A = \begin{pmatrix} a & b/2 \\ b/2 & c \end{pmatrix}.$$

CHAPTER XVII

The Change of
Variables Formula

If you have not already done so, you should now read the section on cross products, Chapter 1, §7 because we are going to use it.

§1. DETERMINANTS AS AREA AND VOLUME

We shall study the manner in which area changes under an arbitrary mapping by approximating this mapping with a linear map. Therefore, first we study how area and volume change under a linear map, and this leads us to interpret the determinant as area and volume according as we are in $\mathbf{R}^2$ or $\mathbf{R}^3$.

Let us first consider $\mathbf{R}^2$. Let

$$A = \begin{pmatrix} a \\ c \end{pmatrix} \quad \text{and} \quad B = \begin{pmatrix} b \\ d \end{pmatrix}$$

be two non-zero vectors in the plane, and suppose that they are not scalar multiples of each other. We have already seen that they span a parallelogram, as shown on Fig. 1.

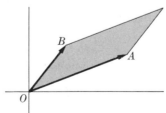

Figure 1

Theorem 1 in $\mathbf{R}^2$. *Let A, B be non-zero elements of $\mathbf{R}^2$, which are not scalar multiples of each other. Then the area of the parallelogram spanned by A and B is equal to the absolute value of the determinant $|D(A, B)|$.*

Proof. We assume known that this area is equal to the product of the lengths of the base times the altitude, and this is equal to

$$\|A\| \, \|B\| \, |\sin \theta|,$$

where θ is the angle between A and B (i.e. between $\overrightarrow{OA}$ and $\overrightarrow{OB}$). This is illustrated on Fig. 2.

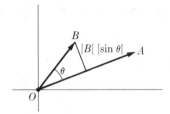

Figure 2

Note that

$$|\sin \theta| = \sqrt{1 - \cos^2 \theta} \, ,$$

and recall from the theory of the dot product that

$$\cos \theta = \frac{A \cdot B}{\|A\| \, \|B\|} \, .$$

We have

$$\text{Area of parallelogram} = \|A\| \, \|B\| \sqrt{1 - \frac{(A \cdot B)^2}{\|A\|^2 \|B\|^2}}$$

$$= \sqrt{\|A\|^2 \|B\|^2 - (A \cdot B)^2} \, .$$

All that remains to be done is to plug in the coordinates of A and B to see what we want come out. Indeed, the above expression is equal to the square root of

$$(a^2 + c^2)(b^2 + d^2) - (ab + cd)^2.$$

If you expand this out, you will find that this last expression is equal to

$$(ad - bc)^2.$$

Consequently, the area of the parallelogram is equal to

$$\sqrt{(ad - bc)^2} = |ad - bc| = |D(A, B)|.$$

This proves our assertion.

Example 1. Let $A = (3, 1)$ and $B = (2, -5)$. Then the area of the parallelogram spanned by A and B is equal to the absolute value of the determinant

$$\begin{vmatrix} 3 & 1 \\ 2 & -5 \end{vmatrix} = -15 - 2 = -17.$$

Hence this area is equal to 17. *Note:* We wrote our vectors horizontally. We get the same determinant as if we write them vertically, namely

$$\begin{vmatrix} 3 & 2 \\ 1 & -5 \end{vmatrix} = -17,$$

because we know that the determinant of the transpose of a matrix is equal to the determinant of the matrix.

We interpret Theorem 1 in terms of linear maps. Given vectors A, B in the plane, we know that there exists a unique linear map

$$L: \mathbf{R}^2 \rightarrow \mathbf{R}^2$$

such that $L(E^1) = A$ and $L(E^2) = B$. In fact, if

$$A = aE^1 + cE^2, \qquad B = bE^1 + dE^2,$$

then the matrix associated with the linear map is

$$\begin{pmatrix} a & b \\ c & d \end{pmatrix}.$$

Let us define the **determinant of a linear map** to be the determinant of its associated matrix. Then

$$\det L = ad - bc,$$

and we see that this is the same thing as the determinant

$$D(A, B).$$

Let S be the unit square, so S consists of all points

$$t_1 E^1 + t_2 E^2$$

with $0 \leq t_1 \leq 1$ and $0 \leq t_2 \leq 1$ as shown on Fig. 3(a).

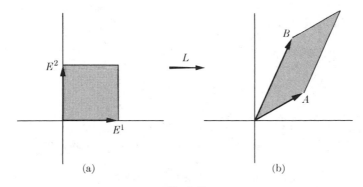

Figure 3

Let P be the parallelogram spanned by A and B. Then P consists of all combinations

$$t_1A + t_2B$$

with $0 \leq t_1 \leq 1$ and $0 \leq t_2 \leq 1$ as shown on Fig. 3 (b). Since

$$L(t_1E^1 + t_2E^2) = t_1L(E^1) + t_2L(E^2) = t_1A + t_2B,$$

we conclude that the image of the unit square by L is precisely that parallelogram. Furthermore,

$$\boxed{\text{Area of } P = |\text{Det}(L)|.}$$

Example 2. The area of the parallelogram spanned by the vectors $(2, 1)$ and $(3, -1)$ (Fig. 4) is equal to the absolute value of

$$\begin{vmatrix} 2 & 1 \\ 3 & -1 \end{vmatrix} = -5,$$

and hence is equal to 5.

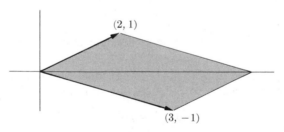

(2, 1)

(3, −1)

Figure 4

We can also obtain a formula showing how the area of an arbitrary parallelogram changes under a linear map.

Theorem 2. *Let P be a parallelogram spanned by two vectors in $\mathbf{R}^2$. Let $L: \mathbf{R}^2 \to \mathbf{R}^2$ be a linear map. Then*

$$\text{Area of } L(P) = |\text{Det } L|(\text{Area of } P).$$

Proof. Suppose that P is spanned by two vectors A, B. Then $L(P)$ is spanned by $L(A)$ and $L(B)$. (Cf. Fig. 5). There is a linear map $L_1: \mathbf{R}^2 \to \mathbf{R}^2$ such that

$$L_1(E^1) = A \quad \text{and} \quad L_1(E^2) = B.$$

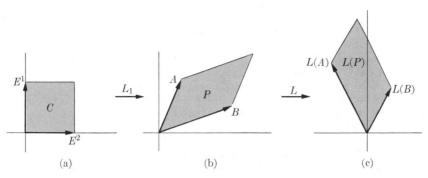

Figure 5

Then $P = L_1(S)$, where S is the unit square, and

$$L(P) = L(L_1(S)) = (L \circ L_1)(S).$$

By what we proved above, we obtain

$$\text{Area } L(P) = |\text{Det}(L \circ L_1)| = |\text{Det}(L)\,\text{Det}(L_1)| = |\text{Det}(L)|\text{Area}(P),$$

thus proving our assertion.

Corollary. *For any rectangle R with sides parallel to the axes, and any linear map*

$$L: \mathbf{R}^2 \to \mathbf{R}^2,$$

we have

$$\text{Area } L(R) = |\text{Det}(L)|\text{Area}(R).$$

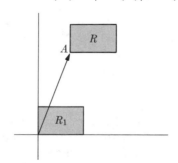

Figure 6

Proof. The rectangle R is equal to the translation of a rectangle R_1 as shown on Fig. 6, with one corner at the origin, that is

$$R = R_1 + A.$$

Then

$$L(R) = L(R_1) + L(A).$$

The area of $L(R_1)$ is the same as the area of $L(R_1) + L(A)$ (i.e. the translation of $L(R_1)$ by $L(A)$). All we have to do is apply Theorem 2 to complete the proof.

Next we consider volumes of boxes in 3-dimensional space. The box spanned by three independent vectors A, B, C in 3-space is also called a **parallelotope**.

Theorem 3. *Let A, B, C be vectors in $\mathbf{R}^3$, and assume that the segments $\overrightarrow{OA}, \overrightarrow{OB}, \overrightarrow{OC}$ do not all lie in a plane. Then the volume of the box spanned by A, B, C is equal to the absolute value of the determinant.*

$$|D(A, B, C)|.$$

Proof. Similar arguments to those which applied in $\mathbf{R}^2$ show us that the area of the base of the box, spanned by A and B, is equal to

(∗) $$\sqrt{\|A\|^2\|B\|^2 - (A \cdot B)^2}\ .$$

Look at Fig. 7.

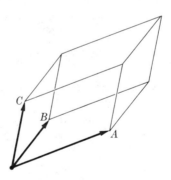

Figure 7

The volume of the box is equal to the area of this base times the altitude, and this altitude is equal to the length of the projection of C along a vector perpendicular to A and B. You should now have read the section on cross products, because the simplest way to handle the present situation is to use the cross product. We know that $A \times B$ is such a vector, perpendicular to A and B. The projection of C on $A \times B$ is equal to

$$\frac{C \cdot (A \times B)}{(A \times B) \cdot (A \times B)} A \times B,$$

where the number in front of $A \times B$ is the component of C along $A \times B$ as studied in Chapter I. Therefore the length of this projection is equal to

(**)
$$\frac{|C \cdot (A \times B)|}{\|A \times B\|}.$$

On the other hand, if you look at property **CP 6** of the cross product in Chapter I, §7 you will find that (*) is equal to $\|A \times B\|$. Therefore, the volume of the box spanned by A, B, C, which is equal to the product of (*) and (**), is seen to be equal to

$$|C \cdot (A \times B)|.$$

All that remains to be done is for you to plug in the coordinates, to see that this is equal to the absolute value of the determinant. You let

$$A = \begin{pmatrix} a_1 \\ a_2 \\ a_3 \end{pmatrix}, \quad B = \begin{pmatrix} b_1 \\ b_2 \\ b_3 \end{pmatrix}, \quad C = \begin{pmatrix} c_1 \\ c_2 \\ c_3 \end{pmatrix},$$

use the definition of the cross product of $A \times B$, and then dot with C. You will find precisely the six terms which give the determinant $D(A, B, C)$, up to a sign, which is killed by the absolute value. This proves Theorem 3.

Example 3. The volume of the box spanned by the vectors

$$(3, 0, 1), \quad (1, 2, 5), \quad (-1, 4, 2)$$

is equal to 42, because the determinant

$$\begin{vmatrix} 3 & 0 & 1 \\ 1 & 2 & 5 \\ -1 & 4 & 2 \end{vmatrix}$$

has the value -42.

Let E^1, E^2, E^3 be the standard unit vectors in the direction of the coordinate axes in 3-space. Then the unit cube S in 3-space consists of all points

$$t_1 E^1 + t_2 E^2 + t_3 E^3$$

with $0 \leq t_i \leq 1$ for $i = 1, 2, 3$. Let $L: \mathbf{R}^3 \to \mathbf{R}^3$ be a linear map such that the vectors

$$A = L(E^1), \quad B = L(E^2), \quad C = L(E^3)$$

do not lie in a plane, i.e. are independent. Then the image of the unit cube under

L is the set of points

$$L(t_1E^1 + t_2E^2 + t_3E^3) = t_1L(E^1) + t_2L(E^2) + t_3L(E^3)$$
$$= t_1A + t_2B + t_3C.$$

This image is therefore the parallelotope spanned by A, B, C. Furthermore, the linear map L is represented by the matrix

$$\begin{pmatrix} a_1 & b_1 & c_1 \\ a_2 & b_2 & c_2 \\ a_3 & b_3 & c_3 \end{pmatrix}.$$

Again we define the **determinant of the linear map** to be the determinant of its matrix. Then we see that:

> *The volume of the parallelotope $L(S)$ is equal to the determinant of L, if S is the unit cube.*

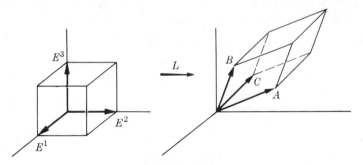

Figure 8

Theorem 4. *Let P be a parallelotope (box) in 3-space, spanned by three vectors. Let $L: \mathbf{R}^3 \to \mathbf{R}^3$ be a linear map. Then*

$$\text{Volume of } L(P) = |\text{Det } L|(\text{Volume of } P).$$

Corollary. *For any rectangular box R in 3-space and any linear map $L: \mathbf{R}^3 \to \mathbf{R}^3$, we have*

$$\text{Vol } L(R) = |\text{Det}(L)|\text{Vol}(R).$$

The proofs are exactly like those in 2-space, drawing 3-dimensional boxes instead of 2-dimensional rectangles.

EXERCISES

1. Find the area of the parallelogram spanned by
   (a) $(-3, 5)$ and $(2, -1)$.   (b) $(2, 3)$ and $(4, -1)$.

2. Find the area of the parallelogram spanned by the following vectors.
   (a) $(2, 1)$ and $(-4, 5)$ (b) $(3, 4)$ and $(-2, -3)$

3. Find the area of the parallelogram such that three corners of the parallelogram are given by the following points
   (a) $(1, 1), (2, -1), (4, 6)$ (b) $(-3, 2), (1, 4), (-2, -7)$
   (c) $(2, 5), (-1, 4), (1, 2)$ (d) $(1, 1), (1, 0), (2, 3)$

4. Find the volume of the parallelotope spanned by the following vectors in 3-space.
   (a) $(1, 1, 3), (1, 2, -1), (1, 4, 1)$ (b) $(1, -1, 4), (1, 1, 0), (-1, 2, 5)$
   (c) $(-1, 2, 1), (2, 0, 1), (1, 3, 0)$ (d) $(-2, 2, 1), (0, 1, 0), (-4, 3, 2)$

§2. DILATIONS

This section will serve as an introduction to the general change of variables formula, and the interpretation of determinants as area and volume.

Let r be a positive number. If A is a vector in $\mathbf{R}^n$ (in practice, $\mathbf{R}^2$ or $\mathbf{R}^3$) we call rA the **dilation** of A by r. Thus dilation by r is a linear mapping,

$$A \mapsto rA.$$

We wish to analyze what happens to area in $\mathbf{R}^2$, and volume in $\mathbf{R}^3$, under a dilation. We start with the simplest case, that of a rectangle. Consider a rectangle whose sides have lengths a, b, as on Fig. 9(a). If we multiply the sides of the rectangle by r, we obtain a rectangle with sides ra, rb as on Fig. 9(b). The area of the dilated rectangle is equal to

$$rarb = r^2ab.$$

Thus dilation by r changes the area of the rectangle by r^2.

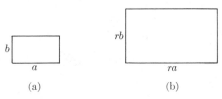

(a)   (b)

Figure 9

In general, let S be an arbitrary region in the plane $\mathbf{R}^2$, whose area can be approximated by the area of a finite number of rectangles. Then the area of S itself changes by r^2 under dilation by r, in other words,

$$\text{Area of } rS = r^2(\text{area of } S).$$

For instance, let D be the disc of radius r, so that D_1 is the disc of radius 1, centered at the origin (Fig. 10). Then $D_r = rD_1$.

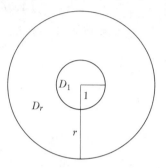

Figure 10

If π is the area of the disc of radius 1, then πr^2 must be the area of the disc of radius r. Of course, we knew this already, but we find this result here again from another point of view. More generally, consider a region S inside a curve as in Fig. 11(a), and let us draw the dilation of S by r in Fig. 11(b). To justify that the area changes by r^2, we draw a grid, approximating the areas by squares.

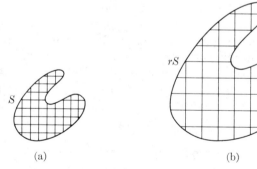

(a)   (b)

Figure 11

Under dilation by r, the area of each square gets multiplied by r^2, and so the sum of the areas of these squares, which approximates the area of S, also gets multiplied by r^2.

The question, of course, arises as to whether the squares lying inside S, and formed by a sufficiently fine grid, actually approximate S. We can see that they do, as follows. Let the sides of the squares in the grid have length c. (Fig. 12a.) Suppose that a square intersects the curve which bounds S. Let Z be this curve. Then any point in the square is at distance at most $c\sqrt{2}$ from the curve Z. This is because the distance between any two points of the square is at most $c\sqrt{2}$ (the length of the diagonal of the square). Let us draw a band of width $c\sqrt{2}$ on each side of the curve, as shown on Fig. 12(b). Then all the squares which intersect the curve must lie within that band. It is very plausible that the area of the band is at most equal to

$$2c\sqrt{2} \text{ times the length of the curve.}$$

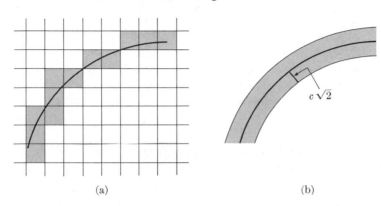

(a)   (b)

Figure 12

Thus if we take c to be very small, i.e. if we take the grid to be a very fine grid, we see that the area of the region S is approximated by the area covered by the squares lying entirely inside the region. This explains why the area of S will get multiplied by r^2 under dilation by r.

We can also make a mixed dilation. Let r, s be two positive numbers. Consider the mapping of $\mathbf{R}^2$ given by

$$(x, y) \longmapsto (rx, sy).$$

Thus we dilate the first coordinate by r and the second by s. If a rectangle R has sides of lengths a, b respectively, then the image of the rectangle under this mapping will be a rectangle with sides of lengths ra, sb. Hence the area of the image will be

$$rasb = rsab.$$

Thus the area changes by a factor of rs under our mapping.

An argument as before shows that if we submit a region S to such a mapping $F_{r,s}$ such that

$$F_{r,s}(x, y) = (rx, sy),$$

then its area will change by a factor of rs.

Example 1. We now have a very easy way of finding the area of an ellipse defined by an equation

$$\frac{x^2}{9} + \frac{y^2}{16} = 1.$$

Indeed, let $u = x/3$ and $v = y/4$. Then

$$u^2 + v^2 = 1,$$

and the ellipse is equal to the image of the circle under the mapping

$$(u, v) \mapsto (3u, 4v).$$

Hence the area of the ellipse is equal to $3 \cdot 4\pi = 12\pi$. Note how we did this without integration! However, the technique of the small grid is of course exactly the same technique which was used in the theory of the integral.

We can also develop the same ideas in 3-space. Consider dilation by r in 3-space; namely consider the mapping

$$(x, y, z) \mapsto (rx, ry, rz).$$

If P is a rectangular box with sides a, b, c, then its dilation by r will be a box with sides ra, rb, rc, and the dilated box will have volume

$$rarbrc = r^3 abc.$$

Thus the volume of a box changes by r^3 under dilation by r.

Similarly, let r, s, t be positive numbers, and consider the linear map

$$F_{r, s, t} \colon \mathbf{R}^3 \mapsto \mathbf{R}^3$$

such that

$$F_{r, s, t}(x, y, z) = (rx, sy, tz).$$

We view this as a mixed dilation. If a rectangular box has sides of lengths a, b, c, then under $F_{r, s, t}$ it gets transformed into a box with sides ra, sb, tc whose volume is

$$rasbtc = rstabc.$$

Thus the volume gets multiplied by rst.

If we approximate an arbitrary region in 3-space by cubes, then we see in a manner analogous to that of 2-space that the volume of the region changes by a factor of r^3 under dilation by r, and changes by a factor of rst under the mixed dilation $F_{r, s, t}$.

Example 2. Find the volume of the region bounded by the equation

$$\frac{x^2}{9} + \frac{y^2}{16} + \frac{z^2}{25} = 1.$$

To do this, let

$$u = \frac{x}{3}, \qquad v = \frac{y}{4}, \qquad w = \frac{z}{5}.$$

The inequality

$$u^2 + v^2 + w^2 \leqq 1$$

defines the unit ball in $\mathbf{R}^3$, and our given region is obtained from this unit ball by the mixed dilation

$$F_{3, 4, 5}.$$

Assuming that the volume of the unit ball in $\mathbf{R}^3$ is equal to $\frac{4}{3}\pi$, we conclude that the volume of our region is equal to

$$3 \cdot 4 \cdot 5 \cdot \tfrac{4}{3}\pi = 80\pi.$$

In the next section, we investigate how area and volume change under general linear maps, not just dilations and mixed dilations.

EXERCISES

1. Find the area of the region bounded by the ellipse

$$\frac{x^2}{a^2} + \frac{y^2}{b^2} = 1.$$

2. Find the volume of the region bounded by the surface

$$\frac{x^2}{a^2} + \frac{y^2}{b^2} + \frac{z^2}{c^2} = 1.$$

In both exercises, a, b, c are positive numbers. Use the ideas of this section.

3. Let A be the region in 3-space defined by the inequalities

$$0 \leq x_i \quad \text{and} \quad x_1^4 + x_2^4 + x_3^4 \leq 1.$$

Let k be the volume of this region.
(a) In terms of k, what is the volume of the region defined by the inequalities

$$0 \leq x_i \quad \text{and} \quad x_1^4 + x_2^4 + x_3^4 \leq 29?$$

(b) Same question if instead of 29 on the right you have a positive number r.

4. Let A be the region in 3-space defined by the inequalities

$$0 \leq x_i \quad \text{and} \quad \sum_{i=1}^{3} x_i^5 \leq 1.$$

Let k be the volume of this region.
(a) In terms of k, what is the volume of the region defined by the inequalities

$$0 \leq x_i \quad \text{and} \quad \sum_{i=1}^{3} x_i^5 \leq 33?$$

(b) Same question if instead of 33 you have an arbitrary positive number r on the right.

§3. CHANGE OF VARIABLES FORMULA IN TWO DIMENSIONS

Let R be a rectangle in $\mathbf{R}^2$ and suppose that R is contained in some open set U. Let

$$G: U \to \mathbf{R}^2$$

be a C^1-map. If G has two coordinate functions,

$$G(u, v) = (g_1(u, v), g_2(u, v)),$$

this means that the partial derivatives of g_1, g_2 exist and are continuous. We let $G(u, v) = (x, y)$, so that

$$x = g_1(u, v) \qquad \text{and} \qquad y = g_2(u, v).$$

Then the Jacobian determinant of the map G is by definition

$$\Delta_G(u, v) = \begin{vmatrix} \dfrac{\partial g_1}{\partial u} & \dfrac{\partial g_1}{\partial v} \\[2mm] \dfrac{\partial g_2}{\partial u} & \dfrac{\partial g_2}{\partial v} \end{vmatrix}.$$

This determinant is nothing but the determinant of the linear map $G'(u, v)$.

Theorem 5. *Assume that G is C^1-invertible on the interior of the rectangle R. Let f be a function on $G(R)$ which is continuous except on a finite number of smooth curves. Then*

$$\iint_R (f \circ G)|\Delta_G| = \iint_{G(R)} f$$

or in terms of coordinates,

$$\iint_R f(G(u, v))|\Delta_G(u, v)| \, du \, dv = \iint_{G(R)} f(x, y) \, dy \, dx.$$

The proof of Theorem 5 is not easy to establish rigorously. However, we can make it plausible in view of Theorem 2.

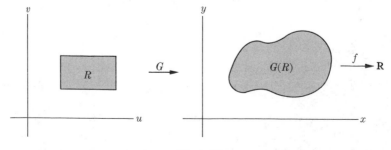

Figure 13

Indeed, suppose first that f is a constant function, say $f(x, y) = 1$ for all (x, y). Then the integral on the right, over $G(R)$, is simply the area of $G(R)$, and our formula reduces to

$$\iint_R |\Delta_G| = \iint_{G(R)} 1.$$

As we pointed out before, Δ_G is the determinant of the approximating linear map to G. If G is itself linear, then $G'(u, v) = G$ for all u, v and in this case, our formula reduces to Theorem 2, or rather its corollary. In the general case, one has to show that when one approximates G by its Jacobian matrix, which depends on (u, v), and then integrates $|\Delta_G|$ one still obtains the same result (Fig. 13). Cf., for instance, my *Introduction to Analysis* for a complete proof. A special case will be proved in the next chapter.

When f is not a constant function, one still has the problem of reducing this case to the case of constant functions. This is done by taking a partition of R into small rectangles S, and then approximating f on each $G(S)$ by a constant function. Again, the details are out of the bounds of this book.

We shall now see how we recover the integral in terms of polar coordinates from the general Theorem 5.

Example 1. Let $x = r \cos \theta$ and $y = r \sin \theta$, $r \geq 0$. Then in this case, we have computed previously the determinant, which is

$$\Delta_G(r, \theta) = r.$$

Thus we find again the formula

$$\iint_R f(r \cos \theta, r \sin \theta) r \, dr \, d\theta = \iint_{G(R)} f(x, y) \, dy \, dx.$$

Of course, we have to take a rectangle for which the map

$$G(r, \theta) = (r \cos \theta, r \sin \theta)$$

is invertible on the interior of the rectangle. For instance, we can take

$$0 \leq r_1 \leq r \leq r_2 \quad \text{and} \quad 0 \leq \theta_1 \leq \theta \leq \theta_2 \leq 2\pi.$$

The image of the rectangle R is the portion $G(R)$ of the sector as shown in Fig. 14.

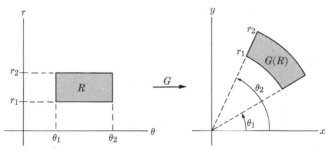

Figure 14

For the next example, we observe that if G is a linear map L, represented by a matrix M, then a Jacobian matrix of G is equal to this matrix M, and hence its Jacobian determinant is the determinant of M.

Example 2. Let T be the triangle whose vertices are $(1, 2)$, $(3, -1)$, and $(0, 0)$. Find the area of this triangle (Fig. 15)

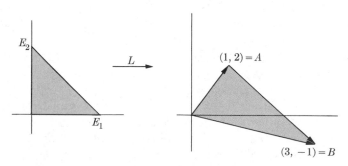

Figure 15

The triangle T is the image of the triangle spanned by 0, E_1, E_2 under a linear map, namely the linear map L such that

$$L(E_1) = (1, 2)$$

and

$$L(E_2) = (3, -1).$$

It is verified at once that $|\text{Det}(L)| = 7$. Since the area of the triangle spanned by O, E_1, E_2 is $\frac{1}{2}$, it follows that the desired area is equal to $\frac{7}{2}$.

Example 3. Let $(x, y) = G(u, v) = (e^u \cos v, e^u \sin v)$. R be the rectangle in the (u, v)-space defined by the inequalities $0 \le u \le 1$ and $0 \le v \le \pi$. It is not difficult to show that G satisfies the hypotheses of Theorem 3, but we shall assume this. The Jacobian matrix of G is given by

$$\begin{pmatrix} e^u \cos v & -e^u \sin v \\ e^u \sin v & e^u \cos v \end{pmatrix}$$

so that its Jacobian determinant is equal to

$$\Delta_G(u, v) = e^{2u}.$$

Let $f(x, y) = x^2$. Then $f^*(u, v) = e^{2u} \cos^2 v$. According to Theorem 5, the integral of f over $G(R)$ is given by the integral

$$\int_0^1 \int_0^\pi e^{4u} \cos^2 v \, du \, dv$$

which can be evaluated very simply by integrating e^{4u} with respect to u and $\cos^2 v$ with respect to v, and taking the product. The final answer is then equal to

$$\pi \frac{(e^4 - 1)}{8}.$$

Example 4. Let S be the region enclosed by ellipse defined by the equation

$$\frac{x^2}{a^2} + \frac{y^2}{b^2} = 1, \qquad\qquad a, b > 0.$$

Its area is πab. (Why?) Let L be the linear map represented by the matrix

$$\begin{pmatrix} 1 & 2 \\ 3 & -5 \end{pmatrix}.$$

Its determinant is equal to -11. Hence the area of the image of S under L is $11\pi ab$.

EXERCISES

In the following exercises, you may assume that the map G satisfies the hypotheses of Theorem 3.

1. Let $(x, y) = G(u, v) = (u^2 - v^2, 2uv)$. Let A be the region defined by $u^2 + v^2 \leq 1$ and $0 \leq u, 0 \leq v$. Find the integral of the function

$$f(x, y) = 1/(x^2 + y^2)^{1/2}$$

   over $G(A)$.

2. (a) Let $(x, y) = G(u, v)$ be the same map as in Exercise 1. Let A be the square $0 \leq u \leq 2$ and $0 \leq v \leq 2$. Find the area of $G(A)$.
   (b) Find the integral of $f(x, y) = x$ over $G(A)$.

3. (a) Let R be the rectangle whose corners are $(1, 2)$, $(1, 5)$, $(3, 2)$, and $(3, 5)$. Let G be the linear map represented by the matrix

$$\begin{pmatrix} 2 & 1 \\ -1 & 3 \end{pmatrix}.$$

   Find the area of $G(R)$.
   (b) Same question if G is represented by the matrix $\begin{pmatrix} 3 & 2 \\ 1 & -6 \end{pmatrix}$.

4. Let $(x, y) = G(u, v) = (u + v, u^2 - v)$. Let A be the region in the first quadrant bounded by the axes and the line $u + v = 2$. Find the integral of the function $f(x, y) = 1/\sqrt{1 + 4x + 4y}$ over $G(A)$.

5. Let R be the unit square in the (u, v)-plane, defined by the inequalities

$$0 \leq u \leq 1 \quad \text{and} \quad 0 \leq v \leq 1.$$

(a) Sketch the image $F(R)$ of R under the mapping F such that

$$F(u, v) = (u, u + v^2).$$

In other words, $x = u$ and $y = u + v^2$.

(b) Compute the integral of the function $f(x, y) = x$ over the region $F(R)$ by using the change of variables formula.

6. Compute the area enclosed by the ellipse, defined by

$$\frac{x^2}{a^2} + \frac{y^2}{b^2} \leq 1.$$

Take $a, b > 0$.

7. Let $(x, y) = G(u, v) = (u, v(1 + u^2))$. Let R be the rectangle $0 \leq u \leq 3$ and $0 \leq v \leq 2$. Find the integral of $f(x, y) = x$ over $G(R)$.

8. Let G be the linear map represented by the matrix

$$\begin{pmatrix} 3 & 0 \\ 1 & 5 \end{pmatrix}.$$

If A is the interior of a circle of radius 10, what is the area of $G(A)$?

9. Let G be the linear map of Exercise 8, and let A be the ellipse defined as in Exercise 6. What is the area of $G(A)$?

10. Let T be the triangle bounded by the x-axis, the y-axis, and the line $x + y = 1$. Let φ be a continuous function of one variable on the interval $[0, 1]$. Let m, n be positive integers. Show that

$$\iint_T \varphi(x + y) x^m y^n \, dy \, dx = c_{m, n} \int_0^1 \varphi(t) t^{m+n+1} \, dt,$$

where $c_{m, n}$ is the constant given by the integral $\int_0^1 (1 - t)^m t^n \, dt$. [*Hint:* Let $x = u - v$ and $y = v$.]

11. Let B be the region bounded by the ellipse $x^2/a^2 + y^2/b^2 = 1$. Find the integral

$$\iint_B y \, dy \, dx.$$

12. Let A be the parallelogram with vertices

$$(0, 0), \quad (1, 1), \quad (1, -1), \quad \text{and} \quad (2, 0).$$

Find

$$\iint_A ((x + y)^2 + (x - y)^2) \, dx \, dy.$$

§4. APPLICATION OF GREEN'S FORMULA TO THE CHANGE OF VARIABLES FORMULA

When a region R is the interior of a closed path, then we can use Green's theorem to prove the change of variables formula in special cases. Indeed, Green's theorem reduces a double integral to an integral over a curve, and change of variables formulas for curves are easier to establish than for 2-dimensional areas. Thus we begin by looking at a special case of change of variables formula for curves.

Let $C: [a, b] \to U$ be a C^1-curve in an open set of $\mathbf{R}^2$. Let $G: U \to \mathbf{R}^2$ be a C^2-map, given by coordinate functions,

$$G(u, v) = (x, y) = (f(u, v), g(u, v)).$$

Thus

$$x = f(u, v) \qquad \text{and} \qquad y = g(u, v).$$

Then the composite $G \circ C$ is a curve. If $C(t) = (\alpha(t), \beta(t))$, then

$$G \circ C(t) = G(C(t)) = (f(\alpha(t), \beta(t)), g(\alpha(t), \beta(t))).$$

In other words, if

$$u = \alpha(t) \qquad \text{and} \qquad v = \beta(t)$$

then

$$x = f(\alpha(t), \beta(t)) \qquad \text{and} \qquad y = g(\alpha(t), \beta(t)).$$

Example 1. Let $G(u, v) = (u, -v)$ be the reflection along the horizontal axis. If $C(t) = (\cos t, \sin t)$, then

$$G \circ C(t) = (\cos t, -\sin t).$$

Thus $G \circ C$ again parametrizes the circle, but observe that the orientation of $G \circ C$ is opposite to that of C, i.e. it is clockwise! (Fig. 16.)

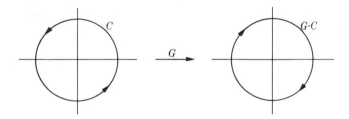

Figure 16

The reason for this reversal of orientation is that the Jacobian determinant of G is negative, namely it is the determinant of

$$\begin{pmatrix} 1 & 0 \\ 0 & -1 \end{pmatrix}.$$

Thus a map G is said to **preserve orientation** if $\Delta_G(u, v) > 0$ for all (u, v) in the domain of definition of G. For simplicity, we only consider such maps G.

Green's theorem leads us to consider the integral

$$\int_{G \circ C} x \, dy.$$

By definition and the chain rule, we have

$$\int_{G \circ C} x \, dy = \int_a^b f(C(t)) \left(\frac{\partial y}{\partial u} \frac{du}{dt} + \frac{\partial y}{\partial v} \frac{dv}{dt} \right) dt$$

$$= \int_C f(u, v) \frac{\partial y}{\partial u} \, du + f(u, v) \frac{\partial y}{\partial v} \, dv.$$

This is true for any curve as above. Hence it remains true for any path, consisting of a finite number of curves.

We are now ready to state and prove the change of variables formula in the case to which Green's theorem applies.

Let U be open in $\mathbf{R}^2$, and let R be a region which is the interior of a closed path C (piecewise C^1 as usual) contained in U. Let

$$G: U \to \mathbf{R}^2$$

be a C^2-map, which is C^1-invertible on U and such that $\Delta_G > 0$. Then $G(R)$ is a region which is the interior of the path $G \circ C$. (Fig. 17.) We then have

$$\boxed{\iint_{G(R)} dy \, dx = \iint_R \Delta_G(u, v) \, du \, dv.}$$

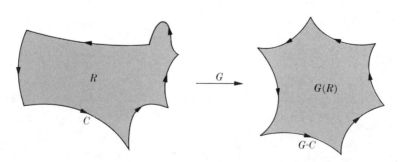

Figure 17

Proof. Let $G(u, v) = (f(u, v), g(u, v))$ be expressed by its coordinates. We have, using Green's theorem:

$$\iint_{G(R)} dy\, dx = \int_{G \circ C} x\, dy = \int_C f\frac{\partial g}{\partial u}\, du + f\frac{\partial g}{\partial v}\, dv$$

$$= \iint_R \left[\frac{\partial}{\partial u}\left(f\frac{\partial g}{\partial v} \right) - \frac{\partial}{\partial v}\left(f\frac{\partial g}{\partial u} \right) \right] du\, dv$$

$$= \iint_R \left[\frac{\partial f}{\partial u}\frac{\partial g}{\partial v} + f\frac{\partial g}{\partial u\, \partial v} - f\frac{\partial g}{\partial u\, \partial v} - \frac{\partial f}{\partial u}\frac{\partial g}{\partial u} \right] du\, dv$$

$$= \iint_R \left[\frac{\partial f}{\partial u}\frac{\partial g}{\partial v} - \frac{\partial g}{\partial u}\frac{\partial f}{\partial v} \right] du\, dv$$

$$= \iint_R \Delta_G(u, v)\, du\, dv,$$

thus proving what we wanted.

EXERCISES

1. Under the same assumptions as the theorem in this section, assume that $\varphi = \varphi(x, y)$ is a continuous function on $G(R)$, and that we can write $\varphi(x, y) = \partial q/\partial x$ for some continuous function q. Prove the more general formula

$$\iint_{G(R)} \varphi(x, y)\, dy\, dx = \iint_R \varphi(G(u, v))\Delta_G(u, v)\, du\, dv.$$

   [*Hint:* Let $p = 0$ and follow the same pattern of proof as in the text.]

2. Let $(x, y) = G(u, v)$ as in the text. We suppose that $G: U \rightarrow \mathbf{R}^2$, and that F is a vector field on $G(U)$. Then $F \circ G$ is a vector field on U. Let C be a curve in U. Show that

$$\int_{G \circ C} F = \int_C (F \circ G) \cdot \frac{\partial G}{\partial u}\, du + (F \circ G) \cdot \frac{\partial G}{\partial v}\, dv.$$

   [Let $F(x, y) = (p(x, y), q(x, y))$ and apply the definitions.]

§5. CHANGE OF VARIABLES FORMULA IN THREE DIMENSIONS

The formula has the same shape as in two dimensions, namely:

Change of Variables Formula. *Let A be a bounded region in $\mathbf{R}^3$ whose boundary consists of finite number of smooth surfaces. Let A be contained in* *some open set U, and let*

$$G: U \to \mathbf{R}^3$$

be a C^1-map, which we assume to be C^1-invertible on the interior of A. Let f be a function on G(A), bounded and continuous except on a finite number of smooth surfaces. Then

$$\iiint_A f(G(u, v, w))|\Delta_G(u, v, w)|\, du\, dv\, dw = \iiint_{G(A)} f(x, y, z)\, dz\, dy\, dx.$$

In the 3-dimensional case, the Jacobian matrix of G at every point is then a 3×3 matrix.

Example 1. Let R be the 3-dimensional rectangle spanned by the three unit vectors E_1, E_2, E_3. Let A_1, A_2, A_3 be three vectors in 3-space, and let

$$G: \mathbf{R}^3 \to \mathbf{R}^3$$

be the linear map such that $G(E_i) = A_i$. Then $G(R)$ is a parallelotope (not necessarily rectangular). (Cf. Fig. 18.)

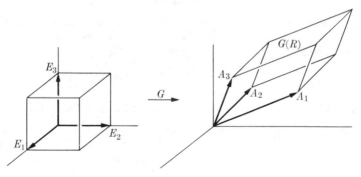

Figure 18

The Jacobian matrix of the map is constant, and is equal to the determinant of the matrix representing the linear map.

The volume of the unit cube is equal to 1. Hence the volume of $G(R)$ is equal to $|\text{Det}(G)|$.

For instance, if

$$A_1 = (3, 1, 2),$$
$$A_2 = (1, -1, 4),$$
$$A_3 = (2, 1, 0),$$

then

$$\text{Det}(G) = \begin{vmatrix} 3 & 1 & 2 \\ 1 & -1 & 4 \\ 2 & 1 & 0 \end{vmatrix} = 2$$

so the volume of $G(R)$ is equal to 2.

Example 2. Find the volume of the tetrahedron spanned by the origin and the three vectors

$$A_1 = (3, 1, 4), \qquad A_2 = (-1, 2, 1), \qquad A_3 = (5, -2, 1).$$

We assume that you have computed the volume of the tetrahedron spanned by the unit vectors, and found $\frac{1}{6}$. There is a unique linear map L which carries E_i on A_i. Hence the volume of our tetrahedron is equal to $\frac{1}{6}$ times the absolute value of the determinant of this linear map, that is to $\frac{1}{6}$ times the absolute value of the determinant

$$\begin{vmatrix} 3 & 1 & 4 \\ -1 & 2 & 1 \\ 5 & -2 & 1 \end{vmatrix} = -14.$$

The answer is $14/6$.

If we are given the four vertices of a tetrahedron and want to find its volume, then we subtract one vertex from the others. This gives us a tetrahedron with one vertex at the origin, whose volume can be found by the above procedure.

Example 3. Consider the cylindrical coordinates map, given by

$$G(r, \theta, z) = (r \cos \theta, r \sin \theta, z).$$

Compute its Jacobian matrix, and its Jacobian determinant. You will easily find

$$\Delta_G(r, \theta, z) = r,$$

so that the general formula for changing variables gives you the same result that was found in Chapter IX by looking at the volume of an elementary region, image of a box under the map G.

Example 4. Let G be the map of spherical coordinates, given by

$$G(\rho, \theta, \varphi) = (\rho \sin \varphi \cos \theta, \rho \sin \varphi \sin \theta, \cos \varphi).$$

Again you should compute the Jacobian matrix and the Jacobian determinant. You will find:

$$\Delta_G(\rho, \theta, \varphi) = \rho^2 \sin \varphi.$$

This gives a justification for the formula of Chapter IX in terms of the change of variables formula, which in the present case reads just like the result of Chapter IX, namely:

$$\iiint_A f(G(\rho, \theta, \varphi))\rho^2 \sin \varphi \; d\rho \; d\varphi \; d\theta = \iiint_{G(A)} f(x, y, z) \; dz \; dy \; dx.$$

Exercise. Carry out in detail the computation of the preceding two examples.

EXERCISES

1. (a) Let $G: \mathbf{R}^3 \rightarrow \mathbf{R}^3$ be the map which sends spherical coordinates (θ, φ, ρ) into cylindrical coordinates (θ, r, z). Write down the Jacobian matrix for this map, and its Jacobian determinant.
   (b) Write down the change of variables formula for this case.

2. Let A be a region in $\mathbf{R}^3$ and assume that its volume is equal to k. Let $G: \mathbf{R}^3 \rightarrow \mathbf{R}^3$ be the map such that $G(x, y, z) = (ax, by, cz)$, where a, b, c are positive numbers. What is the volume of $G(A)$?

3. Find the volume of the ellipsoid

$$\frac{x^2}{a^2} + \frac{y^2}{b^2} + \frac{z^2}{c^2} \leq 1,$$

by the change of variables formula, and by the method of dilations.

4. Find the volume of the solid which is the image of a ball of radius a under the linear map represented by the matrix

$$\begin{pmatrix} 1 & -1 & 1 \\ 0 & 2 & 5 \\ 0 & 0 & 7 \end{pmatrix}.$$

5. (a) Find the volume of the tetrahedron T determined by the inequalities

$$0 \leq x, \; 0 \leq y, \; 0 \leq z \quad \text{and} \quad x + y + z \leq 1.$$

   (b) This tetrahedron can also be written in the form

$$t_1 E_1 + t_2 E_2 + t_3 E_3 \quad \text{with} \quad t_1 + t_2 + t_3 \leq 1, \; 0 \leq t_i.$$

   If L is the linear map such that $L(E_i) = A_i$, show that $L(T)$ is described by similar inequalities. We call it the tetrahedron spanned by $0, A_1, A_2, A_3$.
   (c) Determine the volume of the tetrahedron spanned by the origin and the three vectors $(1, 1, 2), (2, 0, -1), (3, 1, 2)$.
   (d) Using the fact that the volume of a region does not change under translation, determine the volume of the tetrahedron spanned by the four points $(1, 1, 1)$, $(2, 2, 3), (3, 1, 0),$ and $(4, 2, 3)$.

6. (a) Determine the volume of the tetrahedron spanned by the four points $(2, 1, 0)$, $(3, -1, 1)$, $(-1, 1, 2)$, $(0, 0, 1)$.
   (b) Same question for the four points $(3, 1, 2)$, $(2, 0, 0)$, $(4, 1, 5)$, $(5, -1, 1)$.

7. Let $L: \mathbf{R}^3 \rightarrow \mathbf{R}^3$ be the linear map given by

$$L\begin{pmatrix} x \\ y \\ z \end{pmatrix} = \begin{pmatrix} 4x + 4y + 8z \\ 2x + 7y + 4z \\ x + 4y + 3z \end{pmatrix}.$$

   (a) Find the matrix of L.
   (b) Find the determinant of the matrix of L.
   (c) Suppose D is a region in $\mathbf{R}^3$ with volume 5. Find the volume of $L(D)$.

§6. VECTOR FIELDS ON THE SPHERE

Let S be the ordinary sphere of radius 1, centered at the origin. By a **tangent vector field on the sphere**, we mean an association

$$F: S \longrightarrow \mathbf{R}^3$$

which to each point X of the sphere associates a vector $F(X)$ which is **tangent** to the sphere (and hence perpendicular to $\overrightarrow{OX}$). The picture may be drawn as follows (Fig. 19).

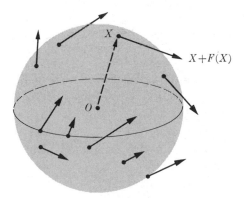

Figure 19

For simplicity of expression, we omit the word tangent, and from now on speak only of vector fields on the sphere. We may think of the sphere as the earth, and we think of each vector as representing the wind at the given point. The wind points in the direction of the vector, and the speed of the wind is the length of the arrow at the point.

We suppose as usual that the vector field is smooth. For instance, the vector field being continuous would mean that if P, Q are two points close by on the sphere, then $F(P)$ and $F(Q)$ are arrows whose lengths are close, and whose directions are also close. As F is represented by coordinates, this means that each coordinate is continuous. We shall actually consider vector fields such that the coordinates are of class C^1, without further repeating this assumption.

Theorem 6. *Given any vector field on the sphere, there exists a point P on the sphere such that $F(P) = O$.*

In terms of the interpretation with the wind, this means that there is some point on earth where the wind is not blowing at all.

To prove Theorem 6, suppose to the contrary that there is a vector field such that $F(X) \neq O$ for all X on the sphere. Define

$$E(X) = \frac{F(X)}{\|F(X)\|},$$

that is, let $E(X)$ be $F(X)$ divided by its norm. Then $E(X)$ is unit vector for each X. Thus from the vector field F we have obtained a vector field E such that all the vectors have norm 1. Such a vector field is called a **unit vector field**. Hence to prove Theorem 6, it suffices to prove:

Theorem 7. *There is no unit vector field on the sphere.*

Until recently, I did not know any relatively simple proof for this classical theorem. The proof which follows is due to Milnor. (*Math. Monthly*, October 1978.)

Suppose that there exists a vector field E on the sphere such that

$$\|E(X)\| = 1$$

for all X. We call this a **unit vector field.** For each small real number t, define

$$G_t(X) = X + tE(X).$$

Geometrically, this means that $G_t(X)$ is the point obtained by starting at X, going in the direction of $E(X)$ with magnitude t. The distance of $X + tE(X)$ from the origin O is then obviously

$$\sqrt{1 + t^2}.$$

Indeed, $E(X)$ is parallel (tangent) to the sphere, and so perpendicular to X itself. Thus

$$\|X + tE(X)\|^2 = (X + tE(X))^2 = X^2 + t^2E(X)^2 = 1 + t^2$$

since both X and $E(X)$ are unit vectors.

Lemma. *For all t sufficiently small, the image $G_t(S)$ of the sphere under G_t is equal to the whole sphere of radius $\sqrt{1 + t^2}$.*

Proof. This amounts to proving a variation of the inverse mapping theorem, and the techniques for such proofs are omitted from this course. Any technique which you would know for proving the inverse mapping theorem would also allow you to prove the present lemma. We shall assume the lemma.

We now extend the vector field E to a bigger region of 3-space, namely the region A between two concentric spheres, defined by the inequalities

$$a \leqq \|X\| \leqq b.$$

This extended vector field is defined by the formula

$$E(rU) = rE(U)$$

for any unit vector U and any number r such that $a \leqq r \leqq b$.

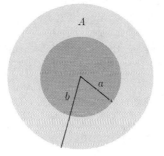

Figure 20

It follows that the formula

$$G_t(X) = X + tE(X),$$

also given in terms of unit vectors U by

$$G_t(rU) = rU + tE(rU) = rG_t(U)$$

defines a mapping which sends the sphere of radius r onto the sphere of radius $r\sqrt{1 + t^2}$ by the lemma, provided that t is sufficiently small. Hence it maps A onto the region between the spheres of radius

$$a\sqrt{1 + t^2} \qquad \text{and} \qquad b\sqrt{1 + t^2}.$$

By the change of volumes under dilations, it is then clear that

$$\text{Volume } G_t(A) = (\sqrt{1 + t^2})^3 \text{ Volume}(A).$$

Observe that taking the cube of $\sqrt{1 + t^2}$ still involves a square root, and is not a polynomial in t.

On the other hand, the Jacobian matrix of G_t is

$$J_{G_t}(X) = I + tJ_E(X),$$

as you can verify easily by writing down the coordinates of $E(X)$, say

$$E(X) = (g_1(x, y, z), g_2(x, y, z), g_3(x, y, z)).$$

Hence the Jacobian determinant has the form

$$\Delta_{G_t}(X) = \det(I + tJ_E(X)),$$

and is therefore a polynomial in t of degree 3, that is we can write

$$\Delta_{G_t}(X) = \varphi_0(X) + \varphi_1(X)t + \varphi_2(X)t^2 + \varphi_3(X)t^3,$$

where $\varphi_0, \ldots, \varphi_3$ are functions. Given the region A, this determinant is then positive for all sufficiently small values of t, by continuity, and the fact that the determinant is 1 when $t = 0$.

For any region A in 3-space, the change of variables formula shows that the volume of $G_t(A)$ is given by the integral

$$\text{Vol } G_t(A) = \iiint_A \Delta_{G_t}(x, y, z) \, dy \, dx \, dz.$$

If we perform the integration, we see that

$$\text{Vol } G_t(A) = c_0 + c_1 t + c_2 t^2 + c_3 t^3$$

where

$$c_i = \iiint_A \varphi_i(x, y, z) \, dy \, dx \, dz.$$

Hence Vol $G_t(A)$ is a polynomial in t of degree 3. Taking for A the region between the spheres yields a contradiction which concludes the proof.

APPENDIX 1

Fourier Series

In this appendix, we discuss a little more systematically the scalar product in the context of spaces of functions. This may be covered at the same time that Chapter I is discussed, but I place the material as an appendix in order not to interrupt the discussion of ordinary vectors after Chapter I.

§1. GENERAL SCALAR PRODUCTS

Let V be the set (also called the space) of continuous functions on some interval, say the interval $[-\pi, \pi]$ which is of interest in Fourier series. We define the **scalar product** of functions f, g in V to be the number

$$\langle f, g \rangle = \int_{-\pi}^{\pi} f(x)g(x)\, dx.$$

This scalar product satisfies conditions analogous to those of Chapter I, namely:

SP 1. *We have* $\langle v, w \rangle = \langle w, v \rangle$ *for all* v, w *in* V.

SP 2. *If* u, v, w *are elements of* V, *then*

$$\langle u, v + w \rangle = \langle u, v \rangle + \langle u, w \rangle.$$

SP 3. *If* x *is a number, then*

$$\langle xu, v \rangle = x\langle u, v \rangle = \langle u, xv \rangle.$$

SP 4. *For all* v *in* V *we have* $\langle v, v \rangle \geq 0$, *and* $\langle v, v \rangle > 0$ *if* $v \neq 0$.

The verification of these properties amounts to recalling simple properties of the integral. For instance, for **SP 1,** we have

$$\langle f, g \rangle = \int_{-\pi}^{\pi} f(x)g(x)\, dx = \int_{-\pi}^{\pi} g(x)f(x)\, dx = \langle g, f \rangle.$$

407

We leave the verification of **SP 2** and **SP 3** as exercises. To prove **SP 4,** suppose that f is a non-zero function. This means that there exists some point c in the interval $[-\pi, \pi]$ such that $f(c) \neq 0$. Then

$$\langle f, f \rangle = \int_{-\pi}^{\pi} f(x)^2 \, dx,$$

and $f(x)^2$ is a function which is always ≥ 0, and such that

$$f(c)^2 > 0.$$

Thus the graph of $f(x)^2$ may look like this.

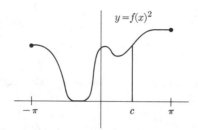

Figure 1

Let $p(x) = f(x)^2$. Geometrically, the integral of $p(x)$ from $-\pi$ to π is the area under the curve $y = p(x)$ between $-\pi$ and π, and this area cannot be 0 since $p(c) > 0$, so the area is > 0. We can give a more formal argument by observing that by continuity, there is an interval of radius r around c and a number $s > 0$ such that

$$p(x) \geq s$$

for all x in this interval. Then by the definition of the integral according to lower sums,

$$\int_{-\pi}^{\pi} p(x) \, dx \geq rs > 0.$$

All the discussion of Chapter I which was carried out using only the four properties **SP 1** through **SP 4** is now seen to be valid in the present context. For instance, we define elements v, w in V to be **orthogonal,** or **perpendicular,** and

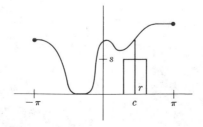

Figure 2

write $v \perp w$, if and only if $\langle v, w \rangle = 0$. We define the **norm** of v to be

$$\|v\| = \sqrt{\langle v, v \rangle} \ .$$

Remark. In analogy with ordinary Euclidean space, elements of V are also sometimes called **vectors**. More generally, one can define the general notion of a vector space, which is simply a set whose elements can be added and multiplied by numbers in such a way as to satisfy the basic properties of addition and multiplication (e.g. associativity and commutativity). Continuous functions on an interval form such a space. In an arbitrary vector space, one can then define the notion of a scalar product satisfying the above four conditions. For our purposes, which is to concentrate on the calculus part of the subject, we work right away in this function space. However, you should observe throughout that all the arguments of this section use only the basic axioms. Of course, when we want to find the norm of a specific function, like $\sin 3x$, then we use specifically the fact that we are working with the scalar product defined by the integral.

We shall now summarize a few properties of the norm.

If c is any number, then we immediately get

$$\|cv\| = |c| \ \|v\|,$$

because

$$\|cv\| = \sqrt{\langle cv, cv \rangle} = \sqrt{c^2 \langle v, v \rangle} = |c| \ \|v\|.$$

Thus we see the same type of arguments as in Chapter I apply here. In fact, any argument given in Chapter I which does not use coordinates applies to our more general situation. We shall see further examples as we go along.

As before, we say that an element $v \in V$ is a **unit vector** if $\|v\| = 1$. If $v \in V$ and $v \neq 0$, then $v/\|v\|$ is a unit vector.

The following two identities follow directly from the definition of the length.

The Pythagoras theorem. *If v, w are perpendicular, then*

$$\|v + w\|^2 = \|v\|^2 + \|w\|^2.$$

The parallelogram law. *For any v, w we have*

$$\|v + w\|^2 + \|v - w\|^2 = 2\|v\|^2 + 2\|w\|^2.$$

The proofs are trivial. We give the first, and leave the second as an exercise. For the first, we have

$$\|v + w\|^2 = \langle v + w, v + w \rangle = \langle v, v \rangle + 2\langle v, w \rangle + \langle w, w \rangle$$
$$= \|v\|^2 + \|w\|^2.$$

Let w be an element of V such that $\|w\| \neq 0$. For any v there exists a unique number c such that $v - cw$ is perpendicular to w. Indeed, for $v - cw$ to be perpendicular to w we must have

$$\langle v - cw, w \rangle = 0,$$

whence $\langle v, w \rangle - \langle cw, w \rangle = 0$ and $\langle v, w \rangle = c\langle w, w \rangle$. Thus

$$c = \frac{\langle v, w \rangle}{\langle w, w \rangle}.$$

Conversely, letting c have this value shows that $v - cw$ is perpendicular to w. We call c the **component of** v **along** w. This component is also called the **Fourier coefficient** of v with respect to w, to fit the applications in the theory of Fourier Series.

In particular, if w is a unit vector, then the component of v along w is simply

$$c = \langle v, w \rangle.$$

Example. Let V be the space of continuous functions on $[-\pi, \pi]$. Let f be the function given by $f(x) = \sin kx$, where k is some integer > 0. Then

$$\|f\| = \sqrt{\langle f, f \rangle} = \left(\int_{-\pi}^{\pi} \sin^2 kx \, dx \right)^{1/2}$$

$$= \sqrt{\pi} .$$

If g is any continuous function on $[-\pi, \pi]$, then the Fourier coefficient of g with respect to f is

$$\frac{\langle g, f \rangle}{\langle f, f \rangle} = \frac{1}{\pi} \int_{-\pi}^{\pi} g(x) \sin kx \, dx.$$

Let c be the component of v along w. As with the case of n-space, we define the **projection** of v along w to be the vector cw, because of our usual picture (Fig. 3):

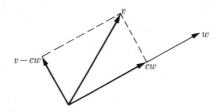

Figure 3

Theorem 1. (Schwarz inequality.) *For all $v, w \in V$ we have*

$$|\langle v, w \rangle| \leq \|v\| \, \|w\|.$$

Proof. If $w = 0$, then both sides are equal to 0 and our inequality is obvious. Next, assume that $w = u$ is a unit vector, that is $u \in V$ and $\|u\| = 1$. If c is the component of v along u, then $v - cu$ is perpendicular to u, and also perpendicular to cu. Hence by the Pythagoras theorem, we find

$$\|v\|^2 = \|v - cu\|^2 + \|cu\|^2$$
$$= \|v - cu\|^2 + c^2.$$

But $\|v - cu\|^2 \geqq 0$. Hence $c^2 \leqq \|v\|^2$, so that $|c| \leqq \|v\|$. Finally, if w is arbitrary $\neq 0$, then

$$u = w/\|w\|$$

is a unit vector, so that by what we just saw,

$$\left|\left\langle v, \frac{w}{\|w\|}\right\rangle\right| \leqq \|v\|.$$

This yields

$$|\langle v, w \rangle| \leqq \|v\| \, \|w\|,$$

as desired.

Theorem 2. *If $v, w \in V$, then*

$$\|v + w\| \leqq \|v\| + \|w\|.$$

Proof. We have:

$$\|v + w\|^2 = \langle v + w, v + w \rangle$$
$$= \langle v, v \rangle + 2\langle v, w \rangle + \langle w, w \rangle$$
$$\leqq \langle v, v \rangle + 2|\langle v, w \rangle| + \langle w, w \rangle$$
$$\leqq \|v\|^2 + 2\|v\| \, \|w\| + \|w\|^2$$
$$= (\|v\| + \|w\|)^2.$$

Taking square roots proves the theorem.

Let $v_1, \ldots, v_n$ be non-zero elements of V which are mutually perpendicular, that is $\langle v_i, v_j \rangle = 0$ if $i \neq j$. Let c_j be the component of v along v_i. Then

$$v - c_1 v_1 - \cdots - c_n v_n$$

is perpendicular to $v_1, \ldots, v_n$. To see this, all we have to do is to take the product with v_j for any j. All the terms involving $\langle v_i, v_j \rangle$ will give 0 if $i \neq j$, and we shall have two remaining terms

$$\langle v, v_j \rangle - c_j \langle v_j, v_j \rangle$$

which cancel. Thus subtracting linear combinations as above orthogonalizes v with respect to $v_1, \ldots, v_n$. The next theorem shows that $c_1 v_1 + \cdots + c_n v_n$ gives the closest approximation to v as a linear combination of $v_1, \ldots, v_n$.

Theorem 3. *Let $v_1, \ldots, v_n$ be vectors which are mutually perpendicular, and such that $\|v_i\| \neq 0$ for all i. Let v be an element of V, and let c_i be the* *component of v along v_i. Let $a_1, \ldots, a_n$ be numbers. Then*

$$\left\| v - \sum_{k=1}^{n} c_k v_k \right\| \leq \left\| v - \sum_{k=1}^{n} a_k v_k \right\|.$$

Proof. We know that

$$v - \sum_{k=1}^{n} c_k v_k$$

is perpendicular to each v_i, $i = 1, \ldots, n$. Hence it is perpendicular to any linear combination of $v_1, \ldots, v_n$. Now we have:

$$\| v - \sum a_k v_k \|^2 = \| v - \sum c_k v_k + \sum (c_k - a_k)v_k \|^2$$
$$= \| v - \sum c_k v_k \|^2 + \| \sum (c_k - a_k)v_k \|^2$$

by the Pythagoras theorem. This proves that

$$\| v - \sum c_k v_k \|^2 \leq \| v - \sum a_k v_k \|^2,$$

and thus our theorem is proved.

The next theorem is known as the **Bessel inequality.**

Theorem 4. *If $v_1, \ldots, v_n$ are mutually perpendicular unit vectors, and if c_i is the Fourier coefficient of v with respect to v_i, then*

$$\sum_{i=1}^{n} c_i^2 \leq \|v\|^2.$$

Proof. We have

$$0 \leq \langle v - \sum c_i v_i, v - \sum c_i v_i \rangle$$
$$= \langle v, v \rangle - \sum 2c_i \langle v, v_i \rangle + \sum c_i^2$$
$$= \langle v, v \rangle - \sum c_i^2.$$

From this our inequality follows.

EXERCISES

1. Prove **SP 2** and **SP 3,** using simple properties of the integral.

2. Let $f_1, \ldots, f_n$ be functions in V which are mutually perpendicular, that is

$$\langle f_i, f_j \rangle = 0 \quad \text{if} \quad i \neq j,$$

and assume that none of the functions f_i is 0. Let $c_1, \ldots, c_n$ be numbers such that

$$c_1 f_1 + \cdots + c_n f_n = 0$$

(the zero function). Prove that all c_i are equal to 0.

3. Let f be a fixed element of V. Let W be the subset of elements h in V such that h is perpendicular to f. Prove that if h_1, h_2 lie in W, then $h_1 + h_2$ lies in W. If c is a number and h is perpendicular to f, prove that ch is also perpendicular to f.

4. Write out the inequalities of Theorem 1 and Theorem 2 explicitly in terms of the integrals. Appreciate the fact that the notation of the text, following that of Chapter I, gives a much neater way, and a more geometric way, of expressing these inequalities.

5. Let m, n be positive integers. Prove that the functions

$$1, \sin nx, \cos mx$$

are mutually orthogonal. Use formulas like

$$\sin A \cos B = \frac{1}{2}[\sin(A + B) + \sin(A - B)],$$

$$\cos A \cos B = \frac{1}{2}[\cos(A + B) + \cos(A - B)].$$

6. Let $\varphi_n(x) = \cos nx$ and $\psi_n(x) = \sin nx$, for a positive integer n. Let φ_0 be the function such that $\varphi_0(x) = 1$, i.e. the constant function 1. Verify by performing the integrals that

$$\|\varphi_n\| = \|\psi_n\| = \sqrt{\pi} \quad \text{and} \quad \|\varphi_0\| = \sqrt{2\pi}.$$

7. Let V be the set of continuous functions on the interval $[0, 1]$. Define the scalar product in V by the integral

$$\langle f, g \rangle = \int_0^1 f(x)g(x)\, dx.$$

   (a) Prove that this satisfies conditions **SP 1** through **SP 4**. How would you define $\|f\|$ in the present context?
   (b) Let $f(x) = x$ and $g(x) = x^2$. Find $\langle f, g \rangle$.
   (c) With f, g as in (b), find $\|f\|$ and $\|g\|$.
   (d) Let $h(x) = 1$, the constant function 1. Find $\langle f, h \rangle$, $\langle g, h \rangle$, and $\|h\|$.

§2. COMPUTATION OF FOURIER SERIES

In the previous section we used continuous functions on the interval $[-\pi, \pi]$. For many applications one has to deal with somewhat more general functions. A convenient class of functions is that of piecewise continuous functions. We say that f is **piecewise continuous** if it is continuous except at a finite number of points, and if at each such point c the limits

$$\lim_{\substack{h \to 0 \\ h > 0}} f(c - h) \quad \text{and} \quad \lim_{\substack{h \to 0 \\ h > 0}} f(c + h)$$

both exist. The graph of a piecewise continuous function then looks like this (Fig. 4):

Figure 4

Let V be the set of functions on the interval $[-\pi, \pi]$ which are piecewise continuous. If f, g are in V, so is the sum $f + g$.

If c is a number, the function cf is also in V, so functions in V can be added and multiplied by numbers, to yield again functions in V. Furthermore, if f, g are piecewise continuous then the ordinary product fg is also piecewise continuous. We can then form the scalar product $\langle f, g \rangle$ since the integral is defined for piecewise continuous functions, and the three properties **SP 1** through **SP 3** are satisfied. However, the scalar product is not positive definite. A function f which is such that $f(x) = 0$ except at a finite number of points has norm 0.

Thus it is convenient, instead of **SP 4**, to formulate a slightly weaker condition:

Weak SP 4. *For all v in V we have $\langle v, v \rangle \geq 0$.*

We then call the scalar product **positive** (not necessarily definite).

We define the **norm** of an element as before, and we ask: For which elements of V is the norm equal to 0? The answer is simple.

Theorem 5. *Let V be the space of functions which are piecewise continuous on the interval $[-\pi, \pi]$. Let f be in V. Then $\|f\| = 0$ if and only if $f(x) = 0$ for all but a finite number of points x in the interval.*

Proof. First, it is clear that if $f(x) = 0$ except for a finite number of x, then

$$\|f\|^2 = \int_{-\pi}^{\pi} f(x)^2 \, dx = 0.$$

(Draw the picture of $f(x)^2$.) Conversely, suppose f is piecewise continuous on $[-\pi, \pi]$ and suppose we have a partition of $[-\pi, \pi]$ into intervals such that f is continuous on each subinterval $[a_i, a_{i+1}]$ except possibly at the end points a_i, $i = 0, \ldots, r - 1$. Suppose that $\|f\| = 0$, so that also $\|f\|^2 = 0 = \langle f, f \rangle$. This means that

$$\int_{-\pi}^{\pi} f(x)^2 \, dx = 0,$$

and the integral is the sum of the integrals over the smaller intervals, so that

$$\sum_{i=0}^{r-1} \int_{a_i}^{a_{i+1}} f(x)^2 \, dx = 0.$$

Each integral satisfies

$$\int_{a_i}^{a_{i+1}} f(x)^2 \, dx \geq 0$$

and hence each such integral is equal to 0. However, since f is continuous on an interval $[a_i, a_{i+1}]$ except possibly at the end points, we must have $f(x)^2 = 0$ for $a_i < x < a_{i+1}$, whence $f(x) = 0$ for $a_i < x < a_{i+1}$. Hence $f(x) = 0$ except at a finite number of points.

The space V of piecewise continuous functions on $[-\pi, \pi]$ is not finite dimensional. Instead of dealing with a finite number of orthogonal vectors, we must now deal with an infinite number.

For each positive integer n we consider the functions

$$\varphi_n(x) = \cos nx, \qquad \psi_n(x) = \sin nx,$$

and we also consider the function

$$\varphi_0(x) = 1.$$

It is verified by easy direct integrations that

$$\|\varphi_n\| = \|\psi_n\| = \sqrt{\pi} \qquad \text{if} \qquad n \neq 0,$$

$$\|\varphi_0\| = \sqrt{2\pi}.$$

Hence the Fourier coefficients of a function f with respect to our functions 1, $\cos nx$, $\sin nx$ are equal to:

$$a_0 = \frac{1}{2\pi} \int_{-\pi}^{\pi} f(x)\, dx, \qquad a_n = \frac{1}{\pi} \int_{-\pi}^{\pi} f(x) \cos nx\, dx,$$

$$b_n = \frac{1}{\pi} \int_{-\pi}^{\pi} f(x) \sin nx\, dx.$$

Furthermore, the functions 1, $\cos nx$, $\sin mx$ are easily verified to be mutually orthogonal. In other words, for any pair of distinct functions f, g among 1, $\cos nx$, $\sin mx$ we have $\langle f, g \rangle = 0$. This means:

If $m \neq n$ and $n \geq 0$, then

$$\int_{-\pi}^{\pi} \cos nx \cos mx\, dx = 0, \qquad \int_{-\pi}^{\pi} \sin nx \sin mx = 0;$$

and for any m, n:

$$\int_{-\pi}^{\pi} \cos nx \sin mx\, dx = 0.$$

The verifications of these orthogonalities are mere exercises in elementary calculus, which you should have already done in §1.

The **Fourier series** of a function f (piecewise continuous) is defined to be the series

$$a_0 + \sum_{k=1}^{\infty} (a_k \cos kx + b_k \sin kx).$$

The partial sum

$$s_n(x) = a_0 + \sum_{k=1}^{n} (a_k \cos kx + b_k \sin kx)$$

is simply the projection of the function f on the space generated by the functions 1, $\cos kx$, $\sin kx$ for $k = 1, \ldots, n$. In the present infinite dimensional case, we write

$$f \sim a_0 + \sum_{k=1}^{\infty} (a_k \cos kx + b_k \sin kx).$$

The sense in which one can replace the sign $\sim$ by an equality depends on various theorems whose proofs go beyond this course. One of these theorems is the following:

Theorem 6. *Assume that the piecewise continuous function f on $[-\pi, \pi]$ is orthogonal to every one of the functions 1, $\cos nx$, $\sin nx$. Then $f(x) = 0$ except at a finite number of x. If f is continuous, then $f = 0$.*

Theorem 6 shows at least that a continuous function is entirely determined by its Fourier series. There is another sense, however, in which we would like f to be equal to its Fourier series, namely we would like the values $f(x)$ to be given by

$$f(x) = a_0 + \sum_{k=1}^{\infty} (a_k \cos kx + b_k \sin kx)$$

$$= a_0 + \lim_{n \to \infty} \sum_{k=1}^{n} (a_k \cos kx + b_k \sin kx).$$

It is false in general that if f is merely continuous then $f(x)$ is given by the series. However, it is true under some reasonable conditions, for instance:

Theorem 7. *Let $-\pi < x < \pi$ and assume that f is differentiable in some open interval containing x, and has a continuous derivative in this interval. Then $f(x)$ is equal to the value of the Fourier series.*

Example 1. Find the Fourier series of the function f such that

$$f(x) = 0 \quad \text{if} \quad -\pi < x < 0,$$
$$f(x) = 1 \quad \text{if} \quad 0 < x < \pi.$$

The graph of f is as follows (Fig. 5).

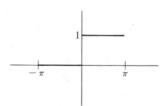

Figure 5

Since the Fourier coefficients are determined by an integral, it does not matter how we define f at $-\pi$, 0, or π. We have

$$a_0 = \frac{1}{2\pi} \int_{-\pi}^{\pi} f(x)\, dx = \frac{1}{2\pi} \int_0^{\pi} dx = \tfrac{1}{2},$$

$$a_n = \frac{1}{\pi} \int_0^{\pi} \cos nx\, dx = 0,$$

$$b_n = \frac{1}{\pi} \int_0^{\pi} \sin nx\, dx = \frac{1}{\pi n}(-\cos nx)\Big|_0^{\pi}$$

$$= \begin{cases} 0 & \text{if } n \text{ is even,} \\ \dfrac{2}{\pi n} & \text{if } n \text{ is odd.} \end{cases}$$

Hence the Fourier series of f is:

$$f(x) \sim \frac{1}{2} + \sum_{m=0}^{\infty} \frac{2}{(2m+1)\pi} \sin(2m+1)x.$$

By Theorem 7, we know that $f(x)$ is actually given by the series except at the points $-\pi$, 0, and π.

Example 2. Find the Fourier series of the function f such that $f(x) = -1$ if $-\pi < x < 0$ and $f(x) = x$ if $0 < x < \pi$.

The graph of f is as follows (Fig. 6).

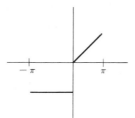

Figure 6

Again we compute the Fourier coefficients. We evaluate the integral over each of the intervals $[-\pi, 0]$ and $[0, \pi]$ since the function is given by different formulas over these intervals. We have

$$a_0 = \frac{1}{2\pi} \int_{-\pi}^{0} (-1) \, dx + \frac{1}{2\pi} \int_{0}^{\pi} x \, dx = \frac{1}{2} + \frac{\pi}{4},$$

$$a_n = \frac{1}{\pi} \int_{-\pi}^{0} (-1) \cos nx \, dx + \frac{1}{\pi} \int_{0}^{\pi} x \cos nx \, dx$$

$$= \begin{cases} 0 & \text{if } n \text{ is even,} \\ -\dfrac{2}{\pi n^2} & \text{if } n \text{ is odd,} \end{cases}$$

$$b_n = \frac{1}{\pi} \int_{-\pi}^{0} (-1) \sin nx \, dx + \frac{1}{\pi} \int_{0}^{\pi} x \sin nx \, dx$$

$$= \begin{cases} -\dfrac{1}{n} & \text{if } n \text{ is even,} \\ \dfrac{2}{\pi n} + \dfrac{1}{n} & \text{if } n \text{ is odd.} \end{cases}$$

Thus we obtain:

$$f(x) = \frac{1}{2} + \frac{\pi}{4} + \sum_{k=1}^{n} (a_k \cos kx + b_k \sin kx).$$

The equality is valid for $-\pi < x < 0$ and $0 < x < \pi$ by Theorem 7.

Example 3. Find the Fourier series of the function $\sin^2 x$. We have

$$\sin^2 x = \frac{1 - \cos 2x}{2} = \frac{1}{2} - \frac{1}{2} \cos 2x.$$

This is already written as a Fourier series, so the expression on the right is the desired Fourier series.

A function f is said to be **periodic** of period 2π if we have

$$f(x + 2\pi) = f(x)$$

for all x. For such a function, we then have by induction $f(x + 2\pi n) = f(x)$ for all positive integers n. Furthermore, letting $t = x + 2\pi$, we see also that

$$f(t - 2\pi) = f(t)$$

for all t, and hence $f(x - 2\pi n) = f(x)$ for all x and all positive integers n.

Given a piecewise continuous function on the interval $-\pi \leqq x < \pi$, we can extend it to a piecewise continuous function which is periodic of period 2π over all of **R**, simply by periodicity.

Example 4. Let $f(x) = x$ on $-\pi \leqq x < \pi$. If we extend f by periodicity, then the graph of the extended function looks like this (Fig. 7):

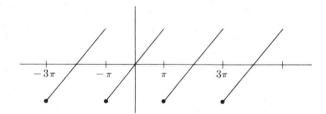

Figure 7

Example 5. Let f be the function on the interval $-\pi \leqq x < \pi$ given by:

$$f(x) = 0 \quad \text{if} \quad -\pi \leqq x \leqq 0,$$
$$f(x) = 1 \quad \text{if} \quad 0 < x < \pi.$$

Then the graph of the function extended by periodicity looks like this (Fig. 8):

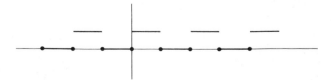

Figure 8

Example 6. Let f be the function on the interval $-\pi \leq x < \pi$ given by $f(x) = e^x$. Then the graph of the extended function looks like this (Fig. 9):

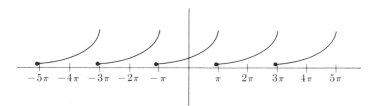

Figure 9

On the other hand, we may also be given a function over the interval $[0, 2\pi]$ and then extend this function by periodicity.

Example 7. Let $f(x) = x$ on the interval $0 \leq x < 2\pi$. The graph of the function extended by periodicity to all of **R** looks like this (Fig. 10):

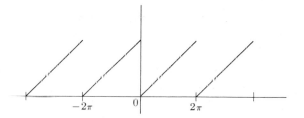

Figure 10

This is different from the function in Example 4, since in the present case, the extended function is never negative. When the function is given on an interval $[0, 2\pi]$, we compute the Fourier coefficients by taking the integral from 0 to 2π. In the present case, we therefore have:

$$a_0 = \frac{1}{2\pi} \int_0^{2\pi} x \, dx = \pi,$$

$$a_n = \frac{1}{\pi} \int_0^{2\pi} x \cos nx \, dx = 0 \qquad \text{for all } n,$$

$$b_n = \frac{1}{\pi} \int_0^{2\pi} x \sin nx \, dx = -\frac{2}{n}.$$

Hence we have, for $0 < x < 2\pi$:

$$x = \pi - 2\left(\sin x + \frac{\sin 2x}{2} + \frac{\sin 3x}{3} + \cdots \right).$$

EXERCISES

1. (a) Let $f(x)$ be the function such that $f(x) = 2$ if $0 \leq x < \pi$ and $f(x) = -1$ if $-\pi \leq x < 0$. Compute $\|f\|$.
   (b) Same question, if $f(x) = x$ for $0 \leq x < \pi$ and $f(x) = -1$ for $-\pi \leq x < 0$.

2. If f is periodic of period 2π and a, b are numbers, show that

$$\int_a^b f(x)\, dx = \int_{a+2\pi}^{b+2\pi} f(x)\, dx = \int_{a-2\pi}^{b-2\pi} f(x)\, dx.$$

   [*Hint:* Change variables, letting $u = x - 2\pi$, $du = dx$.] Also, prove:

$$\int_{-\pi}^{\pi} f(x + a)\, dx = \int_{-\pi}^{\pi} f(x)\, dx = \int_{-\pi+a}^{\pi+a} f(x)\, dx.$$

   [*Hint:* Split the integral over the bounds $-\pi + a$, $-\pi$, π, $\pi + a$.]

3. Let f be an even function, that is $f(x) = f(-x)$, for all x. Assume that f is periodic of period 2π. Show that all its Fourier coefficients with respect to sin nx are 0. Let g be an odd function (that is $g(-x) = -g(x)$). Show that all its Fourier coefficients with respect to cos nx are 0.

4. Compute the Fourier series of the functions, given on the interval $-\pi < x < \pi$ by the following $f(x)$:
   (a) x (b) x^2 (c) $|x|$ (d) $\sin^2 x$
   (e) $|\sin x|$ (f) $|\cos x|$ (g) $\sin^3 x$ (h) $\cos^3 x$

5. Show that the following relations hold:
   (a) For $0 < x < 2\pi$ and $a \neq 0$,

$$\pi e^{ax} = (e^{2a\pi} - 1)\left(\frac{1}{2a} + \sum_{k=1}^{\infty} \frac{a \cos kx - k \sin kx}{k^2 + a^2}\right).$$

   (b) For $0 < x < 2\pi$ and a not an integer,

$$\pi \cos ax = \frac{\sin 2a\pi}{2a} + \sum_{k=1}^{\infty} \frac{a \sin 2a\pi \cos kx + k(\cos 2a\pi - 1) \sin kx}{a^2 - k^2}.$$

   (c) Letting $x = \pi$ in part (b), conclude that

$$\frac{a\pi}{\sin a\pi} = 1 + 2a^2 \sum_{k=1}^{\infty} \frac{(-1)^k}{a^2 - k^2}.$$

   (d) For $0 < x < 2\pi$,

$$\frac{(\pi - x)^2}{4} = \frac{\pi^2}{12} + \sum_{k=1}^{\infty} \frac{\cos kx}{k^2}.$$

APPENDIX 2

Series

This appendix is included for use of the book as a text for a full year, in case series were not discussed during the first year.

§1. CONVERGENT SERIES

Suppose that we are given a sequence of numbers

$$a_1, a_2, a_3, \ldots,$$

i.e. we are given a number a_n for each integer $n \geq 1$. (We picked the starting place to be 1, but we could have picked any integer.) We form the sums

$$s_n = a_1 + a_2 + \cdots + a_n.$$

It would be meaningless to form an infinite sum

$$a_1 + a_2 + a_3 + \cdots$$

because we do not know how to add infinitely many numbers. However, if our sums s_n approach a limit, as n becomes large, then we say that the sum of our sequence **converges**, and we now define its **sum** to be that limit.

The symbols

$$\sum_{n=1}^{\infty} a_n$$

will be called a **series**. We shall say that the **series converges** if the sums s_n approach a limit as n becomes large. Otherwise, we say that it does not converge, or **diverges**. If the series converges, we say that the value of the series is

$$\sum_{n=1}^{\infty} a_n = \lim_{n \to \infty} s_n = \lim_{n \to \infty} (a_1 + \cdots + a_n).$$

The symbols $\lim_{n \to \infty}$ are to be read: "The limit as n becomes large."

Example. Consider the sequence

$$1, \frac{1}{2}, \frac{1}{4}, \frac{1}{8}, \frac{1}{16}, \ldots,$$

and let us form the sums

$$S_n = 1 + \frac{1}{2} + \frac{1}{4} + \cdots + \frac{1}{2^n}.$$

You probably know already that these sums approach a limit and that this limit is 2. To prove it, let $r = \frac{1}{2}$. Then

$$(1 + r + r^2 + \cdots + r^n) = \frac{1 - r^{n+1}}{1 - r} = \frac{1}{1 - r} - \frac{r^{n+1}}{1 - r}.$$

As n becomes large, r^{n+1} approaches 0, whence our sums approach

$$\frac{1}{1 - \frac{1}{2}} = 2.$$

Actually, the same argument works if we take for r any number such that

$$-1 < r < 1.$$

In that case, r^{n+1} approaches 0 as n becomes large, and consequently we can write

$$\sum_{n=0}^{\infty} r^n = \frac{1}{1 - r}.$$

Of course, if $|r| > 1$, then the series Σr^n does not converge. For instance, the partial sums of the series with $r = -3$ are

$$1 - 3 + 3^2 - 3^3 + \cdots + (-1)^n 3^n.$$

Observe that the nth term $(-1)^n 3^n$ does not even approach 0 as n becomes large.

In view of the fact that the limit of a sum is the sum of the limits, and other standard properties of limits, we get the following theorem.

Theorem 1. *Let $\{a_n\}$ and $\{b_n\}$ $(n = 1, 2, \ldots)$ be two sequences and assume that the series*

$$\sum_{n=1}^{\infty} a_n \quad and \quad \sum_{n=1}^{\infty} b_n$$

converge. Then $\Sigma_{n=1}^{\infty}(a_n + b_n)$ also converges, and is equal to the sum of the *two series. If c is a number, then*

$$\sum_{n=1}^{\infty} ca_n = c \sum_{n=1}^{\infty} a_n.$$

Finally, if

$$s_n = a_1 + \cdots + a_n$$

and

$$t_n = b_1 + \cdots + b_n$$

then

$$\sum_{n=1}^{\infty} a_n \sum_{n=1}^{\infty} b_n = \lim_{n\to\infty} s_n t_n.$$

In particular, series can be added term by term. Of course, they cannot be multiplied term by term!

We also observe that a similar theorem holds for the difference of two series.

If a series Σa_n converges, then the numbers a_n must approach 0 as n becomes large. However, there are examples of sequences $\{a_n\}$ for which the series does not converge, and yet

$$\lim_{n\to\infty} a_n = 0.$$

Consider, for instance,

$$1 + \frac{1}{2} + \frac{1}{3} + \cdots + \frac{1}{n} + \cdots.$$

We contend that the partial sums s_n become very large when n becomes large. To see this, we look at partial sums as follows:

$$1 + \frac{1}{2} + \frac{1}{3} + \frac{1}{4} + \frac{1}{5} + \cdots + \frac{1}{8} + \frac{1}{9} + \cdots + \frac{1}{16} + \cdots.$$

In each bunch of terms as indicated, we replace each term by that farthest to the right. This makes our sums smaller. Thus our expression is

$$\geq 1 + \frac{1}{2} + \frac{1}{4} + \frac{1}{4} + \frac{1}{8} + \cdots + \frac{1}{8} + \frac{1}{16} + \cdots + \frac{1}{16} + \cdots$$

$$\geq 1 + \frac{1}{2} + \frac{1}{2} + \frac{1}{2} + \frac{1}{2} + \cdots$$

and therefore becomes arbitrarily large when n becomes large.

§2. SERIES WITH POSITIVE TERMS

Throughout this section, we shall assume that our numbers a_n are ≥ 0. Then the partial sums

$$s_n = a_1 + \cdots + a_n$$

are increasing, i.e.

$$s_1 \leq s_2 \leq s_3 \leq \cdots \leq s_n \leq s_{n+1} \leq \cdots .$$

If they are to approach a limit at all, they cannot become arbitrarily large. Thus in that case there is a number B such that

$$s_n \leq B$$

for all n. The collection of numbers $\{s_n\}$ has therefore a least upper bound, i.e. there is a smallest number S such that

$$s_n \leq S$$

for all n. In that case, the partial sums s_n approach S as a limit. In other words, given any positive number $\varepsilon > 0$, we have

$$S - \varepsilon \leq s_n \leq S$$

for all n sufficiently large (Fig. 1).

Figure 1

This simply expresses the fact that S is the least of all upper bounds for our collection of numbers s_n. We express this as a theorem.

Theorem 2. *Let $\{a_n\}$ $(n = 1, 2, \ldots)$ be a sequence of numbers ≥ 0 and let*

$$s_n = a_1 + \cdots + a_n.$$

If the sequence of numbers $\{s_n\}$ is bounded, then it approaches a limit S, which is its least upper bound.

Example 1. Prove that the series $\sum_{n=1}^{\infty} 1/n^2$ converges.

Let us look at the series:

$$\frac{1}{1^2} + \frac{1}{2^2} + \frac{1}{3^2} + \frac{1}{4^2} + \cdots + \frac{1}{8^2} + \cdots + \frac{1}{16^2} + \cdots + \cdots .$$

We look at the groups of terms as indicated. In each group of terms, if we decrease the denominator in each term, then we increase the fraction. We replace 3 by 2, then 4, 5, 6, 7 by 4, then we replace the numbers from 8 to 15 by 8, and so forth. Our partial sums are therefore less than or equal to

$$1 + \frac{1}{2^2} + \frac{1}{2^2} + \frac{1}{4^2} + \cdots + \frac{1}{4^2} + \frac{1}{8^2} + \cdots + \frac{1}{8^2} \cdots ,$$

and we note that 2 occurs twice, 4 occurs four times, 8 occurs eight times, and so forth. Hence the partial sums are less than or equal to

$$1 + \frac{2}{2^2} + \frac{4}{4^2} + \frac{8}{8^2} + \cdots = 1 + \frac{1}{2} + \frac{1}{4} + \frac{1}{8} + \cdots .$$

Thus our partial sums are less than or equal to those of the geometric series and are bounded. Hence our series converges.

Theorem 2 gives us a very useful criterion to determine when a series with positive terms converges:

Theorem 3. *Let*

$$\sum_{n=1}^{\infty} a_n \qquad and \qquad \sum_{n=1}^{\infty} b_n$$

be two series, with $a_n \geq 0$ for all n and $b_n \geq 0$ for all n. Assume that there is a number $C > 0$ such that

$$a_n \leq C b_n$$

for all n, and that $\sum_{n=1}^{\infty} b_n$ converges. Then $\sum_{n=1}^{\infty} a_n$ converges, and

$$\sum_{n=1}^{\infty} a_n \leq C \sum_{n=1}^{\infty} b_n.$$

Proof. We have

$$a_1 + \cdots + a_n \leq C b_1 + \cdots + C b_n = C(b_1 + \cdots + b_n) \leq C \sum_{n=1}^{\infty} b_n.$$

This means that $C\sum_{n=1}^{\infty} b_n$ is a bound for the partial sums

$$a_1 + \cdots + a_n.$$

The least upper bound of these sums is therefore $\leq C\sum_{n=1}^{\infty} b_n$, thereby proving our theorem.

Theorem 3 has an analogue to show that a series does not converge.

Theorem 3′. *Let*

$$\sum_{n=1}^{\infty} a_n \qquad and \qquad \sum_{n=1}^{\infty} b_n$$

be two series, with a_n and $b_n \geq 0$ for all n. Assume that there is a number $C > 0$ such that

$$a_n \geq Cb_n$$

for all n sufficiently large, and that $\sum_{n=1}^{\infty} b_n$ does not converge. Then $\sum_{n=1}^{\infty} a_n$ diverges.

Proof. Assume $a_n \geq Cb_n$ for $n \geq n_0$. Since Σb_n diverges, we can make the partial sums

$$\sum_{n=n_0}^{N} b_n = b_{n_0} + \cdots + b_N$$

arbitrarily large as N becomes arbitrarily large. But

$$\sum_{n=n_0}^{N} a_n \geq \sum_{n=n_0}^{N} Cb_n = C \sum_{n=n_0}^{N} b_n.$$

Hence the partial sums

$$\sum_{n=1}^{N} a_n = a_1 + \cdots + a_N$$

are arbitrarily large as N becomes arbitrarily large, and hence $\sum_{n=1}^{\infty} a_n$ diverges, as was to be shown.

Example 2. Determine whether the series

$$\sum_{n=1}^{\infty} \frac{n^2}{n^3 + 1}$$

converges.
We write

$$\frac{n^2}{n^3 + 1} = \frac{1}{n + 1/n^2} = \frac{1}{n}\left(\frac{1}{1 + 1/n^3}\right).$$

Then we see that

$$\frac{n^2}{n^3 + 1} \geq \frac{1}{2n}.$$

Since $\Sigma 1/n$ does not converge, it follows that the series of Example 2 does not converge either.

Example 3. The series

$$\sum_{n=1}^{\infty} \frac{n^2 + 7}{2n^4 - n + 3}$$

converges. Indeed, we can write

$$\frac{n^2 + 7}{2n^4 - n + 3} = \frac{n^2(1 + 7/n^2)}{n^4(2 - (1/n)^3 + 3/n^4)} = \frac{1}{n^2} \frac{1 + 7/n^2}{(2 - (1/n)^3 + 3/n^4)}.$$

For n large, the factor

$$\frac{1 + 7/n^2}{2 - (1/n)^3 + 3/n^4}$$

is certainly bounded, and in fact is near $\frac{1}{2}$. Hence we can compare our series with $1/n^2$ to see that it converges, because $\Sigma 1/n^2$ converges, and the factor is bounded.

EXERCISES

1. Show that the series $\Sigma_{n=1}^{\infty} 1/n^3$ converges.

2. (a) Show that the series $\Sigma (\log n)/n^3$ converges. [*Hint:* Estimate $(\log n)/n$.]
   (b) Show that the series $\Sigma (\log n)^2/n^3$ converges.

Test the following series of convergence:

3. $\displaystyle\sum_{n=1}^{\infty} \frac{1}{n^{1/2}}$

4. $\displaystyle\sum_{n=1}^{\infty} \frac{n^2}{n^4 + n}$

5. $\displaystyle\sum_{n=1}^{\infty} \frac{n}{n + 1}$

6. $\displaystyle\sum_{n=1}^{\infty} \frac{n}{n + 5}$

7. $\displaystyle\sum_{n=1}^{\infty} \frac{n^2}{n^3 + n + 2}$

8. $\displaystyle\sum_{n=1}^{\infty} \frac{|\sin n|}{n^2 + 1}$

9. $\displaystyle\sum_{n=1}^{\infty} \frac{|\cos n|}{n^2 + n}$

§3. THE RATIO TEST

We continue to consider only series with terms ≥ 0. To compare such a series with a geometric series, the simplest test is given by the ratio test.

Ratio Test. *Let $\Sigma_{n=1}^{\infty} a_n$ be a series with $a_n > 0$ for all n. Assume that there is a number c with $0 < c < 1$ such that*

$$\frac{a_{n+1}}{a_n} \leq c$$

for all n sufficiently large. Then the series converges.

Proof. Suppose that there exists some integer N such that

$$\frac{a_{n+1}}{a_n} \leq c$$

if $n \geq N$. Then

$$a_{N+1} \leq ca_N$$

$$a_{N+2} \leq ca_{N+1} \leq c^2 a_N$$

and in general, by induction,

$$a_{N+k} \leq c^k a_N.$$

Thus

$$\sum_{n=N}^{N+k} a_n \leq a_N + ca_N + c^2 a_N + \cdots + c^k a_N$$

$$\leq a_N(1 + c + \cdots + c^k) \leq a_N \frac{1}{1-c}.$$

Thus in effect, we have compared our series with a geometric series, and we know that the partial sums are bounded. This implies that our series converges.

The ratio test is usually used in the case of a series with positive terms a_n such that

$$\lim_{n \to \infty} \frac{a_{n+1}}{a_n} = c < 1.$$

Example. Show that the series

$$\sum_{n=1}^{\infty} \frac{n}{3^n}$$

converges.

We let $a_n = n/3^n$. Then

$$\frac{a_{n+1}}{a_n} = \frac{n+1}{3^{n+1}} \frac{3^n}{n} = \frac{n+1}{n} \frac{1}{3}.$$

This ratio approaches $\frac{1}{3}$ as $n \to \infty$, and hence the ratio test is applicable: the series converges.

EXERCISES

Determine whether the following series converge:

1. $\sum n2^{-n}$ 2. $\sum n^2 2^{-n}$ 3. $\sum \dfrac{1}{\log n}$

4. $\sum \dfrac{\log n}{2^n}$ 5. $\sum \dfrac{\log n}{n}$ 6. $\sum \dfrac{n^{10}}{3^n}$

7. $\Sigma \dfrac{1}{\sqrt{n(n+1)}}$ 8. $\Sigma \dfrac{\sqrt{n^3+1}}{e^n}$ 9. $\Sigma \dfrac{n+1}{\sqrt{n^4+n+1}}$

10. $\Sigma \dfrac{n+1}{2^n}$ 11. $\Sigma \dfrac{n}{(4n-1)(n+15)}$ 12. $\Sigma \dfrac{1+\cos(\pi n/2)}{e^n}$

13. $\Sigma \dfrac{1}{(\log n)^{10}}$ 14. $\Sigma n^2 e^{-n^2}$ 15. $\Sigma n^2 e^{-n3}$

16. $\Sigma n^5 e^{-n^2}$ 17. $\Sigma n^4 e^{-n}$ 18. $\Sigma \dfrac{n^n}{n!3^n}$

19. Let $\{a_n\}$ be a sequence of positive numbers, and assume that

$$\frac{a_{n+1}}{a_n} \geqq 1 - \frac{1}{n}$$

for all n. Show that the series Σa_n diverges.

20. A ratio test can be applied in the opposite direction to determine when a series diverges. Prove the following statement: Let a_n be a sequence of positive numbers, and let $c \geqq 1$. If $a_{n+1}/a_n \geqq c$ for all n sufficiently large, then the series Σa_n diverges.

§4. THE INTEGRAL TEST

You must already have felt that there is an analogy between the convergence of an improper integral and the convergence of a series. We shall now make this precise.

Theorem 4. *Let f be a function which is defined and positive for all $x \geqq 1$, and decreasing. The series*

$$\sum_{n=1}^{\infty} f(n)$$

converges if and only if the improper integral

$$\int_1^\infty f(x)\, dx$$

converges.

We visualize the situation in the following diagram (Fig. 2).

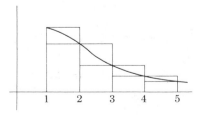

Figure 2

Consider the partial sums

$$f(2) + \cdots + f(n)$$

and assume that our improper integral converges. The area under the curve between 1 and 2 is greater than or equal to the area of the rectangle whose height is $f(2)$ and whose base is the interval between 1 and 2. This base has length 1. Thus

$$f(2) \leq \int_1^2 f(x) \, dx.$$

Again, since the function is decreasing, we have a similar estimate between 2 and 3:

$$f(3) \leq \int_2^3 f(x) \, dx.$$

We can continue up to n, and get

$$f(2) + f(3) + \cdots + f(n) \leq \int_1^n f(x) \, dx.$$

As n becomes large, we have assumed that the integral approaches a limit. This means that

$$f(2) + f(3) + \cdots + f(n) \leq \int_1^\infty f(x) \, dx.$$

Hence the partial sums are bounded, and hence by Theorem 2, they approach a limit. Therefore our series converges.

Conversely, assume that the partial sums

$$f(1) + \cdots + f(n)$$

approach a limit as n becomes large.

The area under the graph of f between 1 and n is less than or equal to the sum of the areas of the big rectangles. Thus

$$\int_1^2 f(x) \, dx \leq f(1)(2 - 1) = f(1)$$

and

$$\int_2^3 f(x) \, dx \leq f(2)(3 - 2) = f(2).$$

Proceeding stepwise, and taking the sum, we see that

$$\int_1^n f(x) \, dx \leq f(1) + \cdots + f(n - 1).$$

The partial sums on the right are less than or equal to their limit. Call this limit L. Then for all positive integers n, we have

$$\int_1^n f(x)\, dx \leqq L.$$

Given any number B, we can find an integer n such that $B \leqq n$. Then

$$\int_1^B f(x)\, dx \leqq \int_1^n f(x)\, dx \leqq L.$$

Hence the integral from 1 to B approaches a limit as B becomes large, and this limit is less than or equal to L. This proves our theorem.

Example. Prove that the series

$$\Sigma \frac{1}{n^2 + 1}$$

converges.
Let

$$f(x) = \frac{1}{x^2 + 1}.$$

Then f is decreasing, and

$$\int_1^B f(x)\, dx = \arctan B - \arctan 1 = \arctan B - \frac{\pi}{4}.$$

As B becomes large, $\arctan B$ approaches $\pi/2$ and therefore has a limit. Hence the integral converges. So does the series, by the theorem.

EXERCISES

1. Show that the following series diverges: $\Sigma_{n+2}^{\infty} 1/(n \log n)$.

2. Show that the following series converges: $\Sigma_{n=1}^{\infty} (n + 1)/(n + 2)n!$.

Test for convergence:

3. $\displaystyle\sum_{n=1}^{\infty} n e^{-n^2}$

4. $\displaystyle\sum_{n=2}^{\infty} \frac{1}{n(\log n)^3}$

5. $\displaystyle\sum_{n=2}^{\infty} \frac{1}{n(\log n)^2}$

6. $\displaystyle\sum_{n=1}^{\infty} \frac{n!}{n^n}$

7. $\displaystyle\sum_{n=1}^{\infty} \frac{n}{e^n}$

8. $\displaystyle\sum_{n=1}^{\infty} \frac{n + 1}{n^3 + 2}$

9. $\displaystyle\sum_{n=1}^{\infty} \frac{1}{n^2 + n - 1}$

10. $\displaystyle\sum_{n=1}^{\infty} \frac{n}{n^3 - n + 5}$

11. Let ε be a number > 0. Show that the series $\Sigma_{n=1}^{\infty} 1/n^{1+\varepsilon}$ converges.

12. Show that the following series converge.

(a) $\displaystyle\sum_{n=1}^{\infty} \frac{\log n}{n^2}$

(b) $\displaystyle\sum_{n=1}^{\infty} \frac{\log n}{n^{3/2}}$

(c) $\displaystyle\sum_{n=1}^{\infty} \frac{\log n}{n^{1+\varepsilon}}$ if $\varepsilon > 0$.

(d) $\displaystyle\sum_{n=1}^{\infty} \frac{(\log n)^2}{n^{3/2}}$

(e) $\displaystyle\sum_{n=1}^{\infty} \frac{(\log n)^3}{n^2}$

13. If $\varepsilon > 0$ show that the series $\sum_{n=2}^{\infty} 1/n(\log n)^{1+\varepsilon}$ converges.

§5. ABSOLUTE AND ALTERNATING CONVERGENCE

We consider a series $\sum_{n=1}^{\infty} a_n$ in which we do not assume that the terms a_n are ≥ 0. We shall say that the series **converges absolutely** if the series

$$\sum_{n=1}^{\infty} |a_n|$$

formed with the absolute values of the terms a_n converges. This is now a series with terms ≥ 0, to which we can apply the tests for convergence given in the two preceding sections. This is important, because we have:

Theorem 5. *Let* $\{a_n\}(n = 1, 2, \dots)$ *be a sequence, and assume that the series*

$$\sum_{n=1}^{\infty} |a_n|$$

converges. Then so does the series $\sum_{n=1}^{\infty} a_n$.

Proof. Let a_n^+ be equal to 0 if $a_n < 0$ and equal to a_n itself if $a_n \geq 0$. Let a_n^- be equal to 0 if $a_n > 0$ and equal to $-a_n$ if $a_n \leq 0$. Then both a_n^+ and a_n^- are ≥ 0. By assumption and comparison with $\sum |a_n|$, we see that each one of the series

$$\sum_{n=1}^{\infty} a_n^+ \qquad \text{and} \qquad \sum_{n=1}^{\infty} a_n^-$$

converges. Hence so does their difference

$$\sum_{n=1}^{\infty} a_n^+ - \sum_{n=1}^{\infty} a_n^-$$

which is equal to

$$\sum_{n=1}^{\infty} (a_n^+ - a_n^-),$$

which is none other than $\sum_{n=1}^{\infty} a_n$. This proves our theorem.

§6. POWER SERIES

Perhaps the most important series are power series. Let x be any number and let $\{a_n\}(n = 0, 1, \ldots)$ be a sequence of numbers. Then we can form the series

$$\sum_{n=0}^{\infty} a_n x^n.$$

The partial sums are

$$a_0 + a_1 x + a_2 x^2 + \cdots + a_n x^n.$$

We have already met such sums when we discussed Taylor's formula.

Example 1. The power series

$$\sum_{n=0}^{\infty} \frac{x^n}{n!}$$

converges for all x, absolutely. Indeed, it will suffice to prove that for any number $R > 0$, the above series converges for $0 < x \leq R$. We use the ratio test. Let

$$b_n = \frac{x^n}{n!}.$$

Then

$$\frac{b_{n+1}}{b_n} = \frac{x^{n+1}}{(n+1)!} \frac{n!}{x^n} = \frac{x}{n+1} \leq \frac{R}{n+1}.$$

When n is sufficiently large, it follows that $R/(n+1)$ is small, and in particular is $< \frac{1}{2}$, so that we can apply the ratio test to prove our assertion.

Similarly, we could prove that the series

$$\sum_{n=0}^{\infty} (-1)^n \frac{x^{2n+1}}{(2n+1)!} \qquad \text{and} \qquad \sum_{n=0}^{\infty} (-1)^n \frac{x^{2n}}{(2n)!}$$

converge absolutely for all x, letting for instance

$$b_n = \frac{x^{2n+1}}{(2n+1)!}$$

for the first one. Then $b_{n+1}/b_n = x^2/[(2n+3)(2n+2)]$, and we can argue as before.

Theorem 6. *Assume that there is a number $r \geq 0$ such that the series*

$$\sum_{n=0}^{\infty} |a_n| r^n$$

converges. Then for all x such that $|x| \leqq r$, the series

$$\sum_{n=0}^{\infty} a_n x^n$$

converges absolutely.

Proof. The absolute value of each term is

$$|a_n|\,|x|^n \leqq |a_n|r^n.$$

Our assertion follows from the comparison Theorem 3.

The least upper bound of all numbers r for which we have the convergence stated in the theorem is called the **radius of convergence** of the series. If there is no upper bound for the numbers r such that the power series above converges, then we say that the radius of convergence is **infinity**.

Suppose that there is an upper bound for the numbers r above, and thus let s be the radius of convergence of the series. Then if $x > s$, the series

$$\sum_{n=0}^{\infty} |a_n| x^n$$

does *not* converge. Thus the radius of convergence s is the number such that the series converges absolutely if $0 < x < s$ but does not converge absolutely if $x > s$.

Theorem 6 allows us to define a function f; namely, for all numbers x such that $|x| < r$, we define

$$f(x) = \lim_{n \to \infty} (a_0 + a_1 x + \cdots + a_n x^n).$$

The proofs that the remainder term in Taylor's formula approaches 0 for various functions now allow us to say that these functions are given by their Taylor series. Thus

$$\sin x = x - \frac{x^3}{3!} + \frac{x^5}{5!} - \cdots$$

$$e^x = 1 + x + \frac{x^2}{2!} + \cdots$$

for all x. Furthermore,

$$\log(1 + x) = x - \frac{x^2}{2} + \cdots$$

is valid for $-1 < x < 1$.

(Here we saw that the series converges for $x = 1$, but it does not converge absolutely; cf. §1.)

However, we can now define functions at random by means of a power series, provided we know the power series converges absolutely, for $|x| < r$.

The ratio test usually gives an easy way to determine when a power series converges, or when it diverges.

Example 2. Prove that the series

$$\sum_{n=2}^{\infty} \frac{\log n}{n^2} x^n$$

converges absolutely for $|x| < 1$, and diverges for $|x| > 1$.

Let $0 < c < 1$ and consider x such that $0 < x \le c$. Let

$$b_n = \frac{\log n}{n^2} x^n.$$

Then

$$\frac{b_{n+1}}{b_n} = \frac{\log(n + 1)}{(n + 1)^2} x^{n+1} \frac{n^2}{\log n} \frac{1}{x^n} = \frac{\log(n + 1)}{\log n} \left(\frac{n}{n + 1}\right)^2 x.$$

Since $\log(n + 1)/\log n$ and $(n/(n + 1))^2$ approach 1 when n becomes very large, it follows that if $c < c_1 < 1$, then for all n sufficiently large

$$\frac{b_{n+1}}{b_n} \le c_1$$

and hence our series converges. This is true for every c such that $0 < c < 1$, and hence the series converges absolutely for $|x| < 1$.

Let $c > 1$. If $x \ge c$ then for all n sufficiently large, it follows that

$$b_{n+1}/b_n \ge 1,$$

whence the series does not converge. This is so for all $c > 1$, and hence the series does not converge if $x > 1$. Hence 1 is the radius of convergence.

If a power series converges absolutely only for $x = 0$, then we agree to say that its radius of convergence is 0. For example, the radius of convergence of the series

$$\sum_{n=1}^{\infty} n! x^n$$

is equal to 0, as one sees by using the ratio test in the divergent case.

Root Test. *Let $\sum a_n x^n$ be a power series and assume that*

$$\lim |a_n|^{1/n} = s,$$

where s is a number. If $s \neq 0$ then the radius of convergence of the series is *equal to* $1/s$. *If* $s = 0$, *then the radius of convergence is infinity. If* $|a_n|^{1/n}$ *becomes arbitrarily large as n becomes large, then the radius of convergence is* 0.

Proof. Without loss of generality, we may assume that $a_n \geq 0$ for all n. Suppose first that s is a number $\neq 0$, and let $0 \leq r < 1/s$. Then $sr < 1$. The numbers $a_n^{1/n}r$ approach sr and hence there is some number $\varepsilon > 0$ such that

$$a_n^{1/n}r < 1 - \varepsilon$$

for all n sufficiently large. Hence the series $\Sigma a_n r^n$ converges by comparison with the geometric series. If on the other hand $r > 1/s$, then $a_n^{1/n}r$ approaches $sr > 1$, and hence we have

$$a_n^{1/n}r \geq 1 + \varepsilon$$

for all n sufficiently large. Comparison from below shows that the series $\Sigma a_n r^n$ diverges. We leave the cases $s = 0$ or $s = \infty$ to the reader.

Example 3. The series $\Sigma x^n/n^2$ has a radius of convergence equal to 1, because

$$\lim\left(\frac{1}{n^2}\right)^{1/n} = \lim \frac{1}{n^{2/n}} = 1,$$

by Corollary 3 of Theorem 11, Chapter VIII, §5 of the *First Course*.

To give other examples, we recall an inequality which you should have worked out in Chapter X, §1, of the *First Course* namely

$$\boxed{(n - 1)! \leq n^n e^{-n} \leq n!.}$$

From it, we prove:

As n becomes large, the expression

$$\frac{(n!)^{1/n}}{n} = \left[\frac{n!}{n^n}\right]^{1/n}$$

approaches $1/e$.

Proof. Take the nth root of the inequality $n^n e^{-n} \leq n!$. We get

$$ne^{-1}e^{1/n} \leq (n!)^{1/n}.$$

Dividing by n yields

$$\frac{1}{e}e^{1/n} \leq \frac{(n!)^{1/n}}{n}.$$

On the other hand, multiply both sides of the inequality

$$(n-1)! \leqq n^n e^{-n} e$$

by n. We get $n! \leqq n^n e^{-n} en$. Take an nth root:

$$(n!)^{1/n} \leqq n e^{-1} e^{1/n} n^{1/n}.$$

Dividing by n yields

$$\frac{(n!)^{1/n}}{n} \leqq \frac{1}{e} e^{1/n} n^{1/n}.$$

But we know that both $n^{1/n}$ and $e^{1/n}$ approach 1 as n becomes large. Thus our quotient is squeezed between two numbers approaching $1/e$, and must therefore approach $1/e$.

Example 4. We have

$$\lim_{n \to \infty} \left[\frac{(3n)!}{n^{3n}} \right]^{1/n} = \frac{27}{e^3}.$$

Proof. We write

$$\frac{(3n)!}{n^{3n}} = \frac{(3n)!}{(3n)^{3n}} 3^{3n}.$$

The $3n$th root of this expression is

$$\left[\frac{(3n)!}{(3n)^{3n}} \right]^{1/3n} 3.$$

We have seen that

$$\left(\frac{m!}{m^m} \right)^{1/m} \qquad \text{approaches} \qquad \frac{1}{e}$$

as m becomes large. We use $m = 3n$. We conclude that

$$\left(\frac{(3n)!}{n^{3n}} \right)^{1/3n} \qquad \text{approaches} \qquad \frac{3}{e}.$$

Hence

$$\left(\frac{(3n)!}{n^{3n}} \right)^{1/n} \qquad \text{approaches} \qquad \frac{27}{e^3},$$

as desired.

EXERCISES

1. Use the abbreviation $\lim_{n \to \infty}$ to mean: limit as n becomes very large. Prove that

(a) $\lim_{n \to \infty} \left[\dfrac{(3n)!}{n^{3n}} \right]^{1/n} = \dfrac{27}{e^3}$

(b) $\lim_{n \to \infty} \left[\dfrac{(3n)!}{n! \, n^{2n}} \right]^{1/n} = \dfrac{27}{e^2}$

2. Find the limit:

(a) $\lim_{n \to \infty} \left[\dfrac{(2n)!}{n^{2n}} \right]^{1/n}$

(b) $\lim_{n \to \infty} \left[\dfrac{(2n)! \, (5n)!}{n^{4n}(3n)!} \right]^{1/n}$

Find the radius of convergence of the following series:

3. (a) $\sum n^n x^n$

(b) $\sum \dfrac{x^n}{n^n}$

4. $\sum \dfrac{n}{n + 5} x^n$

5. $\sum (\log n) x^n$

6. $\sum \dfrac{1}{\log n} x^n$

7. $\sum (\log n)^2 x^n$

8. $\sum 2^n x^n$

9. $\sum 2^{-n} x^n$

10. $\sum (1 + n)^n x^n$

11. $\sum \dfrac{x^n}{n}$

12. $\sum \dfrac{x^n}{\sqrt{n}}$

13. $\sum (1 + (-2)^n) x^n$

14. $\sum (1 + (-1)^n) x^n$

15. $\sum_{n=1}^{\infty} \dfrac{(2n)!}{(n!)^2} x^n$

16. $\sum_{n=1}^{\infty} \dfrac{n^n}{n!} x^n$

17. $\sum_{n=1}^{\infty} \dfrac{(n!)^3}{(3n)!} x^n$

18. $\sum_{n=1}^{\infty} \dfrac{n^{5n}}{(2n)! \, n^{3n}} x^n$

19. $\sum_{n=1}^{\infty} \dfrac{(3n)!}{(n!)^2} x^n$

20. $\sum_{n=1}^{\infty} \dfrac{\sin n\pi/2}{2^n} x^n$

21. $\sum_{n=2}^{\infty} \dfrac{\log n}{2^n} x^n$

22. $\sum_{n=1}^{\infty} \dfrac{1 + \cos 2\pi n}{3^n} x^n$

23. $\sum_{n=2}^{\infty} n x^n$

24. $\sum_{n=1}^{\infty} \dfrac{\sin 2\pi n}{n!} x^n$

25. $\sum_{n=2}^{\infty} n^2 x^n$

26. $\sum_{n=1}^{\infty} \dfrac{\cos n^2}{n^n} x^n$

27. $\sum_{n=2}^{\infty} \dfrac{n}{\log n} x^n$

28. $\sum_{n=2}^{\infty} \dfrac{(-1)^n}{n! - 1} x^n$

29. $\sum_{n=1}^{\infty} \dfrac{n!}{n^n} x^n$

30. $\sum_{n=1}^{\infty} \dfrac{(-1)^n + 1}{n!} x^n$

Note: For some of the above radii of convergence, recall that

$$\lim_{n \to \infty} \left(1 + \frac{1}{n}\right)^n = \lim_{n \to \infty} \left(\frac{1 + n}{n}\right)^n = e.$$

§7. DIFFERENTIATION AND INTEGRATION OF POWER SERIES

If we have a polynomial

$$a_0 + a_1 x + \cdots + a_n x^n$$

with numbers $a_0, a_1, \ldots, a_n$ as coefficients, then we know how to find its derivative. It is $a_1 + 2a_2x + \cdots + na_nx^{n-1}$. We would like to say that the derivative of a series can be taken in the same way, and that the derivative converges whenever the series does.

Theorem 7. *Let r be a number > 0 and let $\Sigma a_n x^n$ be a series which converges absolutely for $|x| < r$. Then the series $\Sigma na_n x^{n-1}$ also converges absolutely for $|x| < r$.*

Proof. Since we are interested in the absolute convergence, we may assume that $a_n \geqq 0$ for all n. Let $0 < x < r$, and let c be a number such that $x < c < r$. Recall that

$$\lim_{n \to \infty} n^{1/n} = 1.$$

We may write

$$na_n x^n = a_n (n^{1/n} x)^n.$$

Then for all n sufficiently large, we conclude that

$$n^{1/n} x < c$$

because $n^{1/n}x$ comes arbitrarily close to x. Hence for all n sufficiently large, we have

$$na_n x^n < a_n c^n.$$

We can then compare the series $\Sigma na_n x^n$ with $\Sigma a_n c^n$ to conclude that $\Sigma na_n x^n$ converges. Since

$$\Sigma na_n x^{n-1} = \frac{1}{x} \Sigma a_n x^n$$

we have proved Theorem 7.

A similar result holds for integration, but trivially. Indeed, if we have a series $\Sigma_{n=1}^\infty a_n x^n$ which converges absolutely for $|x| < r$, then the series

$$\sum_{n=1}^\infty \frac{a_n}{n+1} x^{n+1} = x \sum_{n=1}^\infty \frac{a_n}{n+1} x^n$$

has terms whose absolute value is smaller than in the original series.

The preceding results can be expressed by saying that an absolutely convergent power series can be integrated and differentiated term by term and still yield an absolutely convergent power series.

It is natural to expect that if

$$f(x) = \sum_{n=1}^\infty a_n x^n,$$

then f is differentiable and its derivative is given by differentiating the series term by term. The next theorem proves this.

Theorem 8. *Let*

$$f(x) = \sum_{n=1}^{\infty} a_n x^n$$

be a power series, which converges absolutely for $|x| < r$. *Then* f *is differentiable for* $|x| < r$, *and*

$$f'(x) = \sum_{n=1}^{\infty} n a_n x^{n-1}.$$

Proof. Let $0 < b < r$. Let $\delta > 0$ be such that $b + \delta < r$. We consider values of x such that $|x| < b$ and values of h such that $|h| < \delta$. We have the numerator of the Newton quotient:

$$f(x + h) - f(x) = \sum_{n=1}^{\infty} a_n (x + h)^n - \sum_{n=1}^{\infty} a_n x^n = \sum_{n=1}^{\infty} a_n \left[(x + h)^n - x^n \right].$$

By the mean value theorem, there exists a number x_n between x and $x + h$ such that

$$(x + h)^n - x^n = n x_n^{n-1} h$$

and consequently

$$f(x + h) - f(x) = \sum_{n=1}^{\infty} n a_n x_n^{n-1} h.$$

Therefore

$$\frac{f(x + h) - f(x)}{h} = \sum_{n=1}^{\infty} n a_n x_n^{n-1}.$$

We have to show that the Newton quotient above approaches the value of the series obtained by taking the derivative term by term. We have

$$\frac{f(x + h) - f(x)}{h} - \sum_{n=1}^{\infty} n a_n x^{n-1} = \sum_{n=1}^{\infty} n a_n x_n^{n-1} - \sum_{n=1}^{\infty} n a_n x^{n-1}$$

$$= \sum_{n=1}^{\infty} n a_n \left[x_n^{n-1} - x^{n-1} \right].$$

Using the mean value theorem again, there exists y_n between x_n and x such that the preceding expression is

$$\frac{f(x + h) - f(x)}{h} - \sum_{n=1}^{\infty} n a_n x^{n-1} = \sum_{n=2}^{\infty} (n - 1) n a_n y_n^{n-2} (x_n - x).$$

We have $|y_n| \leqq b + \delta < r$, and $|x_n - x| \leqq |h|$. Consequently,

$$\left| \frac{f(x + h) - f(x)}{h} - \sum_{n=1}^{\infty} na_n x^{n-1} \right| \leqq \sum_{n=2}^{\infty} (n - 1)n|a_n| \, |y_n|^{n-2}|h|$$

$$\leqq |h| \sum_{n=2}^{\infty} (n - 1)n|a_n|(b + \delta)^{n-2}.$$

By Theorem 7 applied twice, we know that the series appearing on the right converges. It is equal to a fixed constant. As h approaches 0, it follows that the expression on the left also approaches 0. This proves that f is differentiable at x, and that its derivative is equal to $\sum_{n=1}^{\infty} na_n x^{n-1}$, for all x such that $|x| < b$. This is true for all b, $0 < b < r$, and therefore concludes the proof of our theorem.

Theorem 9. *Let* $f(x) = \sum_{n=1}^{\infty} a_n x^n$ *be a power series, which converges absolutely for* $|x| < r$. *Then the relation*

$$\int f(x) \, dx = \sum_{n=1}^{\infty} \frac{a_n}{n + 1} x^{n+1}$$

is valid in the interval $|x| < r$.

Proof. We know that the series for f integrated term by term converges absolutely in the interval. By the preceding theorem, its derivative term by term is the series for the derivative of the function, thereby proving our assertion.

EXERCISES

1. Verify in detail that differentiating term by term the series for the sine and cosine given at the end of the section yields

$$S'(x) = C(x) \quad \text{and} \quad C'(x) = -S(x).$$

2. Let

$$f(x) = \sum_{n=0}^{x} \frac{x^{2n}}{(2n)!}.$$

Prove that $f''(x) = f(x)$.

3. Let

$$f(x) = \sum_{n=0}^{\infty} \frac{x^{2n}}{(n!)^2}.$$

Prove that

$$x^2 f''(x) + xf'(x) = 4x^2 f(x).$$

4. Let

$$f(x) = x - \frac{x^3}{3} + \frac{x^5}{5} - \frac{x^7}{7} + \cdots.$$

Show that $f'(x) = 1/(1 + x^2)$.

5. Let

$$J(x) = \sum_{n=0}^{\infty} \frac{(-1)^n}{(n!)^2} \left(\frac{x}{2}\right)^{2n}.$$

Prove that

$$x^2 J''(x) + x J'(x) + x^2 J(x) = 0.$$

6. For any positive integer k, let

$$J_k(x) = \sum_{n=0}^{\infty} \frac{(-1)^n}{n!\,(n+k)!} \left(\frac{x}{2}\right)^{2n+k}.$$

Prove that

$$x^2 J_k''(x) + x J_k'(x) + (x^2 - k^2) J_k(x) = 0.$$

Answers to Exercises

I am much indebted to Anthony Petrello for the answers to the exercises.

Chapter I, §1 *(p. 7)*

	$A + B$	$A - B$	$3A$	$-2B$
1.	$(1, 0)$	$(3, -2)$	$(6, -3)$	$(2, -2)$
2.	$(-1, 7)$	$(-1, -1)$	$(-3, 9)$	$(0, -8)$
3.	$(1, 0, 6)$	$(3, -2, 4)$	$(6, -3, 15)$	$(2, -2, -2)$
4.	$(-2, 1, -1)$	$(0, -5, 7)$	$(-3, -6, 9)$	$(2, -6, 8)$
5.	$(3\pi, 0, 6)$	$(-\pi, 6, -8)$	$(3\pi, 9, -3)$	$(-4\pi, 6, -14)$
6.	$(15 + \pi, 1, 3)$	$(15 - \pi, -5, 5)$	$(45, -6, 12)$	$(-2\pi, -6, 2)$

Chapter I, §2 *(p. 11)*

1. No 2. Yes 3. No 4. Yes 5. No 6. Yes 7. Yes 8. No

Chapter I, §3 *(p. 14)*

1. (a) 5 (b) 10 (c) 30 (d) 14 (e) $\pi^2 + 10$ (f) 245
2. (a) -3 (b) 12 (c) 2 (d) -17 (e) $2\pi^2 - 16$ (f) $15\pi - 10$
4. (b) and (d)

Chapter I, §4 *(p. 25)*

1. (a) $\sqrt{5}$ (b) $\sqrt{10}$ (c) $\sqrt{30}$ (d) $\sqrt{14}$ (e) $\sqrt{10 + \pi^2}$ (f) $\sqrt{245}$
2. (a) $\sqrt{2}$ (b) 4 (c) $\sqrt{3}$ (d) $\sqrt{26}$ (e) $\sqrt{58 + 4\pi^2}$ (f) $\sqrt{10 + \pi^2}$
3. (a) $(\frac{3}{2}, -\frac{3}{2})$ (b) $(0, 3)$ (c) $(-\frac{2}{3}, \frac{2}{3}, \frac{2}{3})$ (d) $(\frac{17}{26}, -\frac{51}{26}, \frac{34}{13})$
   (e) $\dfrac{\pi^2 - 8}{2\pi^2 + 29}(2\pi, -3, 7)$ (f) $\dfrac{15\pi - 10}{10 + \pi^2}(\pi, 3, -1)$

443

4. (a) $(-\frac{6}{5}, \frac{3}{5})$ (b) $(-\frac{6}{5}, \frac{18}{5})$ (c) $(\frac{2}{15}, -\frac{1}{15}, \frac{1}{3})$ (d) $-\frac{17}{14}(-1, -2, 3)$

(e) $\dfrac{2\pi^2 - 16}{\pi^2 + 10}(\pi, 3, -1)$ (f) $\dfrac{3\pi - 2}{49}(15, -2, 4)$

5. (a) $\dfrac{-1}{\sqrt{5}\,\sqrt{34}}$ (b) $\dfrac{-2}{\sqrt{5}}$ (c) $\dfrac{10}{\sqrt{14}\,\sqrt{35}}$ (d) $\dfrac{13}{\sqrt{21}\,\sqrt{11}}$ (e) $\dfrac{-1}{\sqrt{12}}$

6. (a) $\dfrac{35}{\sqrt{41 \cdot 35}}, \dfrac{6}{\sqrt{41 \cdot 6}}, 0$ (b) $\dfrac{1}{\sqrt{17 \cdot 26}}, \dfrac{16}{\sqrt{41 \cdot 17}}, \dfrac{25}{\sqrt{26 \cdot 41}}$

7. Let us dot the sum

$$c_1 A_1 + \cdots + c_r A_r = O$$

with A_i. We find

$$c_1 A_1 \cdot A_i + \cdots + c_i A_i \cdot A_i + \cdots + c_r A_r \cdot A_i = O \cdot A_i = 0.$$

Since $A_j \cdot A_i = 0$ if $j \neq i$ we find

$$c_i A_i \cdot A_i = 0.$$

But $A_i \cdot A_i \neq 0$ by assumption. Hence $c_i = 0$, as was to be shown.

8. (a) $\|A + B\|^2 + \|A - B\|^2 = (A + B) \cdot (A + B) + (A - B) \cdot (A - B)$
$$= A^2 + 2A \cdot B + B^2 + A^2 - 2A \cdot B + B^2$$
$$= 2A^2 + 2B^2 = 2\|A\|^2 + 2\|B\|^2.$$

9. $\|A - B\|^2 = A^2 - 2A \cdot B + B^2 = \|A\|^2 - 2\|A\|\,\|B\|\cos\theta + \|B\|^2$

Chapter I, §5 *(p. 29)*

1. (a) Let $A = P_2 - P_1 = (-5, -2, 3)$. Parametric representation of the line is
$X(t) = P_1 + tA = (1, 3, -1) + t(-5, -2, 3)$.
(b) $(-1, 5, 3) + t(-1, -1, 4)$

2. $X = (1, 1, -1) + t(3, 0, -4)$ 　　　　　**3.** $X = (-1, 5, 2) + t(-4, 9, 1)$

4. (a) $(-\frac{3}{2}, 4, \frac{1}{2})$ (b) $(-\frac{2}{3}, \frac{11}{3}, 0)$, $(-\frac{7}{3}, \frac{13}{3}, 1)$ (c) $(0, \frac{17}{5}, -\frac{2}{5})$ (d) $(-1, \frac{19}{5}, \frac{1}{5})$

5. $P + \frac{1}{2}(Q - P) = \dfrac{P + Q}{2}$

Chapter I, §6 *(p. 34)*

1. The normal vectors $(2, 3)$ and $(5, -5)$ are not perpendicular because their dot product $10 - 15 = -5$ is not 0.

2. The normal vectors are $(-m, 1)$ and $(-m', 1)$, and their dot product is $mm' + 1$. The vectors are perpendicular if and only if this dot product is 0, which is equivalent with $mm' = -1$.

3. $y = x + 8$　**4.** $4y = 5x - 7$　**6.** (c) and (d)

7. (a) $x - y + 3z = -1$　(b) $3x + 2y - 4z = 2\pi + 26$　(c) $x - 5z = -33$

8. (a) $2x + y + 2z = 7$　(b) $7x - 8y - 9z = -29$　(c) $y + z = 1$

9. $(3, -9, -5)$, $(1, 5, -7)$ (Others would be constant multiples of these.)

10. (a) $2(t^2 + 5)^{1/2}$

In Exercises 10 and 11, let $X(t) = P + tA$ be the line. The square of the distance between Q and $X(t)$ is

$$(X(t) - Q)^2 = (X(t) - Q) \cdot (X(t) - Q) = (P - Q + tA) \cdot (P - Q + tA).$$

If you substitute the specific values for P, Q, and A you will find a polynomial of degree 2 in t, i.e. a quadratic polynomial. To minimize the distance, it suffices to minimize the square of the distance. The advantage of taking the square is that the horrible square root sign disappears. By freshman calculus, you can find the minimum of the quadratic polynomial above by setting its derivative equal to 0. In Exercise 10, we get

$$(X(t) - Q)^2 = 20 + 4t^2.$$

The derivative is $8t$, which is 0 when $t = 0$, so P is the point on the line at minimum distance from Q, and this distance is

$$\|P - Q\| = \sqrt{20} = 2\sqrt{5}.$$

11. $(15t^2 + 26t + 21)^{1/2}$, $\sqrt{146/15}$

12. $(-2, 1, 5)$ **13.** $(11, 13, -7)$ **14.** (a) $X = (1, 0, -1) + t(-2, 1, 5)$
   (b) $X = (-10, -13, 7) + t(11, 13, -7)$ or also $(1, 0, 0) + t(11, 13, -7)$

15. (a) $-\frac{1}{3}$ (b) $-\dfrac{2}{\sqrt{42}}$ (c) $\dfrac{4}{\sqrt{66}}$ (d) $-\dfrac{2}{\sqrt{18}}$

16. (a) $(-4, \frac{11}{2}, \frac{15}{2})$ (b) $(\frac{25}{13}, \frac{10}{13}, -\frac{9}{13})$ **17.** $(1, 3, -2)$ **18.** $2/\sqrt{3}$

19. The distance is the norm of the projection of $\overrightarrow{PQ}$ on the vector perpendicular to the plane, namely $\overrightarrow{ON}$. So it is the norm of the projection of $Q - P$ on the unit vector $N/\|N\|$. This is given by the dot product, so the distance is the absolute value of this dot product, namely

$$\left| (Q - P) \cdot \frac{N}{\|N\|} \right|.$$

This is equal to the answer given in the text.

20. (a) $\dfrac{8}{\sqrt{35}}$ (b) $\dfrac{13}{\sqrt{21}}$

Chapter I, §7 *(p. 38)*

1. $(-4, -3, 1)$ **2.** $(-1, 1, -1)$ **3.** $(-9, 6, -1)$ **4.** all zero

5. E_3, E_1, E_2 in that order

7. $(0, -1, 0)$ and $(0, 0, 0)$; no

9. (a) $\sqrt{494}$ (b) $\sqrt{245}$ (c) $\sqrt{470}$ (d) $\sqrt{381}$

11. $\dfrac{d}{dt}[X(t) \times X'(t)] = X(t) \times \dfrac{d}{dt}X'(t) + \dfrac{d}{dt}X(t) \times X'(t)$
$$= X(t) \times X''(t)$$

12. $(t) = \dfrac{dY}{dt} = \dfrac{d}{dt}[X(t) \cdot (X'(t) \times X''(t))]$

$= X'(t) \cdot (X'(t) \times X''(t)) + X(t) \cdot \dfrac{d}{dt}[X'(t) \times X''(t)]$

$= X(t) \cdot (X'(t) \times X'''(t))$

Chapter II, §1 *(p. 50)*

1. $(e^t, -\sin t, \cos t)$ **2.** $\left(2 \cos 2t, \dfrac{1}{1+t}, 1\right)$ **3.** $(-\sin t, \cos t)$

4. $(-3 \sin 3t, 3 \cos 3t)$ **5.** 0 **7.** B

8. $\left(\dfrac{1}{2}, \dfrac{\sqrt{3}}{2}\right) + t\left(\dfrac{1}{2}, \dfrac{\sqrt{3}}{2}\right)$, $(-1, 0) + t(-1, 0)$, or $y = \sqrt{3}\, x, y = 0$

9. $ex + y + 2z = e^2 + 3$ **10.** $x + y = 1$

11. $\sqrt{(X(t) - Q) \cdot (X(t) - Q)}$

If t_0 is a value of t which minimizes the distance, then it also minimizes the square of the distance, which is easier to work with because it does not involve the square root sign. Let $f(t)$ be the square of the distance, so

$$f(t) = (X(t) - Q)^2 = (X(t) - Q) \cdot (X(t) - Q).$$

At a minimum, the derivative must be 0, and the derivative is

$$f'(t) = 2(X(t) - Q) \cdot X'(t).$$

Hence at a minimum, we have $(X(t_0) - Q) \cdot X'(t_0) = 0$, and hence $X(t_0) - Q$ is perpendicular to $X'(t_0)$, i.e. is perpendicular to the curve. If $X(t) = P + tA$ is the parametric equation of a line, then $X'(t) = A$, so we find

$$(P + t_0 A - Q) \cdot A = 0.$$

Solving for t_0 yields $(P - Q) \cdot A + t_0 A \cdot A = 0$, whence

$$t_0 = \frac{(Q - P) \cdot A}{A \cdot A}.$$

13. Differentiate $X'(t)^2 = $ constant to get

$$2X'(t) \cdot X''(t) = 0.$$

14. Let $v(t) = \|X'(t)\|$. To show $v(t)$ is constant, it suffices to prove that $v(t)^2$ is constant, and $v(t)^2 = X'(t) \cdot X'(t)$. To show that a function is constant it suffices to prove that its derivative is 0, and we have

$$\frac{d}{dt} v(t)^2 = 2X'(t) \cdot X''(t).$$

By assumption, $X'(t)$ is perpendicular to $X''(t)$, so the right-hand side is 0, as desired.

15. Differentiate the relation $X(t) \cdot B = t$, you get

$$X'(t) \cdot B = 1,$$

so $\|X'(t)\| \, \|B\| \cos \theta = 1$. Hence $\|X'(t)\| = 1/\|B\| \cos \theta$ is constant. Hence the square $X'(t)^2$ is constant. Differentiate, you get

$$2X'(t) \cdot X''(t) = 0,$$

so $X'(t) \cdot X''(t) = 0$, and $X'(t)$ is perpendicular to $X''(t)$, as desired.

16. (a) $(0, 1, \pi/8) + t(-4, 0, 1)$ (b) $(1, 2, 1) + t(1, 2, 2)$
(c) $(e^3, e^{-3}, 3\sqrt{2}) + t(3e^3, -3e^{-3}, 3\sqrt{2})$ (d) $(1, 1, 1) + t(1, 3, 4)$

18. Let $X(t) = (e^t, e^{2t}, 1 - e^{-t})$ and $Y(\theta) = (1 - \theta, \cos\theta, \sin\theta)$. Then the two curves intersect when $t = 0$ and $\theta = 0$. Also,

$$X'(t) = (e^t, 2e^{2t}, e^{-t}) \qquad \text{and} \qquad Y'(\theta) = (-1, -\sin\theta, \cos\theta)$$

so

$$X'(0) = (1, 2, 1) \qquad \text{and} \qquad Y'(0) = (-1, 0, 1).$$

The angle between their tangents at the point of intersection is the angle between $X'(0)$ and $Y'(0)$, which is $\pi/2$.

19. $(18, 4, 12)$ when $t = -3$ and $(2, 0, 4)$ when $t = 1$.

The points on the curve which also lie on the plane are those for which

$$3(2t^2) - 14(1 - t) + (3 + t^2) - 10 = 0.$$

This is a quadratic equation for t, which you solve by the quadratic formula. You will get the two values $t = -3$ or $t = 1$, which you substitute back in the parametric curve $(2t^2, 1 - t, 3 + t^2)$ to get the two points.

20. (a) Each coordinate of $X(t)$ has derivative equal to 0, so each coordinate is constant, so $X(t) = A$ for some constant A.
(b) $X(t) = tA + B$ for constant vectors $A \neq O$ and B.

21. Let $E = (0, 0, 1)$ be the unit vector in the direction of the z-axis. Then $X'(t) = (-a\sin t, a\cos t, b)$ and

$$\cos\theta(t) = \frac{X'(t) \cdot E}{\|X'(t)\|} = \frac{b}{\sqrt{a^2 + b^2}}.$$

23. Differentiate the relation $X(t) \cdot B = e^{2t}$, you get

$$X'(t) \cdot B = 2e^{2t} = \|X'(t)\| \, \|B\| \cos\theta.$$

Both B and $\cos\theta$ are constant, divide to get (a). Then differentiate $X'(t)^2 = $ constant to get (b), the answer is

$$X'(t) \cdot X''(t) = \frac{8e^{4t}}{\cos^2\theta}.$$

Chapter II, §2 *(p. 53)*

1. $\sqrt{2}$ **2.** (a) $2\sqrt{13}$ (b) $\frac{\pi}{8}\sqrt{17}$

3. (a) $\frac{3}{2}(\sqrt{41} - 1) + \frac{5}{4}\left(\log\frac{6 + \sqrt{41}}{5}\right)$ (b) $e - \frac{1}{e}$

4. (a) 8   (b) $4 - 2\sqrt{2}$

The integral for the length is $L(t) = \int_a^b \sqrt{2 - 2\cos t}\ dt$. Use the formula

$$\sin^2 u = \frac{1 - \cos 2u}{2},$$

with $u = t/2$.

5. (a) $\sqrt{5} - \sqrt{2} + \log \dfrac{2 + 2\sqrt{2}}{1 + \sqrt{5}}$

(b) $\sqrt{26} - \sqrt{10} + \dfrac{1}{2}\log\left(\dfrac{\sqrt{26} - 1}{\sqrt{26} + 1} \cdot \dfrac{\sqrt{10} + 1}{\sqrt{10} - 1}\right) = \sqrt{26} - \sqrt{10} + \log\dfrac{5}{3}\left(\dfrac{1 + \sqrt{10}}{1 + \sqrt{26}}\right)$

6. $\log(\sqrt{2} + 1)$

Chapter III, §1 *(p. 59)*

1.

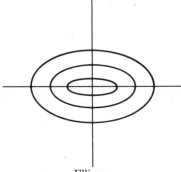

Ellipses

2.

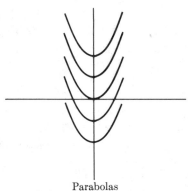

Parabolas

4.

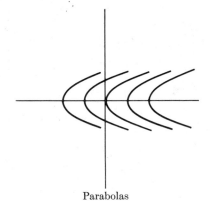

Parabolas

6.

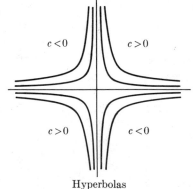

Hyperbolas

13. Using polar coordinates $x = r\cos\theta$ and $y = r\sin\theta$, and identities like $\cos 2\theta = \cos^2\theta - \sin^2\theta$, you should find

$$f(x, y) = r^2 \sin 4\theta.$$

The level curves are then $r^2 \sin 4\theta = $ constant, so

$$r^2 = \frac{c}{\sin 4\theta}.$$

Letting c vary (don't forget negative values for c), the level curves are as follows.

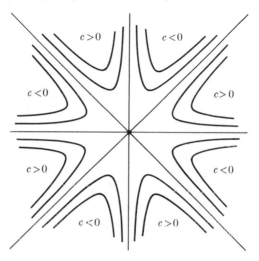

Chapter III, §2 *(p. 65)*

	$\partial f/\partial x$	$\partial f/\partial y$	$\partial f/\partial z$
1.	y	x	1
2.	$2xy^5$	$5x^2y^4$	0
3.	$y \cos(xy)$	$x \cos(xy)$	$-\sin(z)$
4.	$-y \sin(xy)$	$-x \sin(xy)$	0
5.	$yz \cos(xyz)$	$xz \cos(xyz)$	$xy \cos(xyz)$
6.	yze^{xyz}	xze^{xyz}	xye^{xyz}
7.	$2x \sin(yz)$	$x^2z \cos(yz)$	$x^2y \cos(yz)$
8.	yz	xz	xy
9.	$z + y$	$z + x$	$x + y$
10.	$\cos(y - 3z) + \dfrac{y}{\sqrt{1 - x^2y^2}}$	$-x \sin(y - 3z) + \dfrac{x}{\sqrt{1 - x^2y^2}}$	$3x \sin(y - 3z)$

11. (1) $(2, 1, 1)$ (2) $(64, 80, 0)$ (6) $(6e^6, 3e^6, 2e^6)$ (8) $(6, 3, 2)$ (9) $(5, 4, 3)$

12. (4) $(0, 0, 0)$ (5) $(\pi^2 \cos \pi^2, \pi \cos \pi^2, \pi \cos \pi^2)$
(7) $(2 \sin \pi^2, \pi \cos \pi^2, \pi \cos \pi^2)$

13. $(-1, -2, 1)$ **14.** $\dfrac{\partial x^y}{\partial x} = yx^{y-1}$ $\dfrac{\partial x^y}{\partial y} = x^y \log x$

15. $(-2e^{-2} \cos \pi^2, -\pi e^{-2} \sin \pi^2, -\pi e^{-2} \sin \pi^2)$ **16.** $\left(\frac{3}{2}, \frac{1}{2}, -\dfrac{5\sqrt{3}}{2}\right)$

Chapter III, §3 *(p. 70)*

1. $2, -3$ **2.** a, b **3.** a, b, c

5. Select first $H = (h, 0) = hE_1$. Then $A \cdot H = ha_1$ if $A = (a_1, a_2)$. Divide both sides of the relation

$$f(X + H) - f(X) = a_1 h + |h|\, g(H)$$

by $h \neq 0$ and take the limit to see that $a_1 = D_1 f(X)$. Similarly use $H = (0, h) = hE_2$ to see that $a_2 = D_2 f(x, y)$. Similar argument for three variables.

Chapter III, §4 *(p. 74)*

	$\partial^2 f / \partial x^2$	$\partial^2 f / \partial y^2$	$\partial^2 f / \partial x\, \partial y$
1.	$y^2 e^{xy}$	$x^2 e^{xy}$	$yx e^{xy} + e^{xy}$
2.	$-y^2 \sin xy$	$-x^2 \sin xy$	$-xy \sin xy + \cos xy$
3.	$2y^3$	$6x^2 y$	$6xy^2 + 3$
4.	0	2	2
5.	$2e^{x^2+y^2} + 4x^2 e^{x^2+y^2}$	$e^{x^2+y^2}(2 + 4y^2)$	$4xy e^{x^2+y^2}$
6.	$2\cos(x^2 + y)$ $-4x^2 \sin(x^2 + y)$	$-\sin(x^2 + y)$	$-2x \sin(x^2 + y)$
7.	$-(3x^2 + y)^2 \cos(x^3 + xy)$ $-6x \sin(x^3 + xy)$	$-x^2 \cos(x^3 + xy)$	$-(3x^2 + y)x \cos(x^3 + xy)$ $-\sin(x^3 + xy)$

8.
$$\frac{\partial^2 f}{\partial x^2} = \frac{2\left(1 + (x^2 - 2xy)^2\right) - 2(2x - 2y)^2 (x^2 - 2xy)}{\left(1 + (x^2 - 2xy)^2\right)^2}$$

$$\frac{\partial^2 f}{\partial y^2} = \frac{-8x^2(x^2 - 2xy)}{\left[1 + (x^2 - 2xy)^2\right]^2}$$

$$\frac{\partial^2 f}{\partial x\, \partial y} = \frac{-2\left[1 + (x^2 - 2xy)^2\right] - (2x - 2y)(x^2 - 2xy)(-2x)2}{\left[1 + (x^2 - 2xy)^2\right]^2}$$

9. All three $= e^{x+y}$ 10. All three $= -\sin(x + y)$

11. 1 **12.** $2x$ 13. $e^{xyz}(1 + 3xyz + x^2 y^2 z^2)$

14. $(1 - x^2 y^2 z^2) \cos xyz - 3xyz \sin xyz$

15. $\sin(x + y + z)$ 16. $-\cos(x + y + z)$

17. $-\dfrac{48xyz}{(x^2 + y^2 + z^2)^4}$ 18. $6x^2 y$

20. From $D_1 f = -D_2 g$ we get $D_1^2 f = -D_1 D_2 g$. From $D_2 f = D_1 g$ we get $D_2^2 f = D_2 D_1 g$. Adding yields 0.

Chapter IV, §1 *(p. 78)*

1. $\dfrac{d}{dt}(P + tA) = A$, so this follows directly from the chain rule.

2. 5. Indeed, $C'(t) = (2t, -3t^{-4}, 1)$ and $C'(1) = (2, -3, 1)$. Dot this with the given grad $f(1, 1, 1)$ to find 5.

3. $C'(0) = (0, 1)$

Let $C'(0) = (a, b)$. Now grad $f(C(0)) = (9, 2)$ and grad $g(C(0)) = (4, 1)$, so using the chain rule on the functions f and g, respectively, we obtain

$$2 = \frac{d}{dt}f(C(t))\Big|_{t=0} = (9, 2) \cdot (a, b) = 9a + 2b$$

$$1 = \frac{d}{dt}g(C(t))\Big|_{t=0} = (4, 1) \cdot (a, b) = 4a + b.$$

Solving for the above simultaneous equations yields $C'(0) = (0, 1)$.

4. (a) grad $f(tP) \cdot P$.

(b) Use 4(a) and let $t = 0$.

5. Viewing x, y as constant, put $P = (x, y)$ and use Exercise 4(a). Then put $t = 1$. If you expand out, you will find the stated answer.

7. (a) $\partial f/\partial x = x/r$ and $\partial f/\partial y = y/r$ if $r = \sqrt{x^2 + y^2}$.

(b) $\dfrac{\partial f}{\partial x} = \dfrac{x}{(x^2 + y^2 + z^2)^{1/2}}$, $\dfrac{\partial f}{\partial y} = \dfrac{y}{(x^2 + y^2 + z^2)^{1/2}}$, $\dfrac{\partial f}{\partial z} =$ guess what?

8. $\dfrac{\partial r}{\partial x_i} = \dfrac{x_i}{r}$

9. (a) $\partial f/\partial x = (3x^2y + 4x) \cos(x^3y + 2x^2)$

$\partial f/\partial y = x^3 \cos(x^3y + 2x^{\prime})$

(b) $\partial f/\partial x = -(6xy - 4) \sin(3x^2y - 4x)$

$\partial f/\partial y = -3x^2 \sin(3x^2y - 4x)$

(c) $\partial f/\partial x = \dfrac{2xy}{(x^2y + 5y)}$, $\dfrac{\partial f}{\partial y} = \dfrac{x^2 + 5}{x^2y + 5y} = \dfrac{1}{y}$

(d) $\partial f/\partial x = \frac{1}{2}(2xy + 4)(x^2y + 4x)^{-1/2}$

$\partial f/\partial y = \frac{1}{2}x^2(x^2y + 4x)^{-1/2}$

Chapter IV, §2 *(p. 84)*

1.

	Plane	Line
(a)	$6x + 2y + 3z = 49$	$X = (6, 2, 3) + t(12, 4, 6)$
(b)	$x + y + 2z = 2$	$X = (1, 1, 0) + t(1, 1, 2)$
(c)	$13x + 15y + z = -15$	$X = (2, -3, 4) + t(13, 15, 1)$
(d)	$6x - 2y + 15z = 22$	$X = (1, 7, 2) + t(-6, 2, -15)$
(e)	$4x + y + z = 13$	$X = (2, 1, 4) + t(8, 2, 2)$
(f)	$z = 0$	$X = (1, \pi/2, 0) + t(0, 0, \pi/2 + 1)$

2. (a) $(3, 0, 1)$ (b) $X = (\log 3, \dfrac{3\pi}{2}, -3) + t(3, 0, 1)$

(c) $3x + z = 3 \log 3 - 3$

3. (a) $X = (3, 2, -6) + t(2, -3, 0)$ (b) $X = (2, 1, -2) + t(-5, 4, -3)$

(c) $X = (3, 2, 2) + t(2, 3, 0)$

4. $\|C(t) - Q\|$ and see Exercise 11 of Chapter II, §1.

5. (a) $6x + 8y - z = 25$ (b) $16x + 12y - 125z = -75$
   (c) $\pi x + y + z = 2\pi$ 6. $x - 2y + z = 1$

7. (b) $x + y + 2z = 2$ 8. $3x - y + 6z = 14$

9. $(\cos 3)x + (\cos 3)y - z = 3\cos 3 - \sin 3.$

10. $3x + 5y + 4z = 18$ 11. (a) $\dfrac{1}{\sqrt{27}}(5, 1, 1)$ (b) $5x + y + z - 6 = 0$

12. $-5\sqrt{3}/9$

Chapter IV, §3 *(p. 88)*

1. (a) $\frac{5}{3}$ (b) max $= \sqrt{10}$ min $= -\sqrt{10}$

2. (a) $\dfrac{3}{2\sqrt{5}}$ (b) $\frac{48}{13}$ (c) $2\sqrt{145}$

3. Increasing $\left(-\dfrac{9\sqrt{3}}{2}, -\dfrac{3\sqrt{3}}{2}\right)$,   decreasing $\left(\dfrac{9\sqrt{3}}{2}, \dfrac{3\sqrt{3}}{2}\right)$

4. (a) $\left(\dfrac{9}{2\cdot 6^{7/4}}, \dfrac{3}{2\cdot 6^{7/4}}, -\dfrac{6}{2\cdot 6^{7/4}}\right)$ (b) $(1, 2, -1, 1)$

5. (a) $-2/\sqrt{5}$ (b) $\sqrt{116}$ 6. $\dfrac{1}{\sqrt{54}}(5, 2, 5), 6\sqrt{6}$ 7. $\left(\dfrac{-1}{\sqrt{2}}, \dfrac{1}{\sqrt{2}}\right), \sqrt{2}$

8. $(2e - 5)/\sqrt{3}$

9. For any unit vector A, the function of t given by $f(P + tA)$ has a maximum at $t = 0$ (for small values of t), and hence its derivative is 0 at $t = 0$. But its derivative is grad $f(P + tA) \cdot A$, which at $t = 0$ is grad $f(P) \cdot A$. This is true for all A, whence grad $f(P) = O$. (For instance, let A be any one of the standard unit vectors in the directions of the coordinate axes.)

Chapter IV, §4 *(p. 95)*

1. $\dfrac{\partial f}{\partial x} = \dfrac{dg}{dr}\dfrac{\partial r}{\partial x} = \dfrac{dg}{dr}\dfrac{x}{r}$. Replace x by y and z. Square each term and add. You can factor

$$\frac{x^2}{r^2} + \frac{y^2}{r^2} + \frac{z^2}{r^2} = 1.$$

2. (a) $-X/r^3$ (b) $2X$ (c) $-3X/r^5$ (d) $-2e^{-rx}X$ (e) $-X/r^2$
   (f) $-4mX/r^{m+2}$ (g) $-(\sin r)X/r$

3. $F(t)^2 = (\cos t)^2 A^2 + 2(\cos t)(\sin t)A \cdot B + (\sin t)^2 B^2 = 1$
   because $A^2 = B^2 = 1$ since A, B are unit vectors and $A \cdot B = 0$ by assumption. Hence $\|F(t)\| = 1$, so $F(t)$ lies on the sphere of radius 1.

4. Note that $L(t) = (1 - t)P + tQ$. If $L(t) = 0$ for some value of t, then

$$(1 - t)P = -tQ.$$

Square both sides, use $P^2 = Q^2 = 1$ to get $(1 - t)^2 = t^2$. It follows that $t = 1/2$, so $\frac{1}{2}P = \dfrac{-1}{2}Q$, whence $P = -Q$.

5. By Exercise 4, $L(t) \neq O$ if $0 \leq t \leq 1$. Then $L(t)/\|L(t)\|$ is a unit vector, and this expression is composed of differentiable expressions so is differentiable.

6. Let $C(t) = (\cos t)P + (\sin t)A$. Then $C(t)^2 = 1$, see Exercise 3.

7. Let $x = a \cos t$ and $y = b \sin t$.

9. It suffices to prove that the function $f(C(t))$ is constant (as function of t). Take its derivative, get by the chain rule

$$\frac{d}{dt}f(C(t)) = \operatorname{grad} f(C(t)) \cdot C'(t) = h(C(t))C(t) \cdot C'(t).$$

But $C(t)^2 = a^2$ because $C(t)$ is on the sphere of radius a. Differentiating this with respect to t yields $2C(t) \cdot C'(t) = 0$, so $C(t) \cdot C'(t) = 0$, which you plug in above to see that the derivative of $f(C(t)) = 0$. Hence $f(C(t_1)) = f(C(t_2))$ so $f(P) = f(Q)$.

10. $\operatorname{grad} f(X) = \left(g'(r)\dfrac{x}{r}, g'(r)\dfrac{y}{r}, g'(r)\dfrac{z}{r} \right) = \dfrac{g'(r)}{r} X$ (say in three variables), and $g'(r)/r$ is a scalar factor of X, so $\operatorname{grad} f(X)$ and X are parallel.

12. First $\partial f/\partial x = g'(r)x/r$. Using the rule for the derivative of a product, and a quotient, we then get:

$$\frac{\partial^2 f}{\partial x^2} = g'(r)\left[\frac{r - xx/r}{r^2} \right] + g''(r)\frac{x}{r}\frac{x}{r}.$$

Replace x by y to get $\partial^2 f/\partial y^2$. Then add. Things will cancel to give the desired answer.

13. Same method as in Exercise 12.

Chapter IV, §5 *(p. 100)*

1. $\dfrac{\partial z}{\partial r} = \dfrac{\partial f}{\partial x}\dfrac{\partial u}{\partial r} + \dfrac{\partial f}{\partial y}\dfrac{\partial v}{\partial r}$ and $\dfrac{\partial z}{\partial t} = \dfrac{\partial f}{\partial x}\dfrac{\partial u}{\partial t} + \dfrac{\partial f}{\partial y}\dfrac{\partial v}{\partial t}$

2. (a) $\dfrac{\partial f}{\partial x} = 3x^2 + 3yz,$ $\dfrac{\partial f}{\partial y} = 3xz - 2yz$

$\dfrac{\partial f}{\partial s} = (3x^2 + 3yz) + (3xz - 2yz)(-1) + (3xy - y^2)2s$

$\dfrac{\partial f}{\partial t} = (3x^2 + 3yz)2 + (3xz - 2yz)(-1) + (3xy - y^2)2t$

(b) $\dfrac{\partial f}{\partial x} = \dfrac{y^2 + 1}{(1 - xy)^2},$ $\dfrac{\partial f}{\partial y} = \dfrac{x^2 + 1}{(1 - xy)^2}$

$\dfrac{\partial f}{\partial s} = \dfrac{(x^2 + 1) \sin(3t - s)}{(1 - xy)^2}$

$\dfrac{\partial f}{\partial t} = \dfrac{2(y^2 + 1) \cos 2t - 3(x^2 + 1) \sin(3t - s)}{(1 - xy)^2}$

3. 8, because when $u = 1$, $v = 1$ we have $g(1, 1) = f(0, 0, 0)$, so

$$D_1 g(u, v) = D_1 f(x, y, z)\frac{\partial x}{\partial u} + D_2 f(x, y, z)\frac{\partial y}{\partial u} + D_3 f(x, y, z)\frac{\partial z}{\partial u}$$

so

$$D_1 g(1, 1) = D_1 f(0, 0, 0)1 + D_2 f(0, 0, 0)2 + D_3 f(0, 0, 0)0 = 8.$$

4. Differentiate the relation with respect to t to get

$$D_1 f(tx, ty)x + D_2 f(tx, ty)y = mt^{m-1}f(x, y).$$

Then differentiate once more, to get

$$D_1 D_1 f(tx, ty)xx + D_2 D_1 f(tx, ty)yx + D_1 D_2 f(tx, ty)xy + D_2 D_2 f(tx, ty)y^2$$
$$= m(m - 1)t^{m-2}f(x, y).$$

Then put $t = 1$.

5. Put $s = x - y$ and $t = y - x$. Then $\partial s/\partial x = 1$, etc. Use the formulas

$$\frac{\partial u}{\partial x} = \frac{\partial f}{\partial s}\frac{\partial s}{\partial x} + \frac{\partial f}{\partial t}\frac{\partial t}{\partial x} \qquad \text{and} \qquad \frac{\partial u}{\partial y} = \frac{\partial f}{\partial s}\frac{\partial s}{\partial y} + \frac{\partial f}{\partial t}\frac{\partial t}{\partial y}.$$

6. (b) $\dfrac{\partial g}{\partial y} = f'(2x + 7y)7$ and $\dfrac{\partial g}{\partial x} = f'(2x + 7y)2.$

Chapter V, §1 *(p. 105)*

1. $k \log \|X\|$ **2.** $-\dfrac{k}{2r^2}$ **3.** $\begin{cases} \log r, & k = 2 \\ \dfrac{1}{(2 - k)r^{k-2}}, & k \neq 2 \end{cases}$

Chapter V, §2 *(p. 109)*

1. No **2.** No **3.** No **4.** No **5.** No **6.** No

Chapter V, §3 *(p. 113)*

1. No **2.** No **3.** No **4.** No

5. (a) r (b) $\log r$ (c) $\dfrac{r^{n+2}}{n + 2}$ if $n \neq -2$ **6.** $2x^2y$ **7.** $x \sin xy$ **8.** x^3y^2

9. $x^2 + y^4$ **10.** (a) e^{xy} (b) $\sin xy$ (c) $\sin(x^2y)$ **11.** $g(r)$

12. $x^3y + 2y^2x - y + 2$. You have to add the constant at the end to satisfy $\varphi(1, 1) = 4$.

13. (a) $x^2 + \frac{3}{2}y^2 + 2z^2$ (b) $xy + yz + xz$ (c) xe^{y+2z}
  (d) $xy \sin z$ (e) $xyz + z^3y$ (f) xe^{yz}
  (g) $xz^2 + y^2$ (h) $z \sin xy$ (i) $y^3xz + xy + yz$

14. $\left(\dfrac{-y}{x^2 + y^2}, \dfrac{x}{x^2 + y^2} \right)$

15. div curl $F = D_1(D_2 f_3 - D_3 f_2) + D_2(D_3 f_1 - D_1 f_3) + D_3(D_1 f_2 - D_2 f_1)$
  $= D_1 D_2 f_3 - D_1 D_3 f_2 + D_2 D_3 f_1 - D_2 D_1 f_3 + D_3 D_1 f_2 - D_3 D_2 f_1$
  $= 0.$

Chapter V, §5 *(p. 121)*

1. $D_1\psi(x, y) = e^{xy}$, $D_2\psi(x, y) = \dfrac{xe^{xy} - e^y}{y} - \dfrac{e^{xy} - e^y}{y^2}$

2. $D_1\psi(x, y) = \cos(xy),\ D_2\psi(x, y) = \dfrac{x}{y}\cos(xy) - \dfrac{\sin(xy)}{y^2}$

3. $D_1\psi(x, y) = (y + x)^2$
$D_2\psi(x, y) = 2yx - 2y + x^2 - 1$

4. $D_1\psi(x, y) = e^{y+x}$
$D_2\psi(x, y) = e^{y+x} - e^{y+1}$

5. $D_1\psi(x, y) = e^{y-x}$
$D_2\psi(x, y) = -e^{y-x} + e^{y-1}$

6. $D_1\psi(x, y) = x^2 y^3$
$D_2\psi(x, y) = y^2 x^3$

7. $D_1\psi(x, y) = \dfrac{\log(xy)}{x}$

$D_2\psi(x, y) = \dfrac{\log x}{y}$

8. $D_1\psi(x, y) = \sin(3xy)$

$D_2\psi(x, y) = \dfrac{\cos 3xy - \cos 3y}{3y^2} + \dfrac{x\sin 3xy - \sin 3y}{y}$

Chapter V, §6 *(p. 125)*

Let $F = (f_1, f_2, f_3)$ be a vector field on a rectangular box in 3-dimensional space $\mathbf{R}^3$. Let (x_0, y_0, z_0) be a point of the box. Assume that

$$D_j f_i = D_i f_j \qquad \text{for all indices} \qquad i, j = 1, 2, 3.$$

Define

$$\varphi(x, y, z) = \int_{x_0}^{x} f_1(t, y, z)\, dt + \int_{y_0}^{y} f_2(x_0, t, z)\, dt + \int_{z_0}^{z} f_3(x_0, y_0, t)\, dt.$$

We must verify that $D_1\varphi = f_1,\ D_2\varphi = f_2$ and $D_3\varphi = f_3$. The first condition $D_1\varphi = f_1$ follows from the fundamental theorem of calculus and the fact that the second and third integrals do not depend on x, so their derivatives with respect to x are 0. Next, we have

$$
\begin{aligned}
D_2\varphi(x, y, z) &= \int_{x_0}^{x} D_2 f_1(t, y, z)\, dt + f_2(x_0, y, z) + 0 \\
&= \int_{x_0}^{x} D_1 f_2(t, y, z)\, dt + f_2(x_0, y, z) \\
&= f_2(x, y, z) - f_2(x_0, y, z) + f_2(x_0, y, z) \\
&= f_2(x, y, z)
\end{aligned}
$$

as desired. Finally,

$$
\begin{aligned}
D_3\varphi(x, y, z) &= \int_{x_0}^{x} D_3 f_1(t, y, z)\, dt + \int_{y_0}^{y} D_3 f_2(x_0, t, z)\, dt + f_3(x_0, y_0, z) \\
&= \int_{x_0}^{x} D_1 f_3(t, y, z)\, dt + \int_{y_0}^{y} D_2 f_3(x_0, t, z)\, dt + f_3(x_0, y_0, z) \\
&= f_3(x, y, z) - f_3(x_0, y, z) + f_3(x_0, y, z) - f_3(x_0, y_0, z) + f_3(x_0, y_0, z) \\
&= f_3(x, y, z),
\end{aligned}
$$

as was to be shown.

Chapter VI, §1 *(p. 135)*

1. $-\frac{369}{10}$ **2.** $\frac{23}{6}$ **3.** 0 **4.** 0 **5.** 54 **6.** $\sqrt{3c/2}$ **7.** $\frac{4}{3}$ **8.** $-\pi - \frac{8}{3}$

9. $\frac{4}{15}$ **10.** 4π **11.** (a) $3\pi/4$ (b) 2π (c) 2π (d) 2π **12.** $\frac{264}{5} + 6$

Chapter VI, §2 *(p. 138)*

1. 56

Chapter VI, §3 *(p. 143)*

1. The sum of the integrals over the curves $C_1, C_2, \ldots, C_m$ is equal to

$$\varphi(Q_1) - \varphi(P_1) + \varphi(Q_2) - \varphi(P_2) + \cdots + \varphi(Q_m) - \varphi(P_m).$$

But $Q_1 = P_2, Q_2 = P_3, \ldots, Q_{m-1} = P_m$ so all terms cancel except $\varphi(Q_m) - \varphi(P_1)$, as desired.

2. 9/2. There is a potential function

$$\varphi(x, y, z) = x^2 + \frac{3}{2}y^2 + 2z^2.$$

Then $\varphi(1, 1, 1) - \varphi(0, 0, 0) = 9/2$.

In each case of Exercises 3 through 8 there is a potential function and the integral can be evaluated as in Exercise 2.

3. 3 **4.** (a) 9/2 (b) 3 **5.** 8 **7.** $1 - e^{-2\pi}$

8. There exists a potential function, $\varphi(x, y, z) = xyz^3$, so the integral is independent of the curve by Theorem 5.

9. (a) No (b) $\frac{1}{8}$ (c) 0

10. (a) $0 = g(1) - g(1)$ (b) 0 (c) There is a potential function $g(r) = \sin r$, because by the chain rule, if $\varphi(X) = \sin r$, then grad $\varphi(X) = F(x, y)$.

11. (a) 2π (b) No potential function. We can write the vector field F in the form

$$F(x, y) = G(x, y) + \text{grad } \psi(x, y)$$

where $G(x, y)$ is the usual $\left(\dfrac{-y}{x^2 + y^2}, \dfrac{x}{x^2 + y^2} \right)$ and $\psi(X) = \log r$.

The integral of F is the sum of the integrals of G and grad ψ. The integral of grad ψ over a closed curve is 0. If C is the circle of radius 1 centered at the origin, then

$$\int_C F = \int_C G + \int_C \text{grad } \psi = 2\pi + 0 = 2\pi.$$

12. (a) Yes, because the vector field is defined and has continuous partial derivatives on this rectangle, so Theorem 4 of Chapter V applies. (b) 2π (c) No, because there is some closed curve such that the integral of the vector field around this closed curve is not 0. See the comments to Exercise 11.

13. (a) 0 (b) 0 (c) Yes, $\varphi(X) = e^r$, because a direct partial differentiation with respect to x, y, z shows that

$$\text{grad } \varphi(X) = F(x, y).$$

Theorem 4 of Chapter V is not applicable here since the open set is not a rectangle.

14. (a) $e^5 - e^{\sqrt{5}}$ (b) 0 (c) 0 (d) 0

There is a potential function, and the easiest way to evaluate the integrals is by means of the formula $\varphi(Q) - \varphi(P)$.

15. (a) $-\pi/2$ (b) $2\pi/3$. In this case, there is a potential function, namely θ, on an open set containing the stated path, so the formula $\varphi(Q) - \varphi(P)$ can again be used.

16. $16 + 5/6$. There is a potential function, so use the easy way to compute the value of the integral.

Chapter VII, §2 *(p. 166)*

1. (a) 12 (b) $\frac{11}{5}$ (c) $\frac{1}{10}$ (d) $2 + \pi^2/2$ (e) $\frac{5}{6}$ (f) $\pi/4$ (g) $\frac{8}{3}$ (h) $\frac{49}{32}\pi$ (i) 3

2. (d)   (f)

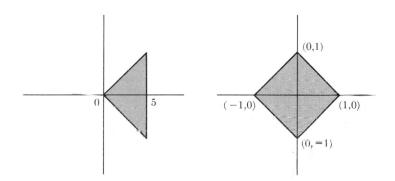

To see (f), suppose first $x \geq 0$ and $y \geq 0$. Then the inequality reads $x + y \leq 1$, which is in the first quadrant, below the line $y = -x + 1$ as shown. But the region is symmetric, in the sense that it does not change if x, y are replaced by $\pm x$ or $\pm y$ because of the absolute values. Hence we get the square as drawn.

3. (a) $-3\pi/2$ (b) $e - 1/e$

To see (b), for instance, we decompose the region of Exercise 2(f) into two pieces as shown.

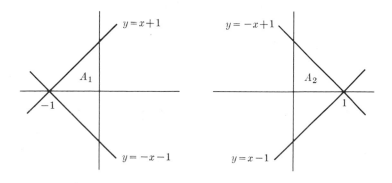

Then we have to sum the integrals

$$\iint_{A_1} e^x e^y \, dy \, dx = \int_{-1}^{0} \int_{-x-1}^{x+1} e^x e^y \, dy \, dx$$

and

$$\iint_{A_2} e^x e^y \, dy \, dx = \int_{0}^{1} \int_{x-1}^{-x+1} e^x e^y \, dy \, dx.$$

In each case, e^x can be taken out of the inner integral with respect to y. You evaluate the inner integral, for instance:

$$\int_{-x-1}^{x+1} e^y \, dy = e^y \Big|_{-x-1}^{x+1} = ee^x - \frac{1}{e} e^{-x}.$$

Then integrate with respect to x.

3. (c) $\pi^2 - \dfrac{40}{9}$ (d) $\dfrac{63}{32}$

4. (a) $\dfrac{1}{20}$ (b) $\dfrac{1}{35}$ (c) 4

6. (a) $\dfrac{49}{20}$ (b) $1 - \cos 2$

For 6(b) watch out. You have to split the integral into

$$\int_{1}^{3} |x - 2| \, dx = \int_{1}^{2} - (x - 2) \, dx + \int_{2}^{3} (x - 2) \, dx$$

because $|t| = t$ if $t \geq 0$ and $|t| = -t$ if $t \leq 0$.
(c) 0. Remember that $\cos(-y) = \cos y$.
(d) 1. You have to split the integral into

$$\int_{-1}^{1} \int_{0}^{|x|} = \int_{-1}^{0} \int_{0}^{-x} + \int_{0}^{1} \int_{0}^{x}.$$

The region of integration is shown on the figure.

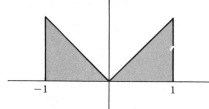

(e) $\dfrac{1}{6}$ (f) $\dfrac{e^2 - 1}{4}$ (g) $\dfrac{2^7}{21} + \dfrac{2^4}{4} + \dfrac{3^7}{21} - \dfrac{3^4}{4} = \dfrac{7895}{84}$

7. $2ka^4/3$

8. (a) $\log 2$ (b) $\frac{1}{3}$ (c) π (d) $-\frac{1}{3}$ (e) $\log\frac{27}{16}$ **9.** $\frac{9}{128}$ **10.** $1/30$ **11.** $3\pi/4$

12. $k\pi/4$

Chapter VII, §3 *(p. 176)*

1. $(e - 1)\pi$ 2. $3\pi/2$ 3. $\pi(1 - e^{-a^2})$ 4. π 5. $3\pi/8$ 6. $3k\pi a^4/2$
7. $\frac{\pi}{2}(e - 1)$ 8. πa^2 9. $\pi a^4/8$ 10. $a^3 \sqrt{2}/6$ 11. $a^2(\pi + 8)/4$
12. $a^3(15\pi + 32)/24$ 13. (a) 1 (b) $2a^2$
14. $\dfrac{2\sqrt{2}\,\pi a^3}{3} - \dfrac{64}{9}a^3 + \dfrac{40\sqrt{2}\,a^3}{9}$ 15. $2\pi\left[-(a^2 + 1)^{-1/2} + 1\right]$
16. $2\pi\left[-\dfrac{1}{2(a^2 + 2)} + \dfrac{1}{4}\right]$. Limit $= \pi/2$
17. $\dfrac{\pi}{2}\left(\dfrac{1}{2^4} - \dfrac{1}{3^4}\right)$ 18. (a) $-5\pi/4$ (b) $\frac{49}{32}\pi a^4$
19. (a) $3\pi/8$ (b) $\frac{2}{3}(2\sqrt{2} - 1)$ (c) 0
20. 0
21. $\dfrac{32\pi}{3} - 4\pi\sqrt{3}$
We work out this problem more fully. The function giving the height of the sphere of radius 2 above a point (x, y) of the disc of radius 1 is given in terms of the polar coordinates by

$$z = \pm\sqrt{4 - r^2}\,.$$

Hence

$$\text{Volume} = 2\int_0^{2\pi}\int_0^1 \sqrt{4 - r^2}\; r\,dr\,d\theta.$$

You can integrate this by substitution, with $u = 4 - r^2$, $du = -2r\,dr$. Multiply and divide the integral by -2.
22. (a) $2\pi\left[\dfrac{b^{-n+2} - a^{-n+2}}{-n + 2}\right]$ if $n \neq 2$
    $2\pi[\log b - \log a]$ if $n = 2$.
    (b) The integral approaches a limit if $n = 0, 1$

Chapter VIII, §1 *(p. 187)*

1. (a) -4 (b) 4 (c) 4π (d) π (e) 8 (f) πab
2. (a) $-5/6$ (b) Directly, let

$$C_1(t) = (t, \sin t) \qquad \text{and} \qquad C_2(t) = (t, 2\sin t)$$

with $0 \leq t \leq \pi$. The integral is

$$\int_C (1 + y^2)\,dx + y\,dy$$

$$= \int_{C_1}(1 + y^2)\,dx + y\,dy - \int_{C_2}(1 + y^2)\,dx + y\,dy$$

$$= \int_0^\pi (1 + \sin^2 t)\,dt + \sin t\cos t\,dt - (1 + 4\sin^2 t)\,dt - 4\sin t\cos t\,dt$$

$$= -3\int_0^\pi \sin^2 t\,dt = -3\pi/2.$$

By Green's theorem,

$$\int_0^\pi \int_{\sin x}^{2 \sin x} -2y \; dy \; dx = \int_0^\pi -3 \sin^2 x \; dx = -3\pi/2.$$

3. (a) -2 (b) $5/3$

4. (a) By Green's theorem, with $p = -y$, $q = x$, we get:

$$\frac{1}{2} \int_C -y \; dx + x \; dy = \frac{1}{2} \iint_A (1 + 1) \; dy \; dx = \iint_A dy \; dx = \text{area of } A.$$

(b) Similar, using $p = 0$ and $q = x$.

5. By Green's theorem the integral is equal to

$$\iint_A \left(-\frac{\partial}{\partial x} \frac{\partial f}{\partial x} - \frac{\partial}{\partial y} \frac{\partial f}{\partial y} \right) dy \; dx = 0.$$

6. 2π, use the method of Example 4, reducing the integral to an integral over the circle of radius 1.

7. You can write $F = G + \text{grad } \psi$, where G is the vector field of Exercise 6 (an old friend), and ψ is a function (which one?). The integral of F over a closed path is therefore equal to the integral of G over a closed path, so no difficulties remain.

Chapter VIII, §2 *(p. 196)*

1. Since $C'(t) = \left(\dfrac{dx}{dt}, \dfrac{dy}{dt} \right)$ is in the direction of the curve, we see that $N(t) \perp C'(t)$ because

$$N(t) \cdot C'(t) = \frac{dy}{dt} \frac{dx}{dt} - \frac{dx}{dt} \frac{dy}{dt} = 0.$$

2. Let $F = (p, q)$. Let G be the vector field $G = (-q, p)$. Apply Green's theorem to G. Then

$$\int_C G = \iint_A \left(\frac{\partial p}{\partial x} + \frac{\partial q}{\partial y} \right) dy \; dx.$$

But

$$\int_C G = \int_C -q \; dx + p \; dy = \int_a^b \left(-q \frac{dx}{dt} + p \frac{dy}{dt} \right) dt.$$

Furthermore,

$$F \cdot N = p \frac{dy}{dt} - q \frac{dx}{dt}.$$

This proves the divergence theorem.

3. The divergence of F is 0 because $\partial y/\partial x = 0$ and $\partial x/\partial y = 0$ also. Hence the divergence theorem implies

$$\int_C F \cdot \mathbf{n} \; ds = 0.$$

4. Let $F = g \text{ grad } f = \left(g \dfrac{\partial f}{\partial x}, g \dfrac{\partial f}{\partial y} \right)$. Take the dot product of F with $\mathbf{n}$, letting $\mathbf{n} =$ (n_1, n_2). You get:

$$F \cdot \mathbf{n} = g \dfrac{\partial f}{\partial x} n_1 + g \dfrac{\partial f}{\partial y} n_2 = g(\text{grad } f \cdot \mathbf{n}) = g D_{\mathbf{n}} f.$$

On the other hand,

$$\text{div } F = \dfrac{\partial}{\partial x}\left(g \dfrac{\partial f}{\partial x} \right) + \dfrac{\partial}{\partial y}\left(g \dfrac{\partial f}{\partial y} \right)$$

$$= g \dfrac{\partial^2 f}{\partial x^2} + \dfrac{\partial g}{\partial x}\dfrac{\partial f}{\partial x} + g \dfrac{\partial^2 f}{\partial y^2} + \dfrac{\partial g}{\partial y}\dfrac{\partial f}{\partial y}$$

$$= (\text{grad } g) \cdot (\text{grad } f) + g \Delta f,$$

by combining the first and third term, and the second and fourth term. Then apply the divergence theorem. Do this also for the vector field $f \text{ grad } g$.

For (b), take the difference of the formulas obtained in part (a) for the two vector fields $g \text{ grad } f$ and $f \text{ grad } g$.

Chapter IX, §1 *(p. 203)*

1. $\dfrac{\pi}{6}$ **2.** 0 **3.** (a) 25 (b) 15/2

  3 − *e*. Solution:

$$\int_{-1}^{0}\int_{-x}^{1}\int_{0}^{-y} e^{x+y-z}\, dz\, dy\, dx = \int_{-1}^{0}\int_{-x}^{1} (e^y - e^{x+y})\, dy\, dx$$

$$= \int_{-1}^{0} (e - e^{-x} - e^{x+1} + 1)\, dx = 3 - e.$$

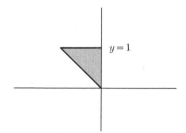

$y = 1$

Chapter IX, §2 *(p. 214)*

1. $\frac{4}{3}\pi a^3$ **2.** $\dfrac{\pi}{3}$ **3.** (a) $\pi k a^4$ (b) $2\pi(1 - a^2)$, 2π **4.** $2\pi k(b^2 - a^2)$

5. $\pi b a^4/4$ **6.** $k\pi a^4/2$ **7.** $\pi/8$

8. $2\pi\left[-\dfrac{1}{3}(1 - r_0^2)^{3/2} + \dfrac{1}{3} - \dfrac{r_0^4}{4} \right]$ where $r_0 = \dfrac{-1 + \sqrt{5}}{2}$

9. $\frac{2}{9}a^3(3\pi - 4)$ **10.** πa^3, center at $(0, 0, a)$, $\cos^2 \varphi = \sin^2 \varphi$

11. (a) $\pi/3$ (b) $2\pi\sqrt{2}/3$ (c) $\pi/2$ (d) $\pi/32$

12. $7a^2b^3/3$ **13.** $64/3$

14. (a) $4\pi\left[\dfrac{b^{3-n} - a^{3-n}}{3 - n}\right]$ if $n \neq 3$ (b) The integral approaches

a limit if $n = 0, 1, 2$.

$\qquad 4\pi[\log b - \log a]$ if $n = 3$.

Chapter IX, §3 *(p.217)*

1. $(1, 5/3)$ **2.** $(5/2, 5)$ **3.** $\left(0, \dfrac{4b}{3\pi}\right)$ **4.** $(1, -\frac{4}{5})$ **5.** $(\pi/2, \pi/8)$

6. $\bar{x} = \dfrac{\pi}{2} + \dfrac{\pi\sqrt{2}}{4} - 1 - \sqrt{2}$, $\bar{y} = \dfrac{\sqrt{2} + 1}{4}$

7. $\bar{x} = \dfrac{2a^2 \log a - a^2 + 1}{4(a \log a - a + 1)}$, $\bar{y} = \dfrac{a(\log a)^2}{2(a \log a - a + 1)} - 1$

8. $(0, 0, \frac{3}{4}h)$ **9.** (a) $\frac{20}{3}k\pi$ (b) $\left(\dfrac{21}{10}, \dfrac{96}{25\pi}\right)$ **10.** (a) $\frac{1}{2}k\pi h^2 r^2$ (b) $\frac{2}{3}h$

11. (a) $\frac{2}{3}ka^3\pi$ (b) $(0, 0)$ (c) $\left(\dfrac{3a}{2\pi}, \dfrac{3a}{2\pi}\right)$ **12.** $\frac{1}{2}ka^4h\pi$ **13.** $\left(0, 0, \dfrac{2h}{5}\right)$

Chapter X, §1 *(p. 225)*

1. $\dfrac{\partial X}{\partial \theta} = (-(a + b \cos \varphi) \sin \theta, (a + b \cos \varphi) \cos \theta, 0)$

$$\left\|\frac{\partial X}{\partial \theta}\right\| = |a + b \cos \varphi|$$

$\dfrac{\partial X}{\partial \varphi} = (- b \sin \varphi \cos \theta, - b \sin \varphi \sin \theta, b \cos \varphi)$

$$\left\|\frac{\partial X}{\partial \varphi}\right\| = |b|$$

2. $\dfrac{\partial X}{\partial \theta} = (-z \sin \alpha \sin \theta, z \sin \alpha \cos \theta, 0)$

$\dfrac{\partial X}{\partial z} = (\sin \alpha \cos \theta, \sin \alpha \sin \theta, \cos \alpha)$

$\dfrac{\partial X}{\partial \theta} \times \dfrac{\partial X}{\partial z} = (z \sin \alpha \cos \theta \cos \alpha, z \sin \alpha \sin \theta \cos \alpha, - z \sin^2 \alpha)$

$$\left\|\frac{\partial X}{\partial \theta} \times \frac{\partial X}{\partial z}\right\| = z \sin \alpha$$

Equation of surface is $x^2 + y^2 = (\tan \alpha)^2 z^2$.

3. $\dfrac{\partial X}{\partial t} = (a \cos \theta, a \sin \theta, 2t)$

$\dfrac{\partial X}{\partial \theta} = (-at \sin \theta, at \cos \theta, 0)$

$\dfrac{\partial X}{\partial t} \times \dfrac{\partial X}{\partial \theta} = (-2at^2 \cos \theta, -2at^2 \sin \theta, a^2 t)$

$\left\| \dfrac{\partial X}{\partial t} \times \dfrac{\partial X}{\partial \theta} \right\| = \sqrt{4a^2 t^4 + a^4 t^2}$

The equation is $x^2 + y^2 = a^2 z$.

4. $\dfrac{\partial X}{\partial \varphi} = (a \cos \varphi \cos \theta, b \cos \varphi \sin \theta, -c \sin \varphi)$

$\dfrac{\partial X}{\partial \theta} = (-a \sin \varphi \sin \theta, b \sin \varphi \cos \theta, 0)$

$\dfrac{\partial X}{\partial \varphi} \times \dfrac{\partial X}{\partial \theta} = (cb \sin^2 \varphi \cos \theta, ac \sin^2 \varphi \sin \theta, ab \sin \varphi \cos \varphi)$

$\left\| \dfrac{\partial X}{\partial \varphi} \times \dfrac{\partial X}{\partial \theta} \right\| = \sqrt{c^2 b^2 \sin^4 \varphi^4 \cos^2 \theta + a^2 c^2 \sin^4 \varphi \sin^2 \theta + a^2 b^2 \sin^2 \varphi \cos^2 \varphi}$

The equation is $\dfrac{x^2}{a^2} + \dfrac{y^2}{b^2} + \dfrac{z^2}{c^2} = 1$.

5. $\dfrac{\partial X}{\partial \theta} = (-a \sin \theta, a \cos \theta, 0)$

$\dfrac{\partial X}{\partial z} = (0, 0, 1)$

$\dfrac{\partial X}{\partial \theta} \times \dfrac{\partial X}{\partial z} = (a \cos \theta, a \sin \theta, 0)$

$\left\| \dfrac{\partial X}{\partial \theta} \times \dfrac{\partial X}{\partial z} \right\| = a$

The equation is $x^2 + y^2 = a^2$.

6. $\dfrac{\partial X}{\partial r} = (\cos \theta, \sin \theta, f'(r))$

$\dfrac{\partial X}{\partial \varphi} = (-r \sin \theta, r \cos \theta, 0)$

$\dfrac{\partial X}{\partial r} \times \dfrac{\partial X}{\partial \theta} = (-f'(r)r \cos \theta, -f'(r)r \sin \theta, r)$

$\left\| \dfrac{\partial X}{\partial r} \times \dfrac{\partial X}{\partial \theta} \right\| = r\sqrt{f'(r)^2 + 1}$

The equation is $z = f(\sqrt{x^2 + y^2})$.

Chapter X, §2 *(p.231)*

1. (a) $\pi\sqrt{2}$ (b) $\dfrac{\sqrt{10}}{9} \pi h^2$ **2.** $\dfrac{\pi}{6}(5\sqrt{5} - 1)$ **3.** $2\pi(\sqrt{3} - \tfrac{1}{3})$

4. $\tfrac{2}{3}\pi(2\sqrt{2} - 1)$ **5.** $2\pi(\tfrac{1}{8}e^2 \text{ arcsinh } 1 - \tfrac{1}{8}e^{-2 \text{ arcsinh } 1} + \tfrac{1}{2} \sinh 1)$

6. $2\pi\sqrt{6}$ **7.** $2\sqrt{2}\,\pi$ **8.** $2\pi(1 - \sqrt{2}/2)$ **9.** $4\pi^2 a$

Chapter X, §3 *(p.238)*

1. (a) $4\pi a^4/3$ (b) $\pi a^5/2$ (c) $4\pi a^6/15$ (d) $\pi a^7/3$
2. $4\pi/3$
3. πa^3
4. $\pi(10\sqrt{5}\ /3 + 2/15)/8$ 5. $\frac{\pi}{60}(25\sqrt{5} - 11)$
6. 0 7. 0 8. πa^3 9. $\pi a^4/2$ 12. $4\pi/3$ 13. $-(\pi^2/4 + 2\pi)$ 14. 2π
15. $104/3$ 16. $2\pi\sqrt{2}$ 17. $\frac{\pi}{12}(8 - 5\sqrt{2}\ )$ 18. $5/12$
19. (a) $2\pi a^2$ (b) $3\pi a^2$ 20. $3/2$ 21. $5\pi/4$ 22. 4π 23. $2\pi/3$

Chapter X, §4 *(p.243)*

1. $\nabla \cdot F = 2x + xz + 2yz$
   $\nabla \times F = (z^2 - xy, 0, yz)$
2. $\nabla \cdot F = \frac{y}{x} + \frac{x}{y} + \frac{xy}{z}$
   $\nabla \times F = (x \log z, -y \log z, \log y - \log x)$
3. $\nabla \cdot F = 2x + x \cos xy + e^x y$
   $\nabla \times F = (e^x z, -e^x yz, y \cos xy)$
4. $\nabla \cdot F = ye^{xy} \sin z + e^{xz} \cos y + ye^{yz} \cos x$
   $\nabla \times F = (ze^{yz} \cos x - xe^{xz} \sin y, e^{xy} \cos z + e^{yz} \sin x, ze^{xz} \sin y - xe^{xy} \sin z)$

Chapter X, §5 *(p.249)*

2. $3/2$ 3. $3\pi/4$ 4. 64π 5. (a) 0 (b) 0 (c) 16 (d) 24
6. $8\pi/3$ 7. $12\pi/5$ 8. 1
9. 24π 10. 48π 11. $243\pi/2$ 12. 135π 13. $11/24$

Chapter X, §6 *(p.259)*

1. 4π 2. $-13/6$ 3. 0 4. 0 5. 0 6. 0 7. $-\pi a^2$
8. (a) Solution of $8 = \pi$.

$$\int_C z\ dx + 2x\ dy + y^2\ dz = \int_0^{2\pi} - \sin^2 t\ dt + 2 \cos^2 t\ dt + \sin^2 t \cos t\ dt$$

$$= -\frac{2\pi}{2} + 2\frac{2\pi}{2} + 0 = \pi.$$

(b) Let $F(x, y, z) = (z, 2x, y^2)$. Then curl $F = (2y, 1, 2)$, and

$$\iint_D (2y, 1, 2) \cdot (0, -1, 1)\ dx\ dy = \iint_D 1\ dx\ dy = \text{Area } (D) = \pi.$$

Chapter XI, §1 *(p.263)*

1. (2, 1), neither max nor min
2. $((2n + 1)\pi, 1)$ and $(2n\pi, -1)$, neither max nor min
3. (0, 0, 0), min, value 0
4. $\pm(\sqrt{2}/2, \sqrt{2}/2)$, neither local max nor min. [*Hint*: Change variables, letting $u = x + y$ and $v = x - y$. Then the critical points are at $\pm(\sqrt{2}, 0)$, and in the (u, v)-plane, near these points, the function increases in one direction and decreases in the other.]
5. All points of form $(0, t, -t)$, neither max nor min.
6. All (x, y, z) with $x^2 + y^2 + z^2 = 2n\pi$ are local max, value 1.
   All (x, y, z) with $x^2 + y^2 + z^2 = (2n + 1)\pi$ are local min, value -1.
7. All points $(x, 0)$ and $(0, y)$ are mins, value 0.
8. (0, 0), min, value 0   9. (t, t), min, value 0
10. $(0, n\pi)$, neither max nor min   11. $(1/2, 0)$, min, value $-1/4$
12. (0, 0, 0), max, value 1   13. (0, 0, 0), min, value 1

Chapter XI, §2 *(p.268)*

1. Min $= -2$ at $(-1, -1)$, max $= 2$ at $(1, 1)$ 2. None
3. Max $\frac{1}{2}$ at $(\sqrt{2}/2, \sqrt{2}/2)$ and $(-\sqrt{2}/2, -\sqrt{2}/2)$, min -1 at $(0, 0)$
4. Max at $(\frac{1}{2}, \frac{1}{3})$, no min
5. Min 0 at (0, 0), max $2/e$ at $(0, \pm 1)$
6. (a) Max 1 at (1, 0), min $1/9$ at (3, 0) (b) Max 1 at (0, 1), min $1/9$ at (0, 3)
7. (a) Both (b) Neither (c) Neither (d) Min (e) Both (f) Max (g) Min
8. $t = (2n + 1)\pi$, so $(-1, 0, 1)$ and $(-1, 0, -1)$
9. Max at (1, 1), value 2
10. Max at (1, 1), value 2
11. Max at (0, 1), (0, 1), and $(-2, 0)$, all with value 0

Chapter XI, §3 *(p.273)*

1. (a) $-1/\sqrt{2}$ (b) $9/8$ 2. $1 + 1/\sqrt{2}$ 3. At $(\frac{5}{3}, \frac{2}{3}, \frac{1}{3})$ min $= 12$
4. $X = \frac{1}{3}(A + B + C)$, min value is $\frac{2}{3}(A^2 + B^2 + C^2 - AB - AC - BC)$
5. 45 at $\pm(\sqrt{3}, \sqrt{6})$ 6. $(\frac{2}{3})^{3/2}$ at $\sqrt{\frac{2}{3}}$ (1, 1, 1)   7. Min 0, max 0
8. Max at $(\pi/8, -\pi/8)$, value $2\cos^2(\pi/8)$; min at $(5\pi/8, 3\pi/8)$ value $\cos^2(5\pi/8) + \cos^2(3\pi/8)$
9. $(0, 0, \pm 1)$ 10. No min, max $= \frac{1}{4}$ at $(\frac{1}{2}, \frac{1}{2})$ 11. 1

12. Max $=\sqrt{3}$ at $\sqrt{3}/3(1, 1, 1)$, min $= -\sqrt{3}$ at $-\sqrt{3}/3(1, 1, 1)$.

13. Values 3 and -3 at $(\frac{1}{3}, -\frac{2}{3}, \frac{2}{3})$ and $(-\frac{1}{3}, \frac{2}{3}, -\frac{2}{3})$

14. $2\sqrt{3}$ at (x, x, x) with $x = \sqrt{4/3}$ **15.** No max or min

16. Max $= 11/6$, min $= 0$ **17.** $25/62$ **18.** $d^2/(a^2 + b^2 + c^2)$

19. Max $= \frac{9}{4}$ at $(-\frac{1}{2}, \sqrt{\frac{3}{4}})$, $(-\frac{1}{2}, -\sqrt{\frac{3}{4}})$, min $= -\frac{1}{4}$ at $(\frac{1}{2}, 0)$

20. $(4-\sqrt{5})/\sqrt{2}$

21. $1/27c^3$ where c is the cost per unit.

22. 8 hours at A and 2 hours at B.

23. 4 units of A and $\frac{16}{3}$ units of B.

24. 8 boxes of candy and 2 cakes.

Chapter XII, §1 *(p.280)*

1. xy **2.** 1   **3.** xy **4.** $x^2 + y^2$

5. $1 + x + y + \dfrac{(x + y)^2}{2!}$ **6.** $1 - \dfrac{y^2}{2}$ **7.** x

8. $y + xy$ **9.** $x + xy + 2y^2$

10. (1) $-\pi(x - 1) - (y - \pi) - (x - 1)(y - \pi)$

     (2) $-1 + \dfrac{\pi^2}{2}(x - 1)^2 + \pi(x - 1)(y - \pi) + \dfrac{(y - 1)^2}{2}$

     (3) $\log 7 + \frac{3}{7}(x - 2) + \frac{2}{7}(y - 3) - \frac{9}{98}(x - 2)^2$
        $+ \frac{1}{49}(x - 2)(y - 3) - \frac{4}{98}(y - 3)^2$

     (4) $2\sqrt{\pi}\,(x - \sqrt{\pi}) + 2\sqrt{\pi}\,(y - \sqrt{\pi}) + (x - \sqrt{\pi})^2 + (y - \sqrt{\pi})^2$

     (5) $e^3 + e^3(x - 1) + e^3(y - 2) + \dfrac{e^3}{2}(x - 1)^2$
                        $+ e^3(x - 1)(y - 2) + \dfrac{e^3}{2}(y - 2)^2$

     (6) $-1 + \frac{1}{2}(y - \pi)^2$

     (7) $1 - \frac{1}{2}(x - \pi/2)^2 - \frac{1}{2}(y - \pi)^2$

     (8) $\dfrac{e^2\sqrt{2}}{2} + \dfrac{e^2\sqrt{2}}{2}(x - 2) + \dfrac{e^2\sqrt{2}}{2}(y - \pi/4)$
               $+ \dfrac{e^2\sqrt{2}}{4}(x - 2)^2 + \dfrac{e^2\sqrt{2}}{2}(x - 2)(y - \pi/4)$
               $- \dfrac{e^2\sqrt{2}}{4}(y - \pi/4)^2$

     (9) $4 + 2(x - 1) + 5(x - 1) + (x - 1)(y - 1) + 2(y - 1)^2$

Chapter XII, §2 *(p.286)*

2. (a) If you take a first partial, each term will have an x or a y left in it, so vanishes at $(0, 0)$.
   (b) $q(x, y)$ is the same as $f(x, y)$.
3. (a) $x^2 + 4xy - y^2$
   (b) At $((2n + 1)\pi, 1)$, $-xy$. At $(2n\pi, -1)$, $+xy$.
   (c) $-\dfrac{1}{\sqrt{2}} e^{-1/2}\left(\dfrac{x^2}{2} + 3xy + \dfrac{y^2}{2}\right)$ at $(\sqrt{2}/2, \sqrt{2}/2)$
   $\dfrac{1}{\sqrt{2}} e^{1/2}\left(\dfrac{x^2}{2} + 3xy + \dfrac{y^2}{2}\right)$ at $(-\sqrt{2}/2, -\sqrt{2}/2)$
   (d) At points $(a, 0)$ we get a^2y^2. At points $(0, b)$, we get b^2x^2.
   (e) y^2 (f) 0
   (g) At the points $(0, n, \pi)$, we get $\pm xy$ according to whether n is odd or even.
   (h) $x^2 + 2y^2$
4. (a) Neither   (b) Min   (c) Max   (d) Neither   (e) Neither
   (f) Neither   (g) Max   (h) Neither

Chapter XII, §3 *(p. 291)*

1 through 5, neither.
6, min   7. Max at $\left(0, \dfrac{1}{\sqrt{3}}\right)$, Min at $\left(0, -\dfrac{1}{\sqrt{3}}\right)$, Saddle points at $(\pm 1, 0)$.
8. Min at $(0, 0)$; Saddle points at $(0, 2/3)$, $(-2/3, 0)$, Max at $(-2/3, 2/3)$

Chapter XII, §4 *(p. 297)*

1. $9D_1^2 + 12D_1D_2 + 4D_2^2$
2. $D_1^2 + D_2^2 + D_3^2 + 2D_1D_2 + 2D_2D_3 + 2D_1D_3$
3. $D_1^2 - D_2^2$ 4. $D_1^2 + 2D_1D_2 + D_2^2$
5. $D_1^3 + 3D_1^2D_2 + 3D_1D_2^2 + D_2^3$
6. $D_1^4 + 4D_1^3D_2 + 6D_1^2D_2^2 + 4D_1D_2^3 + D_2^4$
7. $2D_1^2 - D_1D_2 - 3D_2^2$ 8. $D_1D_2 - D_3D_2 + 5D_1D_3 - 5D_3^2$
9. $\left(\dfrac{\partial}{\partial x}\right)^3 + 12\left(\dfrac{\partial}{\partial x}\right)^2 \dfrac{\partial}{\partial y} + 48 \dfrac{\partial}{\partial x}\left(\dfrac{\partial}{\partial y}\right)^2 + 64\left(\dfrac{\partial}{\partial y}\right)^3$
10. $4\left(\dfrac{\partial}{\partial x}\right)^2 + 4\dfrac{\partial}{\partial x}\dfrac{\partial}{\partial y} + \left(\dfrac{\partial}{\partial y}\right)^2$ 11. $h^2\left(\dfrac{\partial}{\partial x}\right)^2 + 2hk\dfrac{\partial}{\partial x}\dfrac{\partial}{\partial y} + k^2\left(\dfrac{\partial}{\partial y}\right)^2$
12. $h^3\left(\dfrac{\partial}{\partial x}\right)^3 + 3h^2k\left(\dfrac{\partial}{\partial x}\right)^2\dfrac{\partial}{\partial y} + 3hk^2\dfrac{\partial}{\partial x}\left(\dfrac{\partial}{\partial y}\right)^2 + k^3\left(\dfrac{\partial}{\partial y}\right)^3$
13. 8   14. 4   15. 0   16. 4   17. (a) $4!5!xy$ (b) 0   (c) $4!3!$ (d) $10 \cdot 4!3!$
18. (a) 0   (b) $3 \cdot 7!9!$ (c) $11 \cdot 7!9!$ (d) 0
19. (a) $(4!)^2$ (b) $-7 \cdot 9!4!$ (c) 6   (d) 0
20. (a) 0   (b) $4 \cdot 2!3!$ (c) $7 \cdot 6!10!7!$ (d) 0

Chapter XIII, §1 *(p. 308)*

1. $A + B = \begin{pmatrix} 0 & 7 & 1 \\ 0 & 1 & 1 \end{pmatrix}$, $\qquad 3B = \begin{pmatrix} -3 & 15 & -6 \\ 3 & 3 & -3 \end{pmatrix}$

$-2B = \begin{pmatrix} 2 & -10 & 4 \\ -2 & -2 & 2 \end{pmatrix}$, $A + 2B = \begin{pmatrix} -1 & 12 & -1 \\ 1 & 2 & 0 \end{pmatrix}$

$2A + B = \begin{pmatrix} 1 & 9 & 4 \\ -1 & 1 & 3 \end{pmatrix}$, $\qquad A - B = \begin{pmatrix} 2 & -3 & 5 \\ -2 & -1 & 3 \end{pmatrix}$

$A - 2B = \begin{pmatrix} 3 & -8 & 7 \\ -3 & -2 & 4 \end{pmatrix}$, $\qquad B - A = \begin{pmatrix} -2 & 3 & -5 \\ 2 & 1 & -3 \end{pmatrix}$

2. $A + B = \begin{pmatrix} 0 & 0 \\ 2 & -2 \end{pmatrix}$, $\qquad 3B = \begin{pmatrix} -3 & 3 \\ 0 & -9 \end{pmatrix}$ $\quad -2B = \begin{pmatrix} 2 & -2 \\ 0 & 6 \end{pmatrix}$,

$A + 2B = \begin{pmatrix} -1 & 1 \\ 2 & -5 \end{pmatrix}$, $A - B = \begin{pmatrix} 2 & -2 \\ 2 & 4 \end{pmatrix}$, $\qquad B - A = \begin{pmatrix} -2 & 2 \\ -2 & -4 \end{pmatrix}$

3. ${}^tA = \begin{pmatrix} 1 & -1 \\ 2 & 0 \\ 3 & 2 \end{pmatrix}$, $\quad {}^tB = \begin{pmatrix} -1 & 1 \\ 5 & 1 \\ -2 & -1 \end{pmatrix}$

4. ${}^tA = \begin{pmatrix} 1 & 2 \\ -1 & 1 \end{pmatrix}$, $\quad {}^tB = \begin{pmatrix} -1 & 0 \\ 1 & -3 \end{pmatrix}$ 7. Same

8. $\begin{pmatrix} 0 & 2 \\ 0 & -2 \end{pmatrix}$, same 9. $A + {}^tA = \begin{pmatrix} 2 & 1 \\ 1 & 2 \end{pmatrix}$, $\quad B + {}^tB = \begin{pmatrix} -2 & 1 \\ 1 & -6 \end{pmatrix}$

11. Rows of A: $(1, 2, 3), (-1, 0, 2)$

Columns of A: $\begin{pmatrix} 1 \\ -1 \end{pmatrix}, \begin{pmatrix} 2 \\ 0 \end{pmatrix}, \begin{pmatrix} 3 \\ 2 \end{pmatrix}$

Rows of B: $(-1, 5, -2), (1, 1, -1)$

Columns of B: $\begin{pmatrix} -1 \\ 1 \end{pmatrix}, \begin{pmatrix} 5 \\ 1 \end{pmatrix}, \begin{pmatrix} -2 \\ -1 \end{pmatrix}$

12. Rows of A: $(1, -1), (2, 1)$ $\qquad$ Columns of A: $\begin{pmatrix} 1 \\ 2 \end{pmatrix}, \begin{pmatrix} -1 \\ 1 \end{pmatrix}$

Rows of B: $(-1, 1), (0, -3)$ $\qquad$ Columns of B: $\begin{pmatrix} -1 \\ 0 \end{pmatrix}, \begin{pmatrix} 1 \\ -3 \end{pmatrix}$

Chapter XIII, §2 *(p. 312)*

1. $IA = AI = A$ 2. O

3. (a) $\begin{pmatrix} 3 & 2 \\ 4 & 1 \end{pmatrix}$ (b) $\begin{pmatrix} 10 \\ 14 \end{pmatrix}$ (c) $\begin{pmatrix} 33 & 37 \\ 11 & -18 \end{pmatrix}$

5. $AB = \begin{pmatrix} 4 & 2 \\ 5 & -1 \end{pmatrix}$, $\qquad BA = \begin{pmatrix} 2 & 4 \\ 4 & 1 \end{pmatrix}$

6. $AC = CA = \begin{pmatrix} 7 & 14 \\ 21 & -7 \end{pmatrix}$, $\qquad BC = CB = \begin{pmatrix} 14 & 0 \\ 7 & 7 \end{pmatrix}$.

If $C = xI$, where x is a number, then $AC = CA = xA$.

7. $(3, 1, 5)$, first row 8. Second row, third row, i-th row 9. O 10. O

11. (a) $A^2 = \begin{pmatrix} 0 & 0 & 1 \\ 0 & 0 & 0 \\ 0 & 0 & 0 \end{pmatrix}$, $\qquad A^3 = O$ matrix. $\quad$ If $B = \begin{pmatrix} 0 & 1 & 1 & 1 \\ 0 & 0 & 1 & 1 \\ 0 & 0 & 0 & 1 \\ 0 & 0 & 0 & 0 \end{pmatrix}$ then

$$B^2 = \begin{pmatrix} 0 & 0 & 1 & 2 \\ 0 & 0 & 0 & 1 \\ 0 & 0 & 0 & 0 \\ 0 & 0 & 0 & 0 \end{pmatrix}, \quad B^3 = \begin{pmatrix} 0 & 0 & 0 & 1 \\ 0 & 0 & 0 & 0 \\ 0 & 0 & 0 & 0 \\ 0 & 0 & 0 & 0 \end{pmatrix} \text{ and } B^4 = O.$$

(b) $A^2 = \begin{pmatrix} 1 & 2 & 3 \\ 0 & 1 & 2 \\ 0 & 0 & 1 \end{pmatrix}, \quad A^3 = \begin{pmatrix} 1 & 3 & 6 \\ 0 & 1 & 3 \\ 0 & 0 & 1 \end{pmatrix}, \quad A^4 = \begin{pmatrix} 1 & 4 & 10 \\ 0 & 1 & 4 \\ 0 & 0 & 1 \end{pmatrix}$

12. (a) $\begin{pmatrix} 4 \\ 9 \\ 5 \end{pmatrix}$ (b) $\begin{pmatrix} 3 \\ 1 \end{pmatrix}$ (c) $\begin{pmatrix} x_2 \\ 0 \end{pmatrix}$ (d) $\begin{pmatrix} 0 \\ x_1 \end{pmatrix}$

13. (a) $\begin{pmatrix} 2 \\ 4 \end{pmatrix}$ (b) $\begin{pmatrix} 1 \\ 1 \end{pmatrix}$ (c) $\begin{pmatrix} 3 \\ 5 \end{pmatrix}$ (d) $\begin{pmatrix} 4 \\ 6 \end{pmatrix}$

14. (a) $\begin{pmatrix} 3 \\ 1 \\ 2 \end{pmatrix}$ (b) $\begin{pmatrix} 7 \\ -1 \\ 1 \end{pmatrix}$ (c) $\begin{pmatrix} 5 \\ 4 \\ 8 \end{pmatrix}$ (d) $\begin{pmatrix} 12 \\ 3 \\ 9 \end{pmatrix}$

15. Second column of A 16. i-th column of A 17. j-th row of A

18. $\begin{pmatrix} 1 & a+b \\ 0 & 1 \end{pmatrix}, \begin{pmatrix} 1 & na \\ 0 & 1 \end{pmatrix}$ 20. $\begin{pmatrix} 1 & -a \\ 0 & 1 \end{pmatrix}$ 21. $(AB)^{-1} = B^{-1}A^{-1}$

22. $\begin{pmatrix} a & b \\ -a^2/b & -a \end{pmatrix}$ for any $a, b \neq 0$; if $b = 0$, then $\begin{pmatrix} 0 & 0 \\ 0 & 0 \end{pmatrix}$.

23. $A^n = \begin{pmatrix} \cos n\theta & -\sin n\theta \\ \sin n\theta & \cos n\theta \end{pmatrix}$ 24. $\begin{pmatrix} 0 & 1 \\ -1 & 0 \end{pmatrix}$

25. $\begin{pmatrix} 1 & 0 & 0 \\ 0 & 4 & 0 \\ 0 & 0 & 9 \end{pmatrix}, \begin{pmatrix} 1 & 0 & 0 \\ 0 & 8 & 0 \\ 0 & 0 & 27 \end{pmatrix}, \begin{pmatrix} 1 & 0 & 0 \\ 0 & 16 & 0 \\ 0 & 0 & 81 \end{pmatrix}$

26. Diagonal matrix with diagonal $a_1^k, a_2^k, \ldots, a_n^k$ 27. $A^3 = O$

Chapter XIV, §1 *(p. 321)*

1. (a) 11 (b) 13 (c) 6 2. (a) $(e, 1)$ (b) $(1, 0)$ (c) $(1/e, -1)$
3. (a) 1 (b) 11 4. Ellipse $9x^2 + 4y^2 = 36$ 5. Line $x = 2y$
6. Circle $x^2 + y^2 = e^2$, circle $x^2 + y^2 = e^{2c}$
7. Cylinder, radius 1, z-axis = axis of cylinder 8. Circle $x^2 + y^2 = 1$
12. $A = O$

Chapter XIV, §2 *(p. 326)*

1. (a) $\begin{pmatrix} 5 \\ 3 \end{pmatrix}$ (b) $\begin{pmatrix} 5 \\ 0 \end{pmatrix}$ (c) $\begin{pmatrix} 5 \\ 1 \end{pmatrix}$ (d) $\begin{pmatrix} 0 \\ -3 \end{pmatrix}$

2. $\begin{pmatrix} r & 0 & \cdots & 0 \\ 0 & r & & \vdots \\ \vdots & & \ddots & 0 \\ 0 & & \cdots & r \end{pmatrix}$ 3. $\begin{pmatrix} a & 0 \\ 0 & b \end{pmatrix}$ 4. $\begin{pmatrix} a_1 & 0 & 0 \\ 0 & a_2 & 0 \\ 0 & 0 & a_3 \end{pmatrix}$ 5. $\begin{pmatrix} 1 & 0 & 0 \\ 0 & 1 & 0 \end{pmatrix}$

6. $(1, 0, 0)$ 7. $\begin{pmatrix} 1 & 0 & 0 \\ 0 & 0 & 1 \end{pmatrix}$ 8. $\begin{pmatrix} 0 & 1 & 0 \\ 0 & 0 & 1 \end{pmatrix}$ 9. $\begin{pmatrix} 1 & 0 & 0 & 0 \\ 0 & 1 & 0 & 0 \end{pmatrix}$

10. $\begin{pmatrix} 1 & 0 & 0 & 0 \\ 0 & 1 & 0 & 0 \\ 0 & 0 & 1 & 0 \end{pmatrix}$ **11.** Only $A = 0$.   **12.** $\begin{pmatrix} -5 & 3 \\ 7 & 1 \end{pmatrix}$

13. $\begin{pmatrix} -1 & 2 \\ 4 & 6 \end{pmatrix}$ **14.** $\begin{pmatrix} 1 & -2 & 8 \\ 3 & 7 & -5 \\ 4 & 9 & 2 \end{pmatrix}$ **16.** $\begin{pmatrix} -3 & 4 & 5 \\ 5 & 1 & -2 \\ 0 & -7 & 8 \end{pmatrix}$

17. Let $A = L(1)$. For any number t, we have by linearity, $L(t) = L(t \cdot 1) = tL(1) = tA$.

18. $\begin{pmatrix} a_{11} & a_{12} \\ a_{21} & a_{22} \\ a_{31} & a_{32} \end{pmatrix}$ **19.** $\begin{pmatrix} 3 & -5 \\ 1 & 7 \\ -4 & -8 \end{pmatrix}$

Chapter XIV, §3 *(p. 332)*

3. $L(E^1) = \begin{pmatrix} 3 \\ 5 \end{pmatrix}$, $L(E^2) = \begin{pmatrix} -1 \\ 2 \end{pmatrix}$

5. It is the set of all points

$$t_1 A + t_2 B + t_3 C,$$

with numbers t_i satisfying $0 \leq t_i \leq 1$ for $i = 1, 2, 3$. Let S be this parallelepiped. The image of S under L is the set $L(S)$ consisting of all points

$$t_1 L(A) + t_2 L(B) + t_3 L(C),$$

with t_i satisfying the above inequality. Hence it is a parallelepiped if $L(A)$, $L(B)$, $L(C)$ do not all lie in a plane.

6. $\begin{pmatrix} -3 \\ 2 \\ 1 \end{pmatrix}, \begin{pmatrix} 1 \\ 2 \\ -2 \end{pmatrix}, \begin{pmatrix} 4 \\ 1 \\ 5 \end{pmatrix}$

7. The three column vectors of the matrix.

8. It is the set of points $L(P) + tL(A)$ with all t in **R**.

9. (a) $P + t(Q - P)$ (b) $L(P) + tL(Q - P) = L(P) + t[L(Q) - L(P)]$

10. It is the set of points $tL(A) + sL(B)$, with t, s in **R**.

11. It is the set of points $L(P) + tL(A) + sL(B)$ with t, s in **R**.

Chapter XIV, §4 *(p. 340)*

1. Inverse of F is the map G such that $G(X) = (1/7)X$.

2. $G(X) = (-1/8)X$ **3.** $G(X) = c^{-1} X$.

4. $(AB)^{-1} = B^{-1}A^{-1}$; $(ABC)^{-1} = C^{-1}B^{-1}A^{-1}$. Just multiply out

$$ABB^{-1}A^{-1} = I \qquad \text{and} \qquad ABCC^{-1}B^{-1}A^{-1} = I.$$

The same also holds taking the multiplication on the other side.

5. $(I + A)(I - A) = (I - A)(I + A) = I^2 - A^2 = I$ so $I + A$ is an inverse for $I - A$.

6. $I = A(-2I - A)$, so $-(2I + A)$ is an inverse (it commutes with A).

7. We have $(I - A)(I + A + A^2) = (I + A + A^2)(I - A) = I - A^3 = I$, so $I + A + A^2$ is an inverse for $I - A$.

Chapter XV, §1 *(p. 345)*

1. (a) 26 (b) 5 (c) -5 (d) -42 (e) -3 (f) 9
2. 1 3. (a) 1 (b) -1 (c) $-\frac{1}{2}$ (d) 0 5. $D(cA) = c^2 D(A)$.

Chapter XV, §2 *(p. 348)*

2. (a) -20 (b) 5 (c) 4 (d) 5 (e) -76 (f) -14
3. (a) 140 (b) 120 (c) -60
4. abc 5. (a) 3 (b) -24 (c) 16 (d) 14 (e) 0 (f) 8 (g) 8 (h) -10
6. $a_{11}a_{22}a_{33}$ both (a) and (b)

Chapter XV, §3 *(p. 353)*

4. Same, same
5. (a) -20 (b) 5 (c) 4 (d) 5 (e) -76 (f) -14
6. (a) 1 (b) -42 (c) 0 (d) 0 (e) 24 (f) 14 (g) 108 (h) 135 (i) 10
7. $a_{11}a_{22}a_{33}$
8. (a) 0 (b) 24 (c) -12 (d) 0 (e) 27 (f) -54 (g) -25 (h) -3
   (i) 5 (j) 0 (k) -18 (l) 0
9. $D(cA) = c^3 D(A)$ 14. 1 15. $t^2 + 8t + 5$

Chapter XV, §4 *(p. 357)*

1. If a number x is such that $B = xA$, then

$$D(A, B, C) = D(A, xA, C) = xD(A, A, C) = 0,$$

contrary to assumption.

5. Let x, y, z be numbers such that $xA + yB + zC = 0$. Then

$$
\begin{aligned}
0 = D(O, B, C) &= D(xA + yB + zC, B, C) \\
&= xD(A, B, C) + yD(B, B, C) + zD(C, B, C) \\
&= xD(A, B, C).
\end{aligned}
$$

Since $D(A, B, C) \neq 0$ by assumption, it follows that $x = 0$. A similar argument computing $D(A, O, C)$ and $D(A, B, O)$ shows that $y = 0$ and $z = 0$.

Chapter XV, §6 *(p. 360)*

1. (a) $\begin{pmatrix} \frac{2}{9} & \frac{1}{9} \\ -\frac{5}{9} & \frac{2}{9} \end{pmatrix}$ (b) $\begin{pmatrix} \frac{1}{11} & -\frac{4}{11} \\ \frac{2}{11} & \frac{3}{11} \end{pmatrix}$ (c) $\begin{pmatrix} \frac{2}{9} & -\frac{1}{9} \\ -\frac{1}{9} & \frac{5}{9} \end{pmatrix}$ (d) $\begin{pmatrix} -\frac{4}{5} & \frac{1}{5} \\ \frac{3}{5} & -\frac{2}{5} \end{pmatrix}$

2. $\dfrac{1}{ad - bc}\begin{pmatrix} d & -b \\ -c & a \end{pmatrix}$

Chapter XVI, §1 *(p. 364)*

1. (a) $\begin{pmatrix} 1 & 1 \\ 2xy & x^2 \end{pmatrix}$ (b) $\begin{pmatrix} \cos x & 0 \\ y \sin xy & -x \sin xy \end{pmatrix}$ (c) $\begin{pmatrix} ye^{xy} & xe^{xy} \\ 1/x & 0 \end{pmatrix}$

(d) $\begin{pmatrix} z & 0 & x \\ y & x & 0 \\ 0 & z & y \end{pmatrix}$ (e) $\begin{pmatrix} yz & xz & xy \\ 2xz & 0 & x^2 \end{pmatrix}$

(f) $\begin{pmatrix} yz \cos xyz & xz \cos xyz & yx \cos xyz \\ z & 0 & x \end{pmatrix}$

2. (a) $\begin{pmatrix} 1 & 1 \\ 4 & 1 \end{pmatrix}$ (b) $\begin{pmatrix} -1 & 0 \\ -\dfrac{\pi}{2} \sin \dfrac{\pi^2}{2} & -\pi \sin \dfrac{\pi^2}{2} \end{pmatrix}$ (c) $\begin{pmatrix} 4e^4 & e^4 \\ 1 & 0 \end{pmatrix}$

(d) $\begin{pmatrix} -1 & 0 & 1 \\ 1 & 1 & 0 \\ 0 & -1 & 1 \end{pmatrix}$ (e) $\begin{pmatrix} 1 & -2 & -2 \\ -4 & 0 & 4 \end{pmatrix}$ (f) $\begin{pmatrix} 8 & 4\pi & 2\pi \\ 4 & 0 & \pi \end{pmatrix}$

3. (a) $\begin{pmatrix} y & x \\ 2x & 0 \end{pmatrix}$ (b) $\begin{pmatrix} -y \sin xy & -x \sin xy & 0 \\ y \cos xy & x \cos xy & 0 \\ z & 0 & x \end{pmatrix}$

4. $\Delta_F(X) = x^2 - 2xy$. $\Delta_F(X) = 0$ when $x = 0$, y arbitrary, and also at all points with $x = 2y$.

5. $\Delta_F(X) = -x \cos x \sin xy$

6. $\begin{pmatrix} \cos \theta & -r \sin \theta \\ \sin \theta & r \cos \theta \end{pmatrix}$, r; determinant vanishes only for $r = 0$.

7. $\begin{pmatrix} \sin \varphi \cos \theta & -r \sin \varphi \sin \theta & r \cos \varphi \cos \theta \\ \sin \varphi \sin \theta & r \sin \varphi \cos \theta & r \cos \varphi \sin \theta \\ \cos \varphi & 0 & -r \sin \varphi \end{pmatrix}$

Determinant $r^2 \sin \varphi$

8. $\begin{pmatrix} e^r \cos \theta & -e^r \sin \theta \\ e^r \sin \theta & e^r \cos \theta \end{pmatrix}$

Determinant is e^{2r}. $F(r, \theta) = F(r, \theta + 2\pi)$.

Chapter XVI, §4 *(p. 371)*

1. Yes in all cases 2. (a), (b), (c), (d) all locally C^1-invertible
3. $F(x, y) = F(x, y + 2\pi)$

Chapter XVI, §5 *(p. 375)*

2. Letting $y = \varphi(x)$, we have

$$\varphi''(x) = \frac{-1}{D_2 f(x, y)^2} \begin{pmatrix} D_2 f(x, y)(D_1^2 f(x, y) + D_2 D_1 f(x, y)\varphi'(x)) \\ -D_1 f(x, y)(D_1 D_2 f(x, y) + D_2^2 f(x, y)\varphi'(x)) \end{pmatrix}.$$

3. (a) No (b) Yes (c) Yes
4. (a) We have $2x - y - xy' + 2yy' = 0$. This yields $\varphi'(1) = 0$.
   (b) $\varphi'(1) = \dfrac{-\pi}{2}$ (c) $\varphi'(1) = -\frac{1}{3}$ (d) $\varphi'(-1) = \frac{1}{2}$ (e) $\varphi'(0) = -1$
   (f) $\varphi'(2) = \dfrac{-39}{41}$

7. (a) both -1 8. $D_1\varphi = \frac{4}{3}$, $D_2\varphi = -\frac{2}{3}$
   (b) $D_1\varphi(0, 0) = 0$; $D_2\varphi(0, 0) = 0$

(c) $D_1\varphi(1, 1) = \frac{1}{3}$; $D_2\varphi(1, 1) = \frac{1}{3}$

(d) $D_1\varphi(0, \frac{1}{2}) = -\frac{3}{4}$; $D_2\varphi(0, \frac{1}{2}) = -1$

9. For y as an implicit function of (x, z):   (a) both -1

(b) Not possible since $D_2 f(0, 0, 0) = 0$

(c) $D_1\varphi(1, 2) = -1$; $D_2\varphi(1, 2) = 3$

(d) $D_1\varphi(\frac{1}{2}, \frac{1}{2}) = -\frac{3}{4}$; $D_2\varphi(\frac{1}{2}, \frac{1}{2}) = -1$

10. For x as an implicit function of (y, z):   (a) both -1

(b) Not possible since $D_1 f(0, 0, 0) = 0$

(c) $D_1\varphi(1, 2) = -1$; $D_2\varphi(1, 2) = 3$

(d) $D_1\varphi(\frac{1}{2}, \frac{1}{2}) = -\frac{4}{3}$; $D_2\varphi(\frac{1}{2}, \frac{1}{2}) = -\frac{4}{3}$

Chapter XVII, §1 *(p. 387)*

1. (a) 7 (b) 14   2. (a) 14 (b) 1

3. (a) 11 (b) 38 (c) 8 (d) 1   4. (a) 10 (b) 22 (c) 11 (d) 0

Chapter XVII, §2 *(p. 391)*

1. πab 2. $\frac{4}{3}\pi abc$ 3. (a) $29^{3/4}k$ (b) $r^{3/4}k$ 4. (a) $33^{3/5}k$ (b) $r^{3/5}k$

Chapter XVII, §3 *(p. 395)*

1. π 2. (a) $\frac{148}{3}$ (b) 0   3. (a) 42 (b) 120   4. 2   5. $\frac{1}{2}$ 6. πab 7. $\frac{22}{2}$

8. 1500π 9. $15\pi ab$ 11. 0   12. $1/3$

Chapter XVII, §5 *(p. 402)*

1. (a) $\begin{pmatrix} 1 & 0 & 0 \\ 0 & \rho\cos\varphi & \sin\varphi \\ 0 & -\rho\sin\varphi & \cos\varphi \end{pmatrix}$ and determinant is ρ.

(b) $\iiint_A f(G(\theta, \varphi, \rho))\rho \, d\rho \, d\varphi \, d\theta = \iiint_{G(A)} f(\theta, r, z) \, dz \, dr \, d\theta$

2. $abck$ 3. $\frac{4}{3}\pi abc$ 4. $\frac{4}{3}\pi a^3 \cdot 14$

5. (a) $\frac{1}{6}$ (c) $\frac{1}{3}$ (d) $\frac{1}{3}$ 6. (a) $\frac{1}{6}$ (b) $\frac{3}{2}$

7. (a) $\begin{pmatrix} 4 & 4 & 8 \\ 2 & 7 & 4 \\ 1 & 4 & 3 \end{pmatrix}$ (b) 20 (c) 100

Appendix 1, §1 *(p. 412)*

1. $\int_{-\pi}^{\pi} cf(x) \, dx = c \int_{-\pi}^{\pi} f(x) \, dx$

and

$$\langle f, g + h \rangle = \int_{-\pi}^{\pi} f(x)[g(x) + h(x)] \, dx = \int_{-\pi}^{\pi} [f(x)g(x) + f(x)h(x)] \, dx$$
$$= \int_{-\pi}^{\pi} f(x)g(x) \, dx + \int_{-\pi}^{\pi} f(x)h(x) \, dx = \langle f, g \rangle + \langle f, h \rangle.$$

2. Take the scalar product with f_i. We obtain for each i,

$$0 = \langle c_1 f_1 + \cdots + c_n f_n, f_i \rangle = \sum_{k=1}^{n} c_k \langle f_k, f_i \rangle = c_i.$$

3. If $\langle h_1, f \rangle = 0$ and $\langle h_2, f \rangle = 0$, then

$$\langle h_1 + h_2, f \rangle = \langle h_1, f \rangle + \langle h_2, f \rangle = 0.$$

If c is a number and $\langle h, f \rangle = 0$, then $\langle ch, f \rangle = c\langle h, f \rangle = 0$.

4. $\left| \int_{-\pi}^{\pi} f(x)g(x)\, dx \right| \leqq \left(\int_{-\pi}^{\pi} f(x)^2\, dx \right)^{1/2} \left(\int_{-\pi}^{\pi} g(x)^2\, dx \right)^{1/2}$

$\left(\int_{-\pi}^{\pi} [f(x) + g(x)]^2\, dx \right)^{1/2} \leqq \left(\int_{-\pi}^{\pi} f(x)^2\, dx \right)^{1/2} + \left(\int_{-\pi}^{\pi} g(x)^2\, dx \right)^{1/2}$

7. (b) $1/4$ (c) $\|f\| = \dfrac{\sqrt{3}}{3}$ and $\|g\| = \dfrac{\sqrt{5}}{5}$ (d) $1/2, 1/3, 1$

Appendix 1, §2 *(p. 420)*

1. (a) $\sqrt{5\pi}$ (b) $(\pi + \pi^3/3)^{1/2}$

4. (a) $\dfrac{x}{2} = \sin x - \dfrac{\sin 2x}{2} + \cdots + (-1)^{n+1} \dfrac{\sin nx}{n} + \cdots$

   (b) $x^2 = \dfrac{\pi^2}{3} - 4\left(\cos x - \dfrac{\cos 2x}{2^2} + \cdots + (-1)^{n+1} \dfrac{\cos nx}{n^2} + \cdots \right)$

   (c) $|x| = \dfrac{\pi}{2} - \dfrac{4}{\pi}\left(\cos x + \dfrac{\cos 3x}{3^2} + \cdots + \dfrac{\cos(2n+1)x}{(2n+1)^2} + \cdots \right)$

   (f) $|\cos x| = \dfrac{4}{\pi}\left(\dfrac{1}{2} + \dfrac{\cos 2x}{3} + \cdots + (-1)^{n-1} \dfrac{\cos 2nx}{4n^2 - 1} + \cdots \right)$

   (g) $\sin^3 x = \tfrac{3}{4}\sin x - \tfrac{1}{4}\sin 3x$

   (d) $\dfrac{1}{2} - \dfrac{\cos 2x}{2}$

   (e) $|\sin x| = \dfrac{4}{\pi}\left(\dfrac{1}{2} - \dfrac{\cos 2x}{3} - \cdots - \dfrac{\cos 2nx}{4n^2 - 1} - \cdots \right)$

   (h) $\cos^3 x = \tfrac{3}{4}\cos x + \tfrac{1}{4}\cos 3x$

Appendix 2, §2 *(p. 427)*

3. No   **4.** Yes   **5.** No   **6.** No   **7.** No   **8.** Yes   **9.** Yes

Appendix 2, §3 *(p. 428)*

1. Yes   **2.** Yes   **3.** No   **4.** Yes   **5.** No   **6.** Yes   **7.** No   **8.** Yes   **9.** No
10. Yes   **11.** No   **12.** Yes   **13.** No   **14.** Yes   **15.** Yes   **16.** Yes   **17.** Yes
18. Yes

Appendix 2, §4 *(p. 431)*

3. Yes   **4.** Yes   **5.** Yes   **6.** Yes   **7.** Yes   **8.** Yes   **9.** Yes   **10.** Yes

Appendix 2, §6 *(p. 438)*

2. (a) $4/e^2$ (b) $2^2 5^5 e^{-4}/3^3$ **3.** (a) 0   (b) ∞ **4.** 1   **5.** 1   **6.** 1   **7.** 1   **8.** $\frac{1}{2}$

9. 2   **10.** 0   **11.** 1   **12.** 1   **13.** $\frac{1}{2}$ **14.** $\frac{1}{2}$ **15.** $\frac{1}{4}$ **16.** $\dfrac{1}{e}$ **17.** 27   **18.** $\dfrac{4}{e^2}$

19. 0   **20.** ∞ **21.** 2   **22.** 3   **23.** 1   **24.** ∞ **25.** 1   **26.** ∞ **27.** 1   **28.** ∞ **29.** e

30. ∞

INDEX